Statistical Analysis of Questionnaires

A Unified Approach Based on R and Stata

CHAPMAN & HALL/CRC
Interdisciplinary Statistics Series

Series editors: N. Keiding, B.J.T. Morgan, C.K. Wikle, and P. van der Heijden

Published titles

Chapman & Hall/CRC
Interdisciplinary Statistics Series

Statistical Analysis of Questionnaires

A Unified Approach
Based on R and Stata

Francesco Bartolucci

Silvia Bacci

Michela Gnaldi

CRC Press
Taylor & Francis Group
Boca Raton London New York

CRC Press is an imprint of the
Taylor & Francis Group, an **informa** business

A CHAPMAN & HALL BOOK

First published 2016 by CRC Press

Published 2019 by CRC Press
Taylor & Francis Group
6000 Broken Sound Parkway NW, Suite 300
Boca Raton, FL 33487-2742

First issued in paperback 2022

ISBN 13: 978-1-03-247728-2 (pbk)
ISBN 13: 978-1-4665-6849-5 (hbk)

DOI: 10.1201/b18735

Library of Congress Cataloging-in-Publication Data

Bartolucci, Francesco, author.
   Statistical analysis of questionnaires : a unified approach based on R and Stata / Francesco Bartolucci, Silvia Bacci, Michela Gnaldi.
      pages cm. -- (Chapman & Hall/CRC interdisciplinary statistics series)
   Includes bibliographical references and index.
   ISBN 978-1-4665-6849-5 (hardcover : alk. paper) 1. Social sciences--Statistical methods--Computer programs. 2. Social sciences--Graphic methods--Computer programs. 3. Questionnaires--Design. 4. Sampling (Statistics)--Evaluation. 5. R (Computer program language) 6. Stata. I. Bacci, Silvia, author. II. Gnaldi, Michela, author. III. Title.

HA32.B36 2016
001.4'22--dc23   2015026098

Visit the Taylor & Francis Web site at
http://www.taylorandfrancis.com

and the CRC Press Web site at
http://www.crcpress.com

Contents

List of Figures

List of Tables

Preface

Questionnaires are fundamental tools of data collection in many fields and, in particular, in education, psychology, and sociology. They are becoming important tools of data collection even in economics, management, and medicine, especially when an individual characteristic that is not directly observable, such as the level of satisfaction of a customer or the quality of life of a patient, is of main interest.

A questionnaire is typically composed of a series of items that are scored on a binary or ordinal scale and are obviously correlated within each respondent. Therefore, data collected by a questionnaire require special statistical methods to be analyzed. This area has been the object of an increasing interest in the past decades, soliciting a growing number of academic research studies focused on theoretical and practical advances. In particular, in this book, we deal with the foundations of classical test theory (CTT), item response theory (IRT) fundamentals, IRT for dichotomous and polytomous items, up to the most recent IRT extensions, such as multidimensional IRT models, and latent class IRT models. Estimation methods and diagnostics—including graphical diagnostic tools, parametric and nonparametric tests, and differential item functioning—are also described.

The book is intended to have an applicative cut to mirror the growing interest and use of statistical methods and software in several fields of knowledge. Other than covering CTT and IRT basics, the book comprehensively accounts for the most recent theoretical IRT developments and provides Stata and R software codes for each method described within the chapters. For Stata we use version 12 and for R we use version 3.1.2. To enhance comprehension, real datasets are employed in the examples and the resulting software outputs are illustrated in detail.

The book is specifically designed for graduate students in applied statistics and psychometrics, and practitioners involved in measurement tasks within the education, health, and marketing fields, among others. For these figures, the book can be a reference work that will provide both the theoretical framework behind testing and measuring and a practical guide to carry out relevant analyses also based on the most advanced extensions and developments.

The prerequisites to properly understand the content of this book are at the level of a basic course of statistics which one can learn from the main parts of textbooks such as Mood et al. (1974) or Casella and Berger (2006). Moreover, a modest knowledge of Stata is required, which one can learn from the manuals for this software or from books such as Acock (2008). We use the main Stata module for the examples and suggest using gllamm (Rabe-Hesketh et al. 2004) for analyzing data from a questionnaire. Similarly, a basic knowledge of R is required which one may acquire from textbooks such as Dalgaard (2008). In this regard, the main packages we use are `ltm` (Rizopoulos, 2006) and `MultiLCIRT` (Bartolucci et al., 2014), even though we acknowledge that many R packages exist which may be used alternatively.[*] Finally, the datasets used for the illustrative examples in this book as well as the main updates of `Stata` and R codes are available from the authors' webpage.[†]

The book is organized into six chapters. Chapter 1 addresses preliminary aspects related to measurement in the psychological and educational fields and to the logic, development, and use of questionnaires. The chapter also provides a description of the notation and datasets used throughout the book.

Chapter 2 examines the theory of reliability from the perspective of CTT, the procedures used to determine reliability with test data and the estimation of true scores, together with some indices commonly used for item analysis. After a brief description of the conceptual basis of test validity, the remainder of the chapter deals with test bias and with extensions of CTT (i.e., generalizability theory).

Chapter 3 is aimed at illustrating the main assumptions characterizing IRT models for dichotomously scored items and the concept of item characteristic curve. The chapter also describes the three main IRT models—that is, the well-known Rasch model, the two-parameter logistic model, and the three-parameter logistic model—entangling them in the framework of random-effects models under both the normality assumption for the latent trait and the discreteness assumption.

Chapter 4 is devoted to a generalization of IRT models for binary data to account for the categorical nature of item responses. It is very common for psychological and educational tests to be characterized by polytomous items to supply more precise information about the response process. The most relevant case we typically refer to concerns items with ordered categories. Specifically, Chapter 4 illustrates the assumptions of IRT models for polytomous items and the main criteria used to classify these models, with special attention on their statistical properties.

Chapter 5 deals with two fundamental steps of IRT analyses. The chapter's first part is devoted to estimation methods such as the joint maximum likelihood method, the conditional maximum likelihood method, and the marginal maximum likelihood method. The second part is devoted to diagnostic instruments; we distinguish: graphical tools, methods focused on the

---

[*] See `http://cran.r-project.org/web/views/Psychometrics.html` for a list of R packages, many of which contain functions for the methods that are used to analyze data collected by questionnaires.

[†] See `https://sites.google.com/site/questionnairesbook/datasets` for the datasets and `https://sites.google.com/site/questionnairesbook/software-updates` for the software updates.

goodness-of-fit measurement and based on parametric and nonparametric hypothesis tests, and methods focused on differential item functioning.

In Chapter 6, we illustrate several extensions of the models discussed previously. Traditional IRT models assume that the distribution of individual abilities, either continuous or discrete, is the same for all subjects. The first way to extend classical IRT models is to enclose individual covariates in the model so that the distribution of the ability becomes individual specific. Another way to extend classical IRT models is to allow the individual abilities to be dependent as a result of sharing common factors in the case of multi-level or longitudinal data. In addition, according to traditional IRT models, all dependencies among individual responses are accounted for by only one latent trait. However, questionnaires are often made up of several subsets of items measuring different constructs (or dimensions). Therefore, the chapter also deals with multidimensional models. It concludes with an overview of structural equation models in which the relationships between these models and IRT models are outlined.

Acknowledgments

We specially thank Roberto Ricci and Patrizia Falzetti of the Italian Institute for the Evaluation of the Education System (INVALSI) for allowing us to use some of the data they produce in our examples.

Also, we thank Guido Miccinesi of the Institute for Cancer Study and Prevention of Florence and Andrea Bonacchi of the University of Florence for allowing us to use the Hospital Anxiety and Depression Scale (HADS) data in the available examples.

Finally, we thank Phil Bardsley of the University of California and Olga Demidova and Svetlana Sokolova of the National Research University Higher School of Economics of Moscow for allowing us to use, and for their assistance with, the Russian Longitudinal Monitoring Survey (RLMS) data.

We also acknowledge that a relevant part of the research behind this book has been supported by the Grant RBFR12SHVV of the Italian Government (FIRB 2012 Project "Mixture and Latent Variable Models for Causal Inference and Analysis of Socio-Economic Data").*

* See https://sites.google.com/site/latentfirb/.

1

Preliminaries

1.1 Introduction

Measurement is at the heart of the research process involved in any psychological or behavioral science. However, all measurement instruments, including questionnaires, are affected by challenges that can reduce their accuracy.

This chapter addresses important preliminary aspects related to measurement in the psychological and educational fields, and to the logic, development, and use of psychological measures based on questionnaires. On the other hand, the specific activity of questionnaire construction and the subsequent process of testing, that is, the act of administering questionnaires, is outside the scope of this chapter. Rather, we will focus on questionnaires as tools to study individuals with regard to their psychological attributes (e.g., abilities, skills, attitudes). As these latter attributes are not directly observable entities, they can only be inferred from a set of observed behaviors. This issue is one among the major measurement challenges in the psychological field.

Afterward, the chapter will provide a brief overview of other key issues one has to be aware of when dealing with data in the psychological field, such as the lack of a well-defined unit and origin of scale in most behavioral and social measurements.

As psychometric theory and models also provide useful insight for questionnaire development (i.e., in the pretesting process), this chapter will be devoted to a short discussion on the main steps involved in questionnaire construction. Even if these specific activities are not central for this book, we will briefly describe some of them to highlight the practical role of psychometric methods for ensuring the accuracy and sensibility of a measurement instrument.

To conclude, we outline that in the remainder of this book, we will use terms such as *questionnaire*, *test*, *scale*, and *measurement instrument* as synonyms, reserving our preference to the term questionnaire, instead of test, as the latter can be confused with test in the statistical testing acceptation. Similarly, the psychological attribute, which is an object of measurement throughout a questionnaire, will be indifferently named as a *latent variable*, *latent trait*, or *latent construct*, the word latent being replaceable by *unobserved* or *unobservable*.

The chapter is organized as follows. Sections 1.2 and 1.3 provide a general definition and discussion about latent variables representing psychological attributes and their measurements. Sections 1.4 and 1.5 describe the questionnaires, as tools of measurement, and their construction. After a discussion about psychometric theory in Section 1.6, we provide a description, in Section 1.7, of the basic notation used throughout the book. Finally, the datasets used for the examples in the following chapters are described in detail in Section 1.8.

1.2 Psychological Attributes as Latent Variables

When dealing with measurement in the psychological sciences, the first key issue is that psychological attributes cannot be directly measured. Unlike physical attributes, such as weight or height, which can be straight measured (e.g., through a scale or a yardstick), psychological traits are only indirectly observable entities, hence the denomination of latent constructs or unobserved variables. In other words, there is no means to directly measure attributes such as mathematical ability, anxiety, depression, working memory, or alcohol dependence. The degree to which a certain psychological trait characterizes an individual can be merely inferred from overt behaviors, which represent the construct observable manifestation. For instance, it is not possible to directly measure an individual's mathematical ability, but we can measure his/her performance at a number of test items in mathematics. Thus, the performance at a questionnaire or the responses to certain items of an ability test are the observed behaviors, which are assumed to be indicative of the corresponding latent constructs.

In turn, observed behaviors can be seen only as proxies of the associated latent construct (e.g., the responses to the items of a mathematics test can be considered as proxies of the ability under study). Unlike constructs, observable proxies only reflect specific aspects of the construct and, as such, they are not perfect measures of it (Raykov and Marcoulides, 2011). This lack of correspondence between unobservable constructs and observable proxies is primarily related to the fact that defining a construct is not the same as measuring that construct (Crocker and Algina, 1986). Second, when administering a test, there are many factors unrelated to the construct being measured which, however, can affect individuals' test answers (e.g., students' prior fatigue, boredom, guessing, forgetfulness, misrecording, misscoring).

Another way to think about psychological constructs is in terms of latent dimensions along which individuals are positioned and differ from one another. In the psychological sciences, the process of measurement aims essentially at differentiating individuals according to their relative positions along the latent dimensions of interest (Furr and Bacharach, 2008). As latent dimensions are not directly observable and measurable, individuals' exact locations on the latent traits are not known, and therefore, they are not precisely identified. This issue, together with the lack of identity between proxies and associated constructs, represents two among the major measurement problems in the psychological, educational, and behavioral fields.

1.3 Challenges in the Measurement of Latent Constructs

Measurement of latent constructs presents special challenges mainly related to the fact that psychological constructs are unobservable variables. The literature on this matter (Crocker and Algina, 1986; Raykov and Marcoulides, 2011) underlines the following main measurement challenges in the behavioral and social fields:

1. There is no single approach to measurement.
2. Measurements are based only on samples of behaviors.
3. Measurement is always prone to errors.
4. Units of measurement are not well defined.
5. Constructs must demonstrate relationship with other constructs.

The first challenge emerges from the fact that latent constructs are indirectly measured. Thus, there is always the possibility that the same construct is not operationally defined in the same way by different researchers (i.e., selecting different types of observable behavior). It follows that there is no single approach to measurement, which would yield a single satisfactory measuring instrument.

Second, measurement in the behavioral and social sciences is always based on samples. For instance, it is reasonable that a researcher interested in students' mathematics ability will select only a limited number of observable tasks to assess such an ability. Determining the number of items and the content necessary to adequately cover the construct of interest is a crucial issue when developing a sensible measurement instrument (Crocker and Algina, 1986).

Third, measurement of latent constructs is obviously not error free. Errors occur whenever changes in the observed behavior of the studied individuals are not attributable to actual changes in the underlying ability, but to external factors unrelated to the construct (e.g., students' prior fatigue, guessing, misrecording, misscoring). A very relevant issue in behavioral and social research is, thus, how to estimate the degree of measurement error present in a given set of observations.

The fourth challenge refers to the lack of well-defined unit and origin (i.e., zero point) of the scale in most behavioral and social measurements. For instance, is an individual score of 40 obtained on a mathematics questionnaire indicative of half as much mathematics ability than a score of 80? Is a score of zero on the same questionnaire indicative of no mathematical ability at all? It is not possible to answer univocally to the previous questions as units of measurement are not well defined.

Fifth, a construct cannot be defined only in terms of operational definitions: it is also necessary to demonstrate its relationships with other constructs in the same subject-matter domain. This second type of definition, called *nomothetic definition*, helps in interpreting the obtained measurements. When such relationships cannot be demonstrated, then the measurements are of no value.

1.4 What Is a Questionnaire?

The previous discussion leads us to point out that the development of a questionnaire implies the prior conceptualization of the construct of interest (e.g., mathematical ability, depression) and the subsequent definition of sensible rules of correspondence between the theoretical construct and the observable behaviors (i.e., the responses to the items of a questionnaire) that are assumed to be the overt manifestations of that construct. The process of defining a construct is called an *operational definition* (Crocker and Algina, 1986; Furr and Bacharach, 2008).

In psychological measurement, numerals are typically used to represent an individual's level of a latent attribute. Specifically, measurement of psychological constructs occurs when quantitative values are assigned to the sample observed by administering a questionnaire. As known, when a questionnaire is administered to a sample of subjects, every individual receives a score on every single item and on the questionnaire as a whole. It is from these latter scores that one draws inference about the amount of constructs possessed by the individuals. Making correct decisions based on questionnaire scores (e.g., admitting a student to a college, admitting a patient in a certain psychological treatment group) largely depends on the degree to which such scores truly reflect the attributes under study. It is also worth noting that a questionnaire is not expected to provide exhaustive coverage of all possible behaviors that concur to define a latent construct. On the other hand, a questionnaire approximates the underlying construct through a sample of behaviors, and its quality is linked to the degree this sample is representative of the construct.

Generally, the process of assigning numbers to certain events is called a *measurement*. The rules followed during this process define separate levels of measurement, differing from one another for the degree of complexity of the mathematical system by which numbers are associated to different observations.

At the first level, there is the *nominal measurement*, that is, a classification by inclusion into mutually exclusive subsets (Lord and Novick, 1968) with respect to some characteristics of interest (e.g., the choice of a certain brand in a marketing study). Nominal measurement yields a nominal scale and implies an identity relationship that holds among all members of one group (e.g., the subset of prevalent customers of a certain brand).

The second level of measurement, the *ordinal measurement*, yields an ordinal, or rank, scale. As in nominal scales, rank scales imply mutually exclusive subsets. However, different from nominal scales, they also presuppose that an ordering relationship may be established between all pairs of objects. This relationship can be interpreted in the same way as A is greater than B or A is preferred to B. As in nominal scales, also in ordinal scales the assignment of numerical values to subjects belonging to the various groups is arbitrary, that is, any different set of ordered distinct numbers or labels can be chosen.

The *interval measurement*, which yields an interval scale, implies not only the ordinal property of a rank scale but also a one-to-one correspondence between the observed elements of the construct and the real numbers. In fact, an interval scale establishes a distance function over all pairs of elements. In interval scales, the size of the unit of measurement is constant and additive, but multiplicative operations are not sensible. That is, only additive transformations of such scales are admissible because they preserve the equality of numeric differences (e.g., one can add 3° to 30° or to 15°; in both cases, 3° represents the change in the amount of temperature, but it is not possible to claim that 30° is twice as warm as 15°).

At the fourth level of measurement, the *ratio scale* further assumes the existence of a fixed, nonarbitrary, zero point. Therefore, different from interval scales, ratio scales allow multiplicative transformations (i.e., it is appropriate to assume that a distance of 30 km is twice as far as a distance of 15 km).

Data derived from the administration of questionnaires do not have a fixed zero point. That is, zero scores are arbitrary because they do not truly reflect the total absence of the psychological attribute of interest, but rather a nonmeasurable quantity. Thus, generally, data are treated as if they had interval scale properties, even if it is generally agreed that measurement in the psychological field yields only a nominal or, at the best, an ordinal scale; see, for example, Lord and Novick (1968) and Furr and Bacharach (2008). In fact, in treating these data by interval properties, one assumes a specific distance function, which is completely arbitrary because measurement units are unspecified. For instance, what are the units of measurement when measuring a psychological attribute such as mathematical knowledge, depression, shyness, or intelligence? Presumably, they are responses to a number of items within a questionnaire. The overall measure of the psychological attribute of interest is then obtained by counting up the scores obtained on the questionnaire items. However, how do we know if, and to what extent, those responses are related to the psychological attribute of interest? Thus, for instance, it is not fully appropriate to assume that a 10-point difference in the total scores of two individuals is the same as a 10-point difference in the total scores of other two individuals taking the same (or another) questionnaire. However, even if treating data by interval properties is, formally, an arbitrary strengthening, it is generally considered admissible from a pragmatic point of view. The interval scaling is justified to the extent that measurement provides an accurate empirical predictor of a latent attribute.

On the basis of what was discussed earlier, a questionnaire can be composed by a variety of items, mainly distinguishing between quantitative and qualitative (or categorical) items. An item is quantitative when its values result from counting or measuring something. Examples of quantitative items are given by self-reported evaluations of a given condition (e.g., satisfaction, quality of life, quality of health) on an imaginary latent continuum and expressed on a numeric scale (e.g., 0–10 or 0–100). An item is qualitative when its values do not result from measuring or counting, but refer to descriptions.

Qualitative items are categorical items, dichotomously or polytomously scored. Since test theory mainly deals with categorical items, we will especially refer to these types of item in the book. In the dichotomous case, items may assume only two values, such as true/false, yes/no, or satisfied/unsatisfied. For instance, in educational settings, items that assess the correctness of a given answer, giving one point to the correct answer and zero to the incorrect answer, are true/false items. In the case of polytomously scored items, a finite number of categories greater than two may be observed. The latter items may have *unordered categories*: this is the case of items that investigate, for example, electoral or consumers' choices; otherwise the categories are *ordered*. This is the case, for instance, of items that investigate satisfaction with one's job, whose answers may be ordered into categories of the type absolutely satisfied, mostly satisfied, neutral, not very satisfied, and absolutely unsatisfied. As another example, the items of a questionnaire to assess students' abilities may be scored into ordinal categories of the type correct, partially correct, and incorrect.

Besides, in the case of binary items scored as 0 and 1, it is possible to distinguish the case of true dichotomous items from the case of false dichotomies, which applies when originally polytomously scored items are then rescored and one or more modalities are aggregated to obtain a binary scoring (e.g., satisfied and unsatisfied, in place of the original five categories of job satisfaction).

1.5 Main Steps in Questionnaire Construction

The development of a proper measurement instrument is a rather complex process, which starts with the item specification and ends with the validation of the measurement instrument itself (Spaan, 2007). There is a great attention on the use of shared procedures in test development, as demonstrated by the Standards for Educational and Psychological Testing (American Educational Research Association, American Psychological Association and National Council on Measurement in Education, 1999). In Downing and Haladyna (2006), 12 different steps for an effective test construction are described, from the definition of an overall plan to the final test technical report. These specific activities will not be central for this book. However, some of the main steps will be briefly described in the present section and the practical role played by the psychometric methods covered in the following chapters of this text will be highlighted.

One of the earliest steps in *questionnaire construction* is deciding the purposes of the questionnaire scores and, thus, clarifying if they are intended to differentiate among individuals with respect to a given psychological attribute or with respect to an expected score (Furr and Bacharach, 2008). In fact, questionnaires are categorized as either *criterion referenced* or *norm referenced*. Criterion-referenced questionnaires are often employed when a decision must be taken about individuals' skill level. A fixed cutoff score is predetermined and used to cluster individuals into two groups, that is, the group of individuals whose scores exceed a certain cutoff score and the group of individuals whose scores do not. On the other hand, norm-referenced questionnaires are used to compare individuals' questionnaire scores with scores from a reference sample, which are representative of the population of interest. Individuals' scores are compared with the expected average score that would be obtained if the questionnaire were administered to the whole population. In practice, the distinction between criterion-referenced and norm-referenced questionnaires is often blurred. In fact, in some sense, criterion-referenced tests are always normed, because the criterion cutoff score is never determined at random and expresses an expected score level.

At the next stage of questionnaire development, behaviors representing the latent construct of interest need to be identified. This stage implies a content analysis and a critical review of available researches on the same, or related, content areas. As already mentioned, the first activity is named an *operational definition* and involves the creation of a rule of correspondence between the latent constructs and the item responses that are assumed to be the overt manifestations of that construct. The second, called a *nomothetic definition*, further investigates the relationship between the latent trait of interest and already available constructs concerning the domain.

The proper development of items that effectively measure the targeted construct is crucial and needs to be checked in the subsequent phase of analysis of the responses on the pretested items. Once the items have been specified and included in a questionnaire conditionally on content and format constraints, a preliminary validation of the measurement instrument is carried out to analyze test and item properties and highlight any possible issue related to item formulation, reasonableness of distractors, item difficulty and discrimination, and consistency of the whole test. Therefore, before being administered to the targeted individuals, a preliminary version of the questionnaire is submitted to a subsample and responses are analyzed by using psychometric methods. This preliminary tryout can yield the removal or revision of some items or their distractors (i.e., the incorrect answers to an item). Thus, psychometric methods also have applications in the process of pretesting to test the accuracy and sensitivity of questionnaires.

1.6 What Is Psychometric Theory?

We define *psychometric theory* as the discipline concerned with evaluating the attributes of measurement instruments employed in the psychological and behavioral sciences. In more detail, psychometric theory deals with the study of how general issues related to measurement in the psychological field impact the process of quantification of the aspects under investigation, as well as on methods aimed at providing information about the underlying constructs from their indirect observation.

Psychometric theory is based on formal logic and general mathematical and statistical methods and models (McDonald, 1999). As such, it is of importance regardless of the specific test or measurement instrument used in a given context where evaluation of latent attributes is required. Being so general, psychometric theory provides an overall framework for questionnaire development, construction, revision, and modification, which can be useful to any behavioral theory that a researcher might adopt.

Psychometric theory is built on two main branches, *classical test theory* (CTT) and *item response theory* (IRT). CTT has been the basis of psychometrics for over 80 years (Spearman, 1904; Kuder and Richardson, 1937; Guttman, 1945; Novick, 1966; Lord and Novick, 1968). According to CTT, a person's observed score on a test is an additive function of a true score component and a random error component. The true score may be defined as the mean score that the person would get on the same test if he/she had an infinite number of testing sessions. Thus, true scores are unknown quantities because it is not possible to observe an infinite number of testing sessions.

IRT is a psychometric approach emerged as an alternative to CTT (Hambleton and Swaminathan, 1985; Hambleton et al., 1991; Van der Linden and Hambleton, 1997) with important developments being relatively recent. This theory is advocated by its proponents as a theoretical framework, which overcomes some of the drawbacks of CTT. Different from IRT, CTT methods do not explicitly model the way respondents with different latent trait levels perform on questionnaire items. On the other hand, the heart of IRT is a statistical model for the way respondents at different latent trait levels answer each item. In fact, IRT is built on the central idea that the probability to provide a certain answer to an item is a function of the person's position on the latent trait and one or more parameters that characterize the item.

IRT finds its origins in the early work of Thurstone (1925), which introduced the fundamentals of IRT in the educational field in order to measure students' abilities, so that the term latent trait is commonly referred to as *ability*, also in the following of this book. Among the first and most relevant contributions to IRT, we also remind the works of Richardson (1936) and Ferguson (1942), for the specification of the normal ogive model, and the work of Lord (1952), who stated the substantial difference between observed test scores and latent traits. Another milestone is represented by the statement of the specific objectivity principle in the social science field by Rasch (1960), who also introduced the homonymous model for dichotomously scored items. Afterward, the work by Lord and Novick (1968) provided a first unified treatment of CTT and IRT, other than an important contributed chapter by Birnbaum (1968), in which the two-parameter logistic model for dichotomously scored items was defined. The two models by Rasch (1960) and Birnbaum (1968) were then extended for the treatment of polytomously scored items by Andrich (1978b), Masters (1982), and Samejima (1969). Finally, among the first relevant contributions to the estimation of IRT model parameters, we recall the works of Bock and Lieberman (1970), Bock (1972), Bock and Aitkin (1981), and Baker (2001).

1.7 Notation

We assume to observe responses of n individuals to J items, and the generic individual and the generic item are denoted by i and j, respectively, with $i = 1, \ldots, n$ and $j = 1, \ldots, J$. Moreover, we consider that the questionnaire is aimed to measure, for each individual i, the *true score* s_i or the level θ_i of a certain *latent trait*, according to the language of CTT or IRT, respectively.

This notation will be suitably extended to the case of test items measuring more than one dimension or latent trait.

Moreover, y_{ij} denotes the response of individual i to item j; it assumes values $0, 1$ in the case of binary items or values $0, \ldots, l_j - 1$ in the case of polytomous (nominal or ordinal) items. Note that $l_j = l$ for $j = 1, \ldots, J$ when all items have the same number of response modalities. In the language commonly used in educational and psychological fields for dichotomous items, we may indifferently indicate by $y_{ij} = 1$ the right or correct answer to item j or denote that individual i endorses item j. Conversely, $y_{ij} = 0$ denotes that subject i fails or gives the wrong answer to item j. We also denote by $y_i = (y_{i1}, \ldots, y_{iJ})'$ the vector of all responses provided by subject i in the sample.

All quantities previously introduced are denoted in lowercase. This follows the usual convention in statistics according to which the lowercase is for observed values. However, it is convenient to deal with these quantities as realizations of random variables (or vectors), which are instead denoted in uppercase. Therefore, s_i may be seen as a realization of S_i, θ_i as a realization of Θ_i, y_{ij} as a realization of Y_{ij}, and y_i as a realization of $Y_i = (Y_{i1}, \ldots, Y_{iJ})'$. This principle applies throughout the book.

In the remainder of this section, we introduce some descriptive statistics of the observed set of responses y_{ij} provided by a group of examinees to a set of test items. These statistics are easily interpretable and they are also typically used as the basis for estimating parameters of interest in a modeling framework.

Basic descriptive statistics are

$$y_{i\cdot} = \sum_{j=1}^{J} y_{ij} \quad \text{and} \quad y_{\cdot j} = \sum_{i=1}^{n} y_{ij},$$

which correspond to the *total score of examinee i* and to the *total score for item j*, respectively, and are measures of performance and difficulty.* Note that, in the dichotomous case, $y_{i\cdot}$ simply corresponds to the number of items correctly responded by individual i and, similarly, $y_{\cdot j}$ corresponds to the number of examinees who respond correctly to item j. The corresponding averaged quantities, or *mean scores*, are

$$\bar{y}_{i\cdot} = \frac{y_{i\cdot}}{J} \quad \text{and} \quad \bar{y}_{\cdot j} = \frac{y_{\cdot j}}{n},$$

---

* Indeed, this is a measure of *easiness* as it tends to be larger for easy items than for difficult items; however, it is commonly referred to as a difficulty measure.

which are equal, in the case of binary items, to proportions of correct responses. It may also make sense to compute the *mean individual score* as

$$\bar{y}_{..} = \frac{1}{n} \sum_{i=1}^{n} \sum_{j=1}^{J} y_{ij}.$$

Other descriptive statistics are measures of variability that may be referred to an examinee or to a test item. In particular, we consider

$$v_{i.} = \frac{1}{J} \sum_{j=1}^{J} (y_{ij} - \bar{y}_{i.})^2 \quad \text{and} \quad v_{.j} = \frac{1}{n} \sum_{i=1}^{n} (y_{ij} - \bar{y}_{.j})^2, \tag{1.1}$$

corresponding to the *variance score of examinee i* and the *variance score of item j*, respectively. On the other hand, the *variance of individual score* is

$$v_{..} = \frac{1}{n} \sum_{i=1}^{n} (y_{i.} - \bar{y}_{..})^2. \tag{1.2}$$

Another statistic, which can be useful to consider, is the *covariance between the observed scores* on two items, say, items j_1 and j_2, which is defined as

$$\mathrm{cov}_{j_1 j_2} = \frac{1}{n} \sum_{i=1}^{n} (y_{ij_1} - y_{.j_1})(y_{ij_2} - \bar{y}_{.j_2}). \tag{1.3}$$

The corresponding correlation index is defined as

$$\mathrm{cor}_{j_1 j_2} = \frac{\mathrm{cov}_{j_1 j_2}}{\sqrt{v_{.j_1} v_{.j_2}}}.$$

As usual, this correlation index is between -1 and 1, with values greater than 0 corresponding to positive correlation and values smaller than 0 corresponding to a negative correlation.

Though it is more natural to use the indicators at issue with quantitative variables, it is important to recall their definition here, even if we mainly deal with variables that are categorical. As will be made clear in Chapter 2, in the CTT setting it is usual to compute such indicators independently from the nature of the test items.

In the case of T different tests administered to the same examinees, the quantities at issue may be defined for each single test. In this case, we denote by $J^{(t)}$ the number of items of test t, with $t = 1, \ldots, T$, and the response provided by examinee i to the jth item of this test by $y_{ij}^{(t)}$, with $i = 1, \ldots, n$ and

$j = 1, \ldots, J^{(t)}$. The vector of responses provided by this subject is indicated by $\mathbf{y}_i^{(t)} = (y_{i1}^{(t)}, \ldots, y_{iJ^{(t)}}^{(t)})'$ and the corresponding total score and mean score are

$$y_{i\cdot}^{(t)} = \sum_{j=1}^{J} y_{ij}^{(t)} \quad \text{and} \quad \bar{y}_{i\cdot}^{(t)} = \frac{y_{i\cdot}^{(t)}}{J^{(t)}},$$

respectively. Accordingly, we may define the total and mean score of item j in test t; these are denoted by $y_{\cdot j}^{(t)}$ and $\bar{y}_{\cdot j}^{(t)}$, respectively. We also consider the mean individual score for the overall test, indicated by $\bar{y}_{\cdot\cdot}^{(t)}$.

As will be clear in the following, it may also be important to consider the correlation index between two test scores (or two halves of the same test or two scores on the same test repeated twice). This is defined as follows:

$$\text{cor}^{(t_1 t_2)} = \frac{\sum_{i=1}^{n}(y_{i\cdot}^{(t_1)} - \bar{y}_{\cdot\cdot}^{(t_1)})(y_{i\cdot}^{(t_2)} - \bar{y}_{\cdot\cdot}^{(t_2)})}{\sqrt{\sum_{i=1}^{n}(y_{i\cdot}^{(t_1)} - \bar{y}_{\cdot\cdot}^{(t_1)})^2 \sum_{i=1}^{n}(y_{i\cdot}^{(t_2)} - \bar{y}_{\cdot\cdot}^{(t_2)})^2}}, \quad t_1 \neq t_2. \qquad (1.4)$$

Finally, given one or more random variables, we use the following notation throughout the book: $E(\cdot)$ for the expected value, $V(\cdot)$ for the variance, $\text{Cov}(\cdot)$ for the covariance, and $\text{Cor}(\cdot)$ for the correlation coefficient. In particular, for a continuous random variable X with distribution having density function $f(x)$, we remind the reader that

$$\mu_X = E(X) = \int xf(x)\,dx,$$

$$\sigma_X^2 = V(X) = E[(X - \mu_X)^2] = \int (x - \mu_X)^2 f(x)\,dx;$$

the previous integrals become sums when the random variable is discrete with probability mass function $f(x)$.

Considering now two continuous random variables X and Y with joint density function $f(x, y)$, we have

$$\text{Cov}(X, Y) = E[(X - \mu_X)(Y - \mu_Y)] = \int\int (x - \mu_X)(y - \mu_Y)f(x, y)\,dy\,dx,$$

where, again, the integrals become sums in the case of discrete random variables with joint probability mass function $f(x, y)$. Moreover, the correlation index is defined as

$$\rho_{XY} = \text{Cor}(X, Y) = \frac{\text{Cov}(X, Y)}{\sqrt{V(X)V(Y)}},$$

which is always between -1 and 1.

1.8 Datasets Used for Examples

The examples proposed to illustrate the main topics treated throughout the book are based on datasets taken from different fields. As described in the following, the fields of application concern education (Section 1.8.1), job satisfaction (Section 1.8.2), and anxiety and depression (Section 1.8.3).

1.8.1 Italian Institute for the Evaluation of the Education System Dataset

The Italian Institute for the Evaluation of the Education System (INVALSI) is a national institution in charge of developing national tests for the assessment of primary, lower-middle, and high-school students. In the following chapters, we use data drawn from the INVALSI Italian test and mathematics test administered in June 2009, at the end of the pupils' compulsory educational period.

The INVALSI Italian test includes two sections, a reading comprehension section and a grammar section. The first section is based on two texts: a narrative-type text, where readers engage with imagined events and actions, and an informational text, where readers engage with real settings; see INVALSI (2009). The reading comprehension processes are measured by 30 items, which require students to demonstrate a range of abilities and skills in constructing meaning from the two written texts. The grammar section is made up of 10 items, which measure the ability of understanding the morphological and syntactic structure of sentences within a text.

The INVALSI mathematics test consists of 27 items covering four main content domains: numbers, shapes and figures, algebra, and data and previsions (INVALSI, 2009). The number content domain consists of understanding (and operating with) whole numbers, fractions and decimals, proportions, and percentage values. The algebra domain requires students the ability to understand, among others, patterns, expressions, and first-order equations, and to represent them through words, tables, and graphs. The shapes and figures domain covers topics such as geometric shapes, measurement, location, and movement. The data and previsions domain includes three main topic areas: data organization and representation (reading, organizing, and displaying data using tables and graphs), data interpretation (identifying, calculating, and comparing characteristics of datasets, including mean, median, mode), and chance (e.g., judging the chance of an outcome, using data to estimate the chance of future outcomes).

All items of the reading, grammar, and mathematics dimensions of the INVALSI tests are of multiple choice type, with one correct answer and three distractors, and are dichotomously scored (assigning 1 point to correct answers and 0 otherwise). However, the mathematics test contains also two open questions for which a partial score of 1 was assigned to partially correct answers and a score of 2 was given to correct answers. For the purposes of the analyses described in the following, the open questions of the mathematics test were dichotomously rescored, giving 0 point to incorrect answers and 1 point otherwise.

For the applications illustrated in the following chapters, we employ two subsamples of the INVALSI original dataset*: (1) a dataset, denoted in the following as the *INVALSI full dataset*, made up of the item responses to all three involved dimensions (reading, grammar, and mathematics) of 3774 students coming from 171 schools, with at least 20 respondents per school and (2) a random subset, denoted in the following as *INVALSI reduced dataset*, referred to 1786 male students coming from the Center of Italy, which includes the responses to mathematics items only. See Tables 1.1 and 1.2, respectively, for the relative distribution of correct answers.

Overall, we observe that the difficulty distribution of the set of items at issue is not completely satisfactory, as many items within each dimension of the INVALSI tests get a percentage of correct answers higher than 75%. Besides, the reading dimension includes two items (i.e., item b6 and item b13), which receive a percentage of very high correct answers (95.6% and 94.4%, respectively). For these items, the difficulty is low to such an extent that almost all students (even the least skilled) can provide the correct answer, and therefore, they cannot differentiate students in a satisfactory way. A similar pattern is observed for items c1, c5, c8, and c9 of the grammar dimension and for items d4, d5_1, and d5_2 of the mathematics dimension (Table 1.1); see also items Y4, Y5, and Y6 in Table 1.2. In light of this, we conclude that the INVALSI test is rather unbalanced in the overall difficulty, as many items are too easy.

In addition, the INVALSI full dataset provides information about the gender of respondents (gender) and about the geographical area of each school, distinguishing among northwest, northeast (dummy area_2=1), center (dummy area_3=1), south (dummy area_4=1), and islands (dummy area_5=1); note that northwest is identified by the dummies area_2 to area_5 equal to zero. As confirmed by the results in Table 1.3, females and males are equally represented. Besides, students from the south and the islands account for more than one half of the sample, while each of the other geographical areas accounts for a smaller percentage of students.

Finally, Table 1.4 shows the average percentage score per gender and geographic area for the full INVALSI dataset. We observe that the best performance is for female students from schools in the south of Italy, in grammar. On the contrary, the worst performance is for male students from schools in northeast of Italy, in reading comprehension.

---

* Both datasets are downloadable from https://sites.google.com/site/question nairesbook/datasets and are named INVALSI_full and INVALSI_reduced with extension .dta for Stata and .RData for R.

TABLE 1.1

INVALSI Full Dataset (3774 Students): Proportion of Correct Answers to the Items of the Three Dimensions—Reading, Grammar, and Mathematics

Reading		Grammar		Mathematics	
Item	Correct Answers	Item	Correct Answers	Item	Correct Answers
a1	0.7520	c1	0.9356	d1	0.8747
a2	0.8877	c2	0.7843	d2	0.8718
a3	0.8135	c3	0.8574	d3	0.6918
a4	0.6706	c4	0.6532	d4	0.9277
a5	0.8998	c5	0.9208	d5_1	0.9894
a6	0.8980	c6	0.7276	d5_2	0.9340
a7	0.7650	c7	0.7888	d5_3	0.5456
a8	0.6150	c8	0.9486	d6a	0.7780
a9	0.7957	c9	0.9640	d6b	0.8802
a10	0.5363	c10	0.7660	d6c	0.8410
a11	0.8182			d7	0.8932
a12	0.7997			d8	0.8090
a13	0.7660			d9	0.7501
a14	0.8037			d10	0.8151
a15	0.8768			d11	0.8233
a16	0.6017			d12	0.8959
a17	0.8235			d13	0.7973
b1	0.7417			d14a	0.8328
b2	0.3773			d14b	0.8151
b3	0.6812			d15	0.8214
b4	0.8646			d16	0.8005
b5	0.8498			d17	0.8808
b6	0.9560			d18	0.4179
b7	0.6187			d19a	0.8426
b8	0.6704			d19b	0.7340
b9	0.7477			d20	0.7684
b10	0.7806			d21	0.8720
b11	0.8278				
b12	0.8108				
b13	0.9444				

1.8.2 Russian Longitudinal Monitoring Survey Dataset

The Russian Longitudinal Monitoring Survey (RLMS) is a nationally representative longitudinal survey conducted by the Higher School of Economics and ZAO Demoscope together with the Carolina Population Center (University of North Carolina at Chapel Hill) and the Institute of Sociology

TABLE 1.2

INVALSI Reduced Dataset (1786 Students):
Proportion of Correct Answers to the Items
of the Mathematics Dimension

Mathematics	
Item	**Correct Answers**
Y1	0.8600
Y2	0.8561
Y3	0.6445
Y4	0.9065
Y5	0.9826
Y6	0.9384
Y7	0.4877
Y8	0.7587
Y9	0.8583
Y10	0.8343
Y11	0.8365
Y12	0.7800
Y13	0.7223
Y14	0.7721
Y15	0.8191
Y16	0.8903
Y17	0.8024
Y18	0.7951
Y19	0.6081
Y20	0.7665
Y21	0.7777
Y22	0.8606
Y23	0.3679
Y24	0.7878
Y25	0.4966
Y26	0.7144
Y27	0.8124

of Russian Academy of Sciences. The survey at issue is designed to monitor the health and economic welfare of households and individuals in the Russian Federation. Data have been collected 19 times since 1992 and include household, individual, and community data.

Individual-level data pertaining to a person's health, employment status, demographic characteristics, and anthropometry are available. Such data are taken from the adult questionnaire administered in 2010 (round XIX) giving rise to 17,810 observations and 610 variables grouped into four main sections:

TABLE 1.3

INVALSI Full Dataset (3774 Individuals): Distribution of Covariates

Variable	Category	Proportion
Gender (gender)		
	0 = male	0.4966
	1 = female	0.5034
Geographical area		
area_2 to area_5	0 = Northwest	0.1500
area_2	1 = Northeast	0.1187
area_3	1 = Center	0.1659
area_4	1 = South	0.3227
area_5	1 = Islands	0.2427

TABLE 1.4

INVALSI Full Dataset (3774 Students): Average Score per Gender and Geographical Area for the Three Dimensions—Reading, Grammar, and Mathematics

		Geographic Area				
		Northwest	Northeast	Center	South	Islands
Males	Reading	0.7640	0.7136	0.7577	0.7703	0.7452
	Grammar	0.8264	0.7636	0.7914	0.8618	0.8256
	Mathematics	0.8157	0.7610	0.7959	0.8349	0.8289
Females	Reading	0.7739	0.7649	0.7643	0.7928	0.7784
	Grammar	0.8035	0.8099	0.8189	0.8841	0.8529
	Mathematics	0.7815	0.7760	0.7777	0.8305	0.8278

migration, work, medical services, and health evaluation. In particular, the section about work includes items on individuals' job position, satisfaction with work conditions, earnings, opportunities for professional growth, and so on. The section is made up of several kinds of items, from dichotomous and polytomous items to open-ended questions.

Specifically, in this book, the job satisfaction of a random subsample of 1418 young individuals is considered.* Individuals' job satisfaction is assessed through four polytomously scored items, which refer to satisfaction with one's job in general (item Y1), work conditions (item Y2), earnings (item Y3), and opportunity for professional growth (item Y4). Each item has

---

* The dataset is downloadable from https://sites.google.com/site/question nairesbook/datasets and is named RLMS with extension .dta for Stata and .RData for R; also a binary version of this dataset is available, which is named RLMS_bin with extension .dta for Stata and .RData for R.

TABLE 1.5

RLMS Dataset: Distribution of the Item Response Categories

Item	Response Category						
	0	1	2	3	4	*0, 1*	*2, 3, 4*
Y1	0.1685	0.4711	0.1953	0.1269	0.0381	*0.6396*	*0.3604*
Y2	0.1636	0.4549	0.2073	0.1291	0.0451	*0.6185*	*0.3815*
Y3	0.0719	0.2814	0.2179	0.2680	0.1608	*0.3533*	*0.6467*
Y4	0.0994	0.3237	0.2377	0.2250	0.1142	*0.4231*	*0.5769*

Note: Italics indicate the corresponding values of the dichotomized items.

five ordered categories: absolutely satisfied (0), mostly satisfied (1), neutral 2), not very satisfied (3), and absolutely unsatisfied (4). For some purposes of the analyses described in the following, the five categories of each item were dichotomized assigning value 1 to absolutely or mostly satisfied respondents (original categories labeled as 0 and 1), and 0 to neutral, not very satisfied, and absolutely unsatisfied respondents (original categories labeled as 2, 3, and 4).

As shown in Table 1.5, the best evaluation concerns items Y1 and Y2 with more than 60% of individuals who are absolutely or mostly satisfied about their job in general and about the work conditions in particular. On the contrary, opportunities for professional growth (item Y4) and, mostly, earning (item Y3) represent critical aspects of one's job condition, with 33.9% and 42.9% of individuals not very satisfied or absolutely unsatisfied, respectively.

The dataset at issue also includes various covariates. In the examples provided in the present book, we consider the following (see also Table 1.6 for details): marital status (`marital`), which distinguishes between adults living single and living as a couple; educational level (`education`), which is characterized by six increasing levels, from 0 (maximum 6 years of education) to 5 (college and postgraduate); gender (`gender`), being 0 for males and 1 for females; working status (`work`), with 0 for adults on leave and 1 for adults that are currently working; and age in years (`age`).

As shown in Table 1.6, the RLMS dataset is balanced with regard to gender and marital status distributions. In addition, high educational levels (see categories 3 to 5) are mostly represented in the dataset that is considered. Finally, the dataset is characterized by young people, with age equally distributed between a minimum value of 16 years and a maximum value of 26 years (mean = 22.78; median = 23.00; standard deviation = 1.95).

1.8.3 Hospital Anxiety and Depression Scale Dataset

The Italian version of the Hospital Anxiety and Depression Scale (HADS; Zigmond and Snaith, 1983) consists of 14 polytomous items equally divided between the two dimensions of anxiety and depression.

TABLE 1.6

RLMS Dataset: Distribution of the Categorical Covariates

Variable	Category	Proportion
Marital status (`marital`)		
	0 = living single	0.5190
	1 = living in a couple	0.4810
Education (`education`)		
	0 = 0–6 years of education	0.0042
	1 = presecondary education	0.0127
	2 = some unfinished secondary education	0.1192
	3 = secondary education	0.3611
	4 = secondary professional education	0.2764
	5 = college and postgraduate	0.2264
Gender (`gender`)		
	0 = male	0.4986
	1 = female	0.5014
Working status (`work`)		
	0 = on leave	0.0726
	1 = currently working	0.9274

The items of the questionnaires, which are Italian-validated translations (Costantini et al., 1999) of the original HADS questionnaire, are the following:

$Y1$ = I can laugh and see the funny side of things.
$Y2$ = I get a sort of frightened feeling like butterflies in the stomach.
$Y3$ = I have lost interest in my appearance.
$Y4$ = I feel as if I am slowed down.
$Y5$ = I look forward with enjoyment to things.
$Y6$ = I get sudden feelings of panic.
$Y7$ = I get a sort of frightened feeling as if something bad is about to happen.
$Y8$ = Worrying thoughts go through my mind.
$Y9$ = I feel cheerful.
$Y10$ = I can sit at ease and feel relaxed.
$Y11$ = I feel restless and have to be on the move.
$Y12$ = I feel tense or wound up.
$Y13$ = I still enjoy the things I used to enjoy.
$Y14$ = I can enjoy a good book or radio or TV program.

The dataset* is referred to a sample of 201 oncology Italian patients, who were asked to fill in the questionnaire concerning the measurement of these

---

* The dataset is downloadable from https://sites.google.com/site/questionnaires book/datasets and is named HADS with extension .dta for Stata and .RData for R.

TABLE 1.7

HADS Dataset: Distribution of the Item Responses

Item	Response Category			
	0	1	2	3
Y2	0.3532	0.5274	0.0796	0.0398
Y6	0.3980	0.4627	0.0995	0.0398
Y7	0.4627	0.2239	0.2189	0.0945
Y8	0.1940	0.4925	0.2488	0.0647
Y10	0.0697	0.4080	0.4428	0.0796
Y11	0.3085	0.4975	0.1144	0.0796
Y12	0.3433	0.4627	0.1493	0.0448
Anxiety	0.3042	0.4392	0.1933	0.0633
Y1	0.4378	0.3284	0.1642	0.0697
Y3	0.5672	0.2985	0.0896	0.0448
Y4	0.3184	0.5473	0.1194	0.0149
Y5	0.4627	0.3881	0.1343	0.0149
Y9	0.0896	0.2786	0.5522	0.0796
Y13	0.4229	0.4229	0.1144	0.0398
Y14	0.3085	0.3731	0.2886	0.0299
Depression	0.3724	0.3767	0.2090	0.0419

two pathologies. All the items of the HADS questionnaire have four response categories: from 0 corresponding to the lowest perceived level of anxiety or depression to 3 corresponding to the highest level of anxiety or depression. Table 1.7 shows the distribution of the item responses among the four categories, distinguishing between the two dimensions.

Overall, responses are mainly concentrated in categories 0 and 1 for both anxiety and depression, whereas category 3, corresponding to the highest level of psycho-pathological disturbances, is selected less than 10% of the times for each item.

Exercises

1. Download the following datasets from the web:

   - lsat.dta downloadable from http://www.gllamm.org/lsat.dta. It contains in long format the responses of 1000 individuals to five binary items from the Law School Admission Test. Responses are stacked into variable resp, the five items are detected through the dummies i1-i5, and the number of individuals with the same response pattern is provided in variable wt2, whereas variable id identifies the response patterns; for more details, see Rabe-Hesketh et al. (2004, Chapter 4).

- mislevy.dat downloadable from http://www.gllamm.org/ books/mislevy.dat. It contains the responses to four binary items relating to one's ability in mathematics (variables y1-y4); in addition, the number of white males (covariate cwm), the number of white females (covariate cwf), the number of nonwhite males (covariate cbm), and the number of nonwhite females (covariate cbf) are provided for each response pattern; for more details, see Skrondal and Rabe-Hesketh (2004, Chapter 9.4).

- aggression.dat downloadable from http://www.gllamm.org/ aggression.dat. It contains in long format the responses of 316 individuals to 24 ordinal items relating to verbal aggression and rated as 0, 1, and 2. Responses are stacked into the variable Y, the 24 items are detected through the dummies I1-I24, and each individual is identified by variable PERSON. In addition, a discrete score about anger (covariate Anger) and the information about gender (covariate Gender with category 0 for females and 1 for males) are provided; for more details, see De Boeck and Wilson (2004a, Chapter 1.2).

- delinq.txt downloadable from http://www.gllamm.org/ delinq.txt. It contains the responses of 6442 individuals to six ordinal items relating to juvenile delinquency and rated as 0, 1, 2, and 3 (columns 2–6), other than the information about gender (column 1); for more details, see Rabe-Hesketh et al. (2004, Chapter 8.4).

For each of the aforementioned datasets, perform the following operations through Stata:

(a) Discuss the data structure and, if necessary, reshape it in a suitable way in order to perform descriptive analyses (see the Stata help for function reshape).

(b) Build the distribution with the absolute and relative frequencies of the item responses for each item and each response category.

(c) Build the absolute and relative frequency distributions of the covariates (if present).

(d) Build the conditional distributions of the item responses given the covariates (if present).

(e) On the basis of results obtained at point (d), discuss the differences in the item responses among individuals having different characteristics (e.g., males against females).

2. Consider the following datasets:

- Scored available in the R package irtoys
- Science available in the R package ltm

We recall that, once a package has been installed in R, it is possible to load it by the command `require()` and to extract the data there contained by the command `data()`.

Then, with reference to the dataset Scored, containing binary responses, perform the following operations in R:

(a) Find the number of respondents and of test items.

(b) Compute the number of correct responses provided by each respondent.

(c) Compute the sum of correct responses to each item.

(d) Compute the variance indices illustrated in this chapter.

Regarding the dataset Science:

(a) Indicate the number of response categories for each item.

(b) Indicate the number of respondents and of test items.

(c) Find the distribution of the response categories for each item (with observed percentages of responses for each category).

(d) Show how it is possible to obtain a dichotomized version of this dataset by scoring the first two response categories as 0 and the other two categories as 1.

2

Classical Test Theory

2.1 Introduction

In this chapter, we first examine the theory of reliability from the perspective of classical test theory (CTT). In fact, CTT relies on a few basic ideas that have important implications regarding the variability between test scores and the way reliability of a test is conceptualized.

After a discussion on the foundations of CTT and the theoretical basis of reliability, this chapter focuses on the procedures actually used to determine reliability with real test data. Our illustration goes from the alternate forms and test–retest methods to the split-half method and to the various measures employed within the internal consistency approach, such as the raw α-coefficient, the standardized α-coefficient, and the related coefficients for binary items.

Next, within the same classical perspective on reliability, procedures used to estimate true scores are described, together with some indices commonly used in item analysis to describe the distribution of the responses to a single item and the relationship between responses to an item and a criterion of interest.

The remainder of this chapter is devoted to test validity. An in-depth description of the issues related to test validity is outside the focus of this book, as evidence on this aspect often relies on the expert's subjective interpretation rather than on statistical measures. Hence, the presentation of test validity will be restricted to a brief description of its conceptual bases and will be focused on the statistical procedures employed to assess it. This chapter also deals with test bias and the procedures typically employed to detect it in the context of CTT, which also include factor analysis and regression analysis.

Finally, we deal with extensions of CTT aimed at overcoming some of its drawbacks. In particular, the theoretical framework of generalizability theory is introduced and the steps of a generalizability analysis are described.

It is important to note that, throughout this chapter, we refer to indicators and methods that have their usual statistical application with quantitative variables. However, in the CTT setting, they are applied to any type of variable, such as scores to dichotomous and polytomous items.

This chapter is organized as follows. The following section illustrates basic concepts about CTT and Section 2.3 introduces some CTT models. Concepts about reliability are introduced in Sections 2.4 and 2.5, whereas related estimation methods are discussed in Section 2.6. Then, methods to estimate the individual true score, which is one of the main quantities of interest in CTT, are outlined in Section 2.7, whereas item analysis is discussed in Section 2.8. Validity, item bias, and generalizability are dealt with in Sections 2.9 through 2.11, respectively. Finally, examples in Stata and R are illustrated in Section 2.12.

2.2 Foundation

The basic concepts of CTT are those of the observed score, measurement error, and true score. *Observed scores* are simply the values obtained from the measurement of some characteristics of a person (e.g., mathematical ability of a student, intelligence quotient of a respondent) as opposed to the true amounts of those characteristics. *Measurement errors* are due to temporary factors affecting the measurement quality of physical and psychological attributes. Sources of error include guessing, administration problems, wrong scoring, and temporary physical and/or psychological respondents' conditions. Moreover, the *true score* of a subject on a test may be conceptualized as the average score obtained by that subject in hypothetical independent repeated measurements with a test specified in advance (Lord and Novick, 1968; Raykov and Marcoulides, 2011).

As an example, consider a test administered many times to the same subject in such a way that a large number of statistically independent and identically distributed measurements are obtained. In addition, assume that such measurements are performed within a short time period so that no change in the underlying construct occurs. The true score is the average of the measurements made in this way (Lord and Novick, 1968; Zimmerman, 1975; Raykov and Marcoulides, 2011). The resulting distribution is called a *propensity distribution*, and it is both subject and test specific. It is important to outline that the theory of true and error scores developed over multiple sampling of the same person also holds (Allen and Yen, 1979) for a single administration of an instrument over multiple persons. This last interpretation of true score allows us to consider it as a random variable instead of a fixed quantity.

According to CTT, a person's observed score on a test is an additive function of the true score and the measurement error (or error score; Gulliksen, 1950; Lord and Novick, 1968; Furr and Bacharach, 2008), expressed in the form

$$Y_{i\cdot} = S_i + \eta_i, \quad i = 1, \ldots, n. \tag{2.1}$$

Since categorical (binary or polytomous) items are treated here, then observed score $Y_{i\cdot}$ consists in counts. Nevertheless, the measurement error η_i, as well as the true score S_i, is usually assumed to have a continuous distribution (typically normal) in the CTT framework. This assumption is introduced for simplicity reasons and may be easily motivated in the presence of many test items.

As mentioned previously, random errors affect an individual observed score because of pure chance, making his or her observed scores either higher or lower in comparison to the true score. As errors may affect an examinee's score in either a positive or negative direction, the average effect of errors across test administrations is zero, that is, $E(\eta_i) = 0$ for all i, so that

$$E(Y_{i\cdot}) = E(S_i), \quad i = 1, \ldots, n.$$

Also, it follows logically from the definition of true score that the process of taking the mean (or expectation) of the propensity distribution cancels out the effect of random measurement errors. In other terms, true scores are independent of errors, and thus, the correlation between the corresponding random variables is zero, $\mathrm{Cor}(S_i, \eta_i) = 0$. Consequently, we have

$$V(Y_{i\cdot}) = V(S_i) + V(\eta_i), \quad i = 1, \ldots, n. \tag{2.2}$$

2.3 Models

The concept of true score postulates that for each test of a set of, say, T tests, any subject has a true score. This, in turn, implies the existence of the expectation used to define this score. To stress that we are concerned with a set of tests instead of a single test, Equation 2.1 can be rewritten as follows:

$$Y_{i\cdot}^{(t)} = S_i^{(t)} + \eta_i^{(t)}, \quad i = 1, \ldots, n, \ t = 1, \ldots, T, \tag{2.3}$$

where $S_i^{(t)}$ refers to the true score at test t, $Y_{i\cdot}^{(t)}$ to the corresponding observed score, and $\eta_i^{(t)}$ to the measurement error in this test.

The way true scores in different tests are related is of interest. Even if based on Equation 2.3, the model of parallel tests, the model of true score equivalent tests (τ-equivalent test), and the model of congeneric tests rely on different assumptions.

The *model of parallel tests* asserts that tests are equivalent when they measure the same underlying construct and, thus, for all T tests, the examinee has the same true score. As a consequence, all differences between their observed scores come from differences in their error scores that are also assumed with constant variance. Therefore, two tests, say, test 1 and test 2, are considered parallel when the following assumptions hold:

1. The tests measure the same latent construct, and hence, the respondents' true scores in one test are the same as the true scores on the other test, that is, $p(S_i^{(1)} = S_i^{(2)}) = 1, i = 1, \ldots, n$.

2. The tests have the same level of error variance, that is, $V(\eta_i^{(1)}) = V(\eta_i^{(2)}), i = 1, \ldots, n$.

We observe that unlike Equation 2.3, which is always true (as long as the mathematical expectation of the observed score exists), the last two conditions are indeed assumptions. These assumptions define the model of parallel tests.

The model of parallel tests is based on two strong and restrictive assumptions (the tests measure the same true score with the same precision), which limit its applicability in empirical works, making it difficult to construct tests that fulfill such assumptions. A generalization of the model of parallel tests, the *model of true score equivalent tests*, also called the τ-equivalent test model, assumes that the T tests assess the same true score but with different error variances and, thus, different precisions.

The assumption of equal true scores of a set of tests shared by the model of parallel tests and the model of true score equivalent tests is still strong as it requires that the tests measure the same construct with the same unit of measurement. However, units of measurement are often arbitrary, because they do not have any particular meaning, and they are also test specific, being distinct across tests. The assumption of equality of measurement units is relaxed within the *model of congeneric tests*, which, different from the previous models, assumes that the true scores at different tests are linearly related to one another so that any test true score can be expressed as a linear combination of any other test true score.

To summarize, all three of the aforementioned CTT models assume a single underlying construct or latent dimension that is assessed by the set of T tests. However, these models are nested, with the model of parallel tests as the most restrictive and the model of congeneric test as the least restrictive one.

2.4 Conceptual Approaches of Reliability

CTT conceptual ideas describe properties of true scores and error scores that have important consequences for the study of reliability. According to CTT (Lord and Novick, 1968), *reliability* ρ_{ys}^2 is the most important concept related to the general quality of test scores. It represents the overall consistency of a measurement. In other words, a measure has a high reliability if it produces similar results under similar conditions. Thus, reliability depends on the extent to which differences in respondents' observed scores are consistent with differences in their true scores. More specifically, the degree of reliability of a test depends on (1) the extent to which differences in observed test scores can be attributed to real differences and (2) the extent to which such differences depend on measurement errors (Gulliksen, 1950; Zimmerman and Williams, 1982; Furr and Bacharach, 2008). In principle, there is no perfectly reliable measure since there are many possible sources of error that might affect the observed scores, inflating or deflating them and obscuring, at least partially, real differences between the true scores.

In CTT, there are two ways of thinking about reliability, that is, in terms of proportion of variances or correlation. We refer to them as theoretical frameworks because both rely on the concepts of true score and measurement error that, by definition, are unknown.

The first way considers reliability as the ratio between true score variance and observed score variance, that is,

$$\rho_{ys}^2 = \frac{V(S_i)}{V(Y_{i\cdot})} = 1 - \frac{V(\eta_i)}{V(Y_{i\cdot})}. \tag{2.4}$$

Note that the index i is left unspecified because we have the same value of reliability for all i. This is obviously true when the random variables S_i, $Y_{i\cdot}$, and η_i are identically distributed and this is compatible with the assumption that the observed data correspond to a random sample of units independently drawn from the same population. The same considerations apply throughout this chapter.

Commenting on Equation 2.4, reliability is seen as the proportion of the observed score variance that corresponds to the true score variance. As such, it ranges between 0 and 1. If ρ_{ys}^2 equals zero, all respondents have the same true score, they do not differ in the characteristics being assessed by the questionnaire, and the test reliability is null. On the contrary, when the true score variance and observed score variance are equal, then $\rho_{ys}^2 = 1$ and reliability is perfect. Even though a perfectly reliable test does not exist in practice as measurement errors affecting the observed scores always occur, one may expect a good test to have a reliability of at least 0.70.

According to the second approach, based on correlation, reliability is the degree to which differences in the observed score reflect differences in the true score. This is expressed in terms of the squared correlation between the observed and true scores or in terms of the complement to 1 of the squared correlation between the observed and error scores, that is,

$$\rho_{ys}^2 = \text{Cor}(Y_{i\cdot}, S_i)^2 = 1 - \text{Cor}(Y_{i\cdot}, \eta_i)^2. \tag{2.5}$$

A reliability of 1, as defined in (2.5), would indicate that differences between respondents' observed test scores are perfectly consistent with respondents' true test scores. We may also say that the test is unreliable to the degree that differences in observed test scores fully reflect differences in error scores. When observed scores moderately depend on error scores, reliability is high.

It is important to outline that a test with a high reliability does not necessarily imply that it is measuring what it is supposed to measure. This issue is explored by the validity concept; see Section 2.9.

2.5 Reliability of Parallel and Nonparallel Tests

As illustrated in Section 2.3, the model of parallel tests asserts that all T tests have the same true score. Thus, all differences among the observed scores for the same person i come from differences in the error scores that are also assumed to have the same variance. Consequently, observed scores on two parallel tests, denoted by 1 and 2, will have the same mean, $E(Y_{i\cdot}^{(1)}) = E(Y_{i\cdot}^{(2)})$, and variance, $V(Y_{i\cdot}^{(1)}) = V(Y_{i\cdot}^{(2)})$, for all i, and it will be possible to estimate reliability coefficients through the correlation between the observed scores of two parallel tests. In fact, the correlation between two parallel tests is equal to the reliability. This correlation is defined as

$$\text{Cor}\left(Y_{i\cdot}^{(1)}, Y_{i\cdot}^{(2)}\right) = \frac{\text{Cov}\left(Y_{i\cdot}^{(1)}, Y_{i\cdot}^{(2)}\right)}{\sqrt{V\left(Y_{i\cdot}^{(1)}\right) V\left(Y_{i\cdot}^{(2)}\right)}}. \tag{2.6}$$

If two tests are parallel, their observed scores have equal variance. Besides, $\text{Cov}(Y_{i\cdot}^{(1)}, Y_{i\cdot}^{(2)})$, the covariance between two composite measures, can be expressed as the sum of the covariances of the components within each measure. Hence, Equation 2.6 can be rewritten as

$$\frac{\text{Cov}\left(S_i^{(1)}, S_i^{(2)}\right) + \text{Cov}\left(S_i^{(1)}, \eta_i^{(2)}\right) + \text{Cov}\left(S_i^{(2)}, \eta_i^{(1)}\right) + \text{Cov}\left(\eta_i^{(1)}, \eta_i^{(2)}\right)}{V\left(Y_{i\cdot}^{(t)}\right)},$$

where t may be either equal to 1 or 2, because $V\left(Y_{i\cdot}^{(1)}\right) = V\left(Y_{i\cdot}^{(2)}\right)$. Given that error scores on one test are random, then they are uncorrelated with true scores and with error scores on the other test. Thus, we have that

$$\text{Cor}\left(Y_{i\cdot}^{(1)}, Y_{i\cdot}^{(2)}\right) = \frac{\text{Cov}\left(S_i^{(1)}, S_i^{(2)}\right)}{\text{V}\left(Y_{i\cdot}^{(t)}\right)}, \quad t = 1, 2. \tag{2.7}$$

Provided that the expected true scores of parallel tests are the same and hence the covariance between the true scores equals the variance of true scores (i.e., the variance is the covariance between a variable and itself), Equation 2.7 can be expressed as

$$\text{Cor}\left(Y_{i\cdot}^{(1)}, Y_{i\cdot}^{(2)}\right) = \frac{\text{V}\left(S_i^{(t)}\right)}{\text{V}\left(Y_{i\cdot}^{(t)}\right)}, \quad t = 1, 2, \tag{2.8}$$

which is our first definition of reliability, that is, $\text{Cor}\left(Y_{i\cdot}^{(1)}, Y_{i\cdot}^{(2)}\right) = \rho_{ys}^2$; see Equation 2.4. Therefore, under the assumptions of parallel tests, the correlation between total parallel test scores is equal to the reliability of the two tests.

When tests are not parallel, the correlation between the observed scores on the two measures, $\text{Cor}\left(Y_{i\cdot}^{(1)}, Y_{i\cdot}^{(2)}\right)$, is a function of two quantities: (1) the correlation between the true scores on the two measures, $\text{Cor}\left(S_i^{(1)}, S_i^{(2)}\right)$, and (2) the reliability of the two measures, $\rho_{ys(1)}^2$ and $\rho_{ys(2)}^2$, that is,

$$\text{Cor}\left(Y_{i\cdot}^{(1)}, Y_{i\cdot}^{(2)}\right) = \text{Cor}\left(S_i^{(1)}, S_i^{(2)}\right)\sqrt{\rho_{ys(1)}^2 \rho_{ys(2)}^2}, \tag{2.9}$$

where $\rho_{ys(t)}^2$ is the measure of reliability for test t, that is, the squared correlation between $Y_{i\cdot}^{(t)}$ and $S_i^{(t)}$; see definition (2.5).

Equation 2.9 has many important consequences for applied measurement. One of the most important implications for research is that observed correlations are always lower or weaker than true correlations (i.e., correlations between true scores) because measures are always affected by measurement errors and, therefore, are never perfectly reliable.

Another consequence is that it is possible to estimate true correlations through observed correlations and the reliability coefficients of two measures, rearranging Equation 2.9, so that

$$\text{Cor}\left(S_i^{(1)}, S_i^{(2)}\right) = \frac{\text{Cor}\left(Y_{i\cdot}^{(1)}, Y_{i\cdot}^{(2)}\right)}{\sqrt{\rho_{ys(1)}^2 \rho_{ys(2)}^2}}.$$

This equation provides a correction for attenuation (McDonald, 1999). If two measures are perfectly reliable, observed correlations would equal true correlations. Thus, it is possible to estimate the correlation that would be obtained if perfectly reliable measures had been used (Furr and Bacharach, 2008).

2.6 Procedures for Estimating Reliability

Since within CTT reliability is seen as the degree to which differences in observed scores are consistent with differences in true scores, one readily understands that, empirically, it cannot be determined because true scores (and their variances) are unknown. A way to overcome this limitation is based on determining the common reliability of two parallel tests by correlating them because the correlation between such tests equals reliability (Equation 2.8). However, as clarified previously, ensuring that two measures are truly parallel is very difficult in empirical research. Moreover, instead of trying to determine reliability, one can estimate it using data referred to a representative sample of subjects from a population of interest.

The reliability of a test can be estimated according to two main types of method, that is, methods based on repeated measures, such as the alternate form method and the test–retest method (Guttman, 1945), and methods based on a single measure, such as the split-half method and the internal consistency reliability method. These methods are described in more detail in the following.

2.6.1 Alternate Form Method

The *alternate form method* requires constructing two similar forms of a test, say, form 1 and form 2, with the same content, and administering them to the same group of examinees within a short period of time (Crocker and Algina, 1986). Half of the class receives the forms in the order first 1 then 2 and the other half in reverse order. The correlation between the observed scores on form 1 and form 2, called the *coefficient of equivalence*, is taken as an estimate of test reliability, that is,

$$\hat{\rho}_{ys}^2 = \mathrm{cor}^{(1,2)}, \tag{2.10}$$

where $\mathrm{cor}^{(1,2)}$ is defined as in Equation 1.4. The higher the coefficient of equivalence, the more confident a researcher can be that scores from the different forms are usable interchangeably. A value equal or higher than 0.70 is acceptable.

There are a number of drawbacks related to this method that limits its applicability. First, one can never know whether two forms of a test are exactly equivalent because different forms might include different contents and imply different constructs.

Second, if two forms of a test are administered simultaneously, some of the errors affecting the observed scores on form 1 could carry over and affect the observed scores on form 2 (e.g., temporary physical and psychological conditions of the respondents being assessed might not change over two testing occasions that are close in time). In such cases, error scores from the two forms would be correlated and this would violate the assumption of randomness of errors.

Each of the limitations here discussed implies that the observed score correlation between two forms of a test is not an accurate estimate of reliability.

2.6.2 Test–Retest Method

A different method to estimate reliability consists in administering the same test to the same group of respondents at more than one occasion. In fact, test users might be interested in knowing how consistently examinees respond to the same test at different time occasions, assessing their scores at a first test occasion and retesting them at a second occasion.

The *test–retest method* avoids the problem of differing content that arises with the alternate form procedure. However, it assumes that respondents' true scores remain stable across the two testing occasions. Remember that, according to CTT, true scores of respondents taking the test at the first occasion must be the same as true scores of respondents taking the test at the second occasion. There are some factors affecting the sensibility of this stability assumption, such as the length of the test–retest interval. Thus, true scores are likely to change across long periods of time. On the contrary, short intervals suffer from carryover effects. In addition, changes are more likely to occur in certain periods in an individual's life. For example, changes in cognitive skills and knowledge are more likely to occur during the school-age period than in later periods.

To summarize, if the stability assumption holds, then the test–retest correlation (i.e., the correlation between the observed scores at the two occasions) is an appropriate estimate of reliability. In practice, $\hat{\rho}^2_{ys} = \text{cor}^{(1,2)}$ as in (2.10), where $\text{cor}^{(1,2)}$ is defined as in Equation 1.4. Therefore, an imperfect correlation (not equal to one) between the observed scores at the two testing occasions would only point to the degree to which measurement errors affect observed scores. On the other hand, when the stability assumption is not reasonable, an imperfect correlation would be the result of two factors that could not be disentangled: the degree to which the measurement error affects the observed scores and the amount of change in the true score.

2.6.3 Split-Half Method

As we have seen, the alternate form and the test–retest methods rely on strong assumptions that may not be appropriate. More importantly, they suffer from several practical difficulties as they are expensive and time consuming.

Further practical alternatives to estimating reliability are through the split-half and the internal consistency reliability methods. Such procedures, designed to estimate reliability, require respondents to complete the test at one occasion and treat the different parts of the test (i.e., items or group of items) as if they were different tests.

The basic idea behind the methods at issue is that if the differences among the observed scores in a part of the test (or even a single item) are consistent with the differences among the observed scores in the other parts, then it is likely that the observed scores on the whole test are consistent with the true scores.

More precisely, the *split-half method* consists of dividing the items of a test into two subsets of the same dimension, with the aim of creating two parallel subsets. There is a number of conventional methods that can be used for dividing a test into halves (Crocker and Algina, 1986), such as

1. Assigning even-numbered items to form 1 and odd-numbered items to form 2 (or vice versa)

2. Ranking the items on the basis of their difficulty levels (see Section 2.8) and then assigning items with odd-numbered ranks to form 1 and items with even-numbered ranks to form 2 (or vice versa)

3. Assigning items to the two subsets randomly

The most common measure used to estimate reliability of the full test (i.e., the test composed by the two halves) within the split-half procedure is the *Spearman–Brown formula* (Spearman, 1904; Lord and Novick, 1968), that is,

$$\tilde{\rho}_{ys}^2 = \frac{2\mathrm{cor}^{(1,2)}}{1 + \mathrm{cor}^{(1,2)}}, \tag{2.11}$$

where, as usual, $\mathrm{cor}^{(1,2)}$ is defined as in Equation 1.4.

The Spearman–Brown formula allows us to overcome the limits of using a simple correlation, correcting for the full-length test; see Raykov and Marcoulides (2011), Section 6.2, for details. However, the appropriateness of the use of the split-half method to estimate reliability rests on the assumption that the two halves are parallel (e.g., the subsets should have the same means and variances of observed scores). Therefore, the greater the violation of this assumption, the less accurate the results will be. Furthermore, there are many possible ways of dividing a test into halves. Thus, the method does not yield to a unique reliability estimate.

2.6.4 Internal Consistency Reliability Method

An alternative to prevent the drawbacks of the split-half method is represented by the *internal consistency reliability method*. This method is conventionally defined as an item-level approach, because it considers each item of a test as a separate test. The complete test is administered once to a sample of examinees. Subsequently, covariances are calculated between all the pairs of items. The most well-known measures of internal consistency reliability are the raw α-coefficient (Cronbach's α) and the standardized α-coefficient (generalized Spearman–Brown formula).

The *raw α-coefficient* or *Cronbach's α* (Cronbach and Meehl, 1955) is given by the following formula:

$$\alpha = \frac{J}{J-1} \frac{\sum_{j_1=1}^{J} \sum_{j_2=1, j_2 \neq j_1}^{J} \text{Cov}(Y_{ij_1}, Y_{ij_2})}{V(Y_{i\cdot})}, \tag{2.12}$$

where $\text{Cov}(Y_{ij_1}, Y_{ij_2})$ is the pairwise covariance between the pair of items (j_1, j_2), with $j_1 \neq j_2$, that is,

$$\text{Cov}(Y_{ij_1}, Y_{ij_2}) = \text{E}\left[(Y_{ij_1} - \mu_{j_1})(Y_{ij_2} - \mu_{j_2})\right],$$

where $\mu_j = \text{E}(Y_{ij})$ and, as clarified earlier, does not depend on i, as the random variables are identically distributed across individuals.

When a test measures one construct, then the items of a test would positively covary with each other. In other terms, if two items were good measures of the same construct, they should have a positive covariance. Thus, the sum of the interim covariances reflects the degree to which responses to all the items are generally consistent with each other.

Equation 2.12 can also be expressed in the following form:

$$\alpha = \frac{J}{J-1} \left(1 - \frac{\sum_{j=1}^{J} V(Y_{ij})}{V(Y_{i\cdot})}\right), \tag{2.13}$$

where $V(Y_{ij})$ denotes the variance of scores on item j, that is, $V(Y_{ij}) = \text{E}[(Y_{ij} - \mu_j)^2]$, which is constant with respect to i. Equations 2.12 and 2.13 provide the same piece of information, leading to the same numerical value for the Cronbach's α-coefficient, and differ only with respect to the way they are computed, requiring to calculate the covariances $\text{Cov}(Y_{ij_1}, Y_{ij_2})$ or, alternatively, the variances $V(Y_{ij})$.

In practice, the coefficient at issue is estimated as

$$\hat{\alpha} = \frac{J}{J-1} \frac{\sum_{j_1=1}^{J} \sum_{j_2=1, j_2 \neq j_1}^{J} \text{cov}_{j_1 j_2}}{v_{\cdot\cdot}},$$

where the numerator is based on the covariance between any pair of items and the denominator is the sample variance of the individual scores, as defined in expressions (1.3) and (1.2), respectively. An equivalent expression, based on (2.13), is

$$\hat{\alpha} = \frac{J}{J-1}\left(1 - \frac{\sum_{j=1}^{J} v_{.j}}{v_{..}}\right),$$

with $v_{.j}$ defined in (1.1).

Note that, in the case of binary items, the item variance $V(Y_{ij})$ simplifies to $\mu_j(1 - \mu_j)$ and, therefore, the raw α-coefficient defined in (2.13) reduces to the so-called *Kuder–Richardson 20 coefficient* (KR20; Kuder and Richardson, 1937; Lord and Novick, 1968):

$$\text{KR20} = \frac{J}{J-1}\left[1 - \frac{\sum_{j=1}^{J} \mu_j(1 - \mu_j)}{V(Y_{i.})}\right],$$

which is computed by substituting the corresponding sample quantities.

Another reliability measure is the *standardized α-coefficient*. This measure, also called the *generalized Spearman–Brown formula*, is defined as in expression (2.12) and applied to standardized items, that is, items with a mean score equal to 0 and a variance equal to 1. After some simple algebra, we have

$$\alpha^{st} = \frac{J\overline{\text{Cor}}}{1 + (J-1)\overline{\text{Cor}}}, \tag{2.14}$$

where $\overline{\text{Cor}}$ is the average of the correlations between any pair of items, that is,

$$\overline{\text{Cor}} = \frac{1}{J(J-1)} \sum_{j_1=1}^{J} \sum_{\substack{j_2=1 \\ j_2 \neq j_1}}^{J} \text{Cor}(Y_{ij_1}, Y_{ij_2}).$$

If the average correlation is null, then the numerator of (2.14) is zero and $\alpha^{st} = 0$, whereas the closer α^{st} to 1, the higher the test reliability. If $\alpha^{st} = 1$, all test items perfectly measure the same latent trait. Conventionally, a test is considered reliable when α^{st} is greater than 0.70. Note that such measures of internal consistency can be artificially inflated by the presence of a high number of items and the presence of very similar items (i.e., redundant items). Therefore, measures of internal consistency above 0.90 should be regarded with suspicion. On the other hand, low values (<0.60) indicate that items are very different from each other or are ambiguously defined.

Practically, the generalized Spearman–Brown coefficient is estimated as

$$\hat{\alpha}^{\text{st}} = \frac{J \overline{\text{cor}}}{1 + (J - 1)\overline{\text{cor}}},$$

where

$$\overline{\text{cor}} = \frac{1}{J(J - 1)} \sum_{j_1=1}^{J} \sum_{\substack{j_2=1 \\ j_2 \neq j_1}}^{J} \text{cor}_{j_1 j_2}$$

and $\text{cor}_{j_1 j_2}$ means the estimated correlation between the score on item j_1 and the score on item j_2, that is,

$$\text{cor}_{j_1 j_2} = \frac{\text{cov}_{j_1 j_2}}{\sqrt{v_{\cdot j_1} v_{\cdot j_2}}}.$$

The generalized Spearman–Brown formula provides similar estimates as the raw α-coefficient, but it is often preferred to the latter because it relies only on correlations instead of relying on variances and covariances, which might be nontrivial concepts for a nontechnical audience.

Note that the α-coefficients and the KR20 represent very popular methods to estimate reliability, essentially because they are easier to apply and provide more accurate estimates of reliability than other methods. Indeed, it can be shown that the α-coefficients provide a lower bound for the reliability ρ_{ys}^2 under the usual assumption of uncorrelated errors. Moreover, under certain further conditions, defined by items having the same true score variance and item true scores perfectly correlated, the α-coefficients equal ρ_{ys}^2. These two conditions are known as the *essentially τ-equivalence* (Feldt and Brennan, 1989), and they differ from the parallel tests condition for not requiring the equality of error variances. Therefore, if test items are essentially τ-equivalent but not parallel, the internal consistency methods give accurate estimates of reliability, while split-half method does not.

2.7 True Score Estimation

Observed scores are estimates of the true score due to the presence of measurement errors. Individual observed scores can be used to define two kinds of true score point estimates: one-point estimate and second-point estimate, also called, the adjusted true score estimate.

The *one-point estimate* is the test score observed at one testing occasion, that is,

$$\hat{s}_i = y_{i\cdot}, \quad i = 1, \ldots, n,$$

usually reported along with the true score confidence interval that allows us to account for estimation accuracy. Under the normality assumption of observed scores, this confidence interval around an individual estimated true score can be expressed as

$$y_{i\cdot} \pm z_{\alpha/2}\, \hat{se}(\eta_i), \quad i = 1, \ldots, n, \tag{2.15}$$

where $z_{\alpha/2}$ stands for the quantile of the standard normal distribution, $1 - \alpha$ denotes the confidence level, and $\hat{se}(\eta_i)$ is the estimated standard deviation of the measurement error, which is given by

$$\hat{se}(\eta_i) = \sqrt{v_{..}\left(1 - \hat{\rho}_{ys}^2\right)}, \quad i = 1, \ldots, n.$$

Note that $\hat{\rho}_{ys}^2$ is the reliability estimate obtained according to one of the previously illustrated methods and it may also be substituted with $\tilde{\rho}_{ys}^2$.

The standard error of measurement represents the average size of the error scores or, in other terms, the average size of the deviation between respondents' observed scores from their true scores. Thus, the standard error of measurement is closely linked to its reliability. The higher the reliability of a test, the lower the standard error of measurement, and when $\hat{\rho}_{ys}^2 = 1$, then we have $\hat{se}(\eta_i) = 0$ for all i. Consequently, the reliability also affects the width of a confidence interval and the accuracy and precision of the test score estimator (i.e., highly reliable tests produce narrower confidence intervals than less reliable tests).

Different from the one-point estimate, the *second-point estimate*, also called the *adjusted true score estimate*, accounts for fluctuant factors (i.e., temporary physical or psychological conditions of respondents) that may affect an individual observed score at an occasion, acknowledging that an individual observed score at that occasion can be higher or lower than his or her score on the same test at a second testing occasion.

The adjusted true score estimate takes into account a specific effect, called the *regression to the mean*, according to which the difference between an individual score and the mean score is likely to be smaller upon a second testing occasion than upon a first testing occasion. That is, if an observed score is above (below) the mean at a first testing occasion, then the observed score on a second occasion is expected to be lower (upper) and closer to the mean.

Thus, it is possible to calculate the adjusted true score estimate, $\hat{s}_i^{\text{adj}}$, by using the individual observed score on a single testing occasion as follows:

$$\hat{s}_i^{\text{adj}} = \bar{y}_{..} + \hat{\rho}_{ys}^2 (y_{i.} - \bar{y}_{..}), \qquad (2.16)$$

where $\hat{\rho}_{ys}^2$ may be again substituted with $\tilde{\rho}_{ys}^2$. Note that the difference between the adjusted true score estimate $\hat{s}_i^{\text{adj}}$ and the one-point estimate $\hat{s}_i = y_{i.}$ depends on the estimated reliability of the test, so that when the reliability decreases, the difference between the two true score estimates increases.

2.8 Item Analysis

In constructing a test, the final set of items is usually selected through a process known as *item analysis* (Lord and Novick, 1968; Crocker and Algina, 1986). This term defines the analysis of the psychometric properties of each test item.

While reliability summarizes test quality by a single number, it does not provide information for evaluating each item. Therefore, other indices have to be considered, usually falling into two groups: (1) indices describing the response distribution to a single item and (2) indices describing the relationship between responses to an item and a criterion of interest, that is, internal consistency.

Indices that fall into the first group are, mostly, the item mean and variance, that is, $\bar{y}_{.j}$ and $v_{.j}$, respectively. We remind that, in the case of binary items, the item mean is also known as *item difficulty*, as it corresponds to the proportion of examinees who endorsed an item (see Section 1.7), so that the higher the item mean, the lower the item difficulty. Moreover, in this case the item variance is obtained as $\bar{y}_{.j}(1 - \bar{y}_{.j})$.

Internal consistency reflects the degree to which differences between individual responses to an item are consistent (and are correlated) with differences between responses to the other items. In test construction, internal consistency is usually evaluated, among other measures, through the *inter-item correlation matrix*, that is, a $J \times J$ matrix reporting the correlation coefficient between the scores of each pair of items.

Item discrimination is another common measure employed for evaluating the internal consistency of a test. Item discrimination is the degree to which an item differentiates respondents who obtain a high score on the whole test from those who score low. Thus, high discrimination values are preferred.

There are several methods to measure item discrimination. In the following, we remind the reader of the most commonly used:

- *α-if-item-omitted coefficient* (or *α-if-item-deleted*): It is obtained omitting an item at a time from the Cronbach's α formula. If α-if-item-omitted coefficient is greater than Cronbach's α calculated for the whole test, then removing that item from the test improves the test internal consistency; hence, the opportunity to keep that item in the final version of the test will be questioned.
- *Extreme group method*: For each item, the difficulty for the group of the best performers (e.g., all individuals with a total score in the upper 25–30%) is compared with the difficulty for the group of the worst performers (e.g., all individuals with a total score in the lower 25–30%). The larger the difference in the difficulties, the greater is the discrimination of that item.
- *Item-to-total correlations*: They represent a class of indices that measure the correlation between each item j score and the total test score or the corrected total test score (i.e., the total score calculated omitting item j). In the case of polytomous items (in particular with ordinal scoring), the Pearson product-moment correlation coefficient may be used. It is estimated as

$$\text{cor}_{\text{pm}}^{(j)} = \frac{1}{n} \frac{\sum_{i=1}^{n} \left(y_{ij} - \bar{y}_{\cdot j} \right) \left(y_{i\cdot} - \bar{y}_{\cdot\cdot} \right)}{\sqrt{v_{\cdot j} v_{\cdot\cdot}}}.$$

In the case of binary items (scored as 0 and 1), following Kline (2005), we suggest to distinguish the case of true dichotomous items (i.e., each item has only two response modalities) from the case of "false" dichotomies (i.e., items are originally polytomously scored and then one or more modalities are aggregated to obtain a binary scoring). In the first case, the item-to-total correlation is measured by the point biserial correlation coefficient (Lord and Novick, 1968), given by

$$\text{cor}_{\text{pbis}}^{(j)} = \frac{\bar{y}_{(j)} - \bar{y}_{\cdot\cdot}}{\sqrt{v_{\cdot\cdot}}} \sqrt{\bar{y}_{\cdot j}(1 - \bar{y}_{\cdot j})}, \tag{2.17}$$

where $\bar{y}_{(j)}$ is the mean score on the whole test for respondents who answered 1 on item j. In the case of "false" dichotomously scored items, the item-to-total correlation is measured by biserial correlation coefficient, obtained by dividing Equation 2.17 by $\phi(z_{(j)})$ and removing the second squared root, that is,

$$\text{cor}_{\text{bis}}^{(j)} = \frac{\bar{y}_{(j)} - \bar{y}_{\cdot\cdot}}{\sqrt{v_{\cdot\cdot}}} \frac{\bar{y}_{\cdot j}(1 - \bar{y}_{\cdot j})}{\phi(z_{(j)})}. \tag{2.18}$$

In this expression, $\phi(z_{(j)})$ is the value of the standard normal density function computed at the normal quantile of level $\bar{y}_{\cdot j}$.

All the mentioned item-to-total correlation coefficients range from −1 to +1. Large positive values indicate that respondents who answer 1 to an item tend to obtain also high scores on the overall test and vice versa.

The indices at issue may be accompanied by a plot of the item difficulties against the percentiles of the observed scores (Kline, 2005). This type of plot highlights how the performance on each item relates with the overall performance on the entire test. An example of such a plot is provided in Figure 2.3 with reference to the items of the Italian Institute for the Evaluational of the Education System (INVALSI) questionnaire.

2.9 Validity

The *validity* of a test is the extent to which a test measures what it is supposed to measure; therefore, a test is considered valid when its results are congruent with the objectives pursued by its use (Messick, 1989). Another perspective on validity considers this property as "the degree to which evidence and theory support the interpretations of test scores entailed by the proposed uses" (American Educational Research Association, 1999). Thus, validity is a matter of interpretation and degree in the sense that (1) validity concerns the interpretation of a measurement instrument and not the measurement instrument itself (i.e., a measure itself is neither valid nor invalid) and (2) a measure can be considered strongly or weakly valid (i.e., the alternative is not between fully valid or fully invalid measures).

We may distinguish between two main types of validity that concur to the psychometric judgment about the adequacy of a questionnaire:

- *Construct validity*: It concerns whether a test is truly measuring the construct for which the test is being used.
- *Criterion validity*: It concerns how well test scores predict a certain type of outcome (e.g., future performance or other types of score on the same construct).

Construct validity relies on the judgment of test developers and experts in the field, and therefore, it has not to be confused with the so-called face validity, which represents a subjective impression of a test content given by nonexperts, such as test takers. Construct validity is the degree to which test scores can be interpreted as reflecting a particular psychological construct (American Educational Research Association, 1999; Westen and Rosenthal, 2003). According to the standards for educational and psychological testing, the content of the test, its internal structure, and the association between test scores and other measures are among the most important factors affecting construct validity. More precisely, we may distinguish three main aspects of construct validity: (1) content validity, (2) convergent validity, and (3) discriminant validity.

First, validity evidence based on test content—also called *content validity*—stresses that, to be interpreted as valid, a test actual content should reflect the important facets of the construct being measured. On the other hand, validity evidence based on the test internal structure underlines that the actual structure of the test should match the structure that the test should have.

A different type of evidence requires that the association between the test score and other measures of the construct is close to an expected level of this association. A common distinction related to this type of evidence is between convergent validity and discriminant validity. *Convergent validity* refers to the degree to which test scores are associated with the scores of other tests with similar constructs. On the contrary, *discriminant validity* is the degree to which test scores are uncorrelated with tests measuring different constructs. In other words, with discriminant validity, one evaluates if a test is not accidentally measuring what it should not measure.

Finally, another relevant concept is that of criterion validity. In this regard, a useful distinction is between concurrent validity and predictive validity. *Concurrent validity* is the degree to which test scores are correlated with other variables (usually called *criterion variables*) measured at the same time as the first test. It compares test scores to other measures of the same construct (e.g., in the educational field, mathematical test results are sometimes compared to supervisor ratings of each candidate on the same construct). On the other hand, *predictive validity* is the degree to which test scores are correlated with other variables measured at a future point in time. It focuses on the predictive power of a test for future performance, such as school success and job success.

Many statistical procedures used to evaluate the association between test and criterion variables rely on correlation coefficients. As we have already seen, the correlation between two test scores can be expressed as a function of the true correlation between the two constructs and the reliabilities of the two tests; see Equation 2.9. Thus, to evaluate convergent and discriminant validity, experts rely on the correlation coefficient between the measure of interest and some variable referred to a similar construct. For example, one might be interested in assessing the validity of a mathematical test through the correlation between the test score itself and some other measures of this ability obtained by asking solution of specific problems.

It is clear that the procedures employed to evaluate test validity rest on the subjective judgment of experts and on the correlation coefficients between the test score of interest and criterion variables. Among the most important factors affecting these correlation coefficients, we recall measurement errors, which attenuate this correlation, test reliability, and criterion variable reliability. We have to clarify that a low correlation level is not always a sign of poor validity. This may happen when the relationship between the construct being measured by a test and the chosen criterion is not linear. For instance, consider the relationship between motivation and achievement: high levels of motivation may cause stress and, thus, adversely affect achievement levels, giving rise to a nonlinear relation between the two constructs.

Within CTT, the internal structure of a test is assessed through factor analysis (Lord and Novick, 1968; Gorsuch, 1983; Netemeyer et al., 2003). *Factor analysis* is a common statistical procedure that, in the present framework, can be used to assess the internal structure and dimensionality of a test. In fact, it allows us to identify the number of factors corresponding to different subsets of internally correlated items. Thus, when a test measures only one factor, all items within the test are correlated with each other, and hence, they reflect only one psychological attribute or construct. In this case, the test is said to be unidimensional because its items measure only one latent trait. On the other hand, whenever factor analysis gives evidence of multidimensionality, then it is useful (1) to study associations between dimensions and (2) to identify which items are linked to which factors.

In factor analysis, any item score Y_{ij} can be expressed as a function of latent factors and corresponding loadings (Gorsuch, 1983; Crocker and Algina, 1986; Netemeyer et al., 2003). This model is typically formulated for quantitative responses and it relies on the Bravais–Pearson correlation matrix. Therefore, to use it with categorical items, we need to adopt the appropriate correlation matrix: in the case of binary items, the tetrachoric correlation matrix may be used, whereas the polychoric correlation coefficients have to be computed in the case of polytomously scored items (Drasgow, 1988; Lee et al., 1995).

A common factor is a factor corresponding to a subset of correlated items. A unique factor is a factor corresponding to only one item, so that unique factors do not account for the correlation between item responses. In factor analysis, the choice of the number of factors, or dimensions, is conventionally based on the eigenvalues of the correlation or variance–covariance matrix between items. A common procedure is to examine a screeplot, that is, a graphical representation of eigenvalues (for an example see Figure 2.4). The number of factors is identified as the value on the x-axis corresponding to the largest drop in eigenvalue values (y-axis). Another common rule is to choose the number of factors on the basis of the values of the eigenvalues themselves, such as 1, when these eigenvalues are based on a correlation matrix.

2.10 Test Bias

Test bias is conventionally defined in terms of construct bias and predictive bias. When the relationship between true and observed scores on a single test is systematically different for different groups of respondents (e.g., males and females), then the test is said to show a *construct bias*. On the other hand, a *predictive bias* is about the relationship between the scores attained at two tests: an outcome or criterion test and a predictive test (i.e., an instrument that can be used to predict scores on the outcome test). When the relationship between true scores of a predictive test and the observed scores of an outcome test differs systematically for two or more groups, the first one suffers from a predictive bias.

The existence of differences between observed scores of different groups is a necessary but not sufficient condition to state that the test is biased. A second condition for a test to be biased is that differences in the observed responses given by examinees belonging to different groups are not related to the psychological construct measured by the test, that is, the true score (Jensen, 1980; Furr and Bacharach, 2008). As such, test bias is a theoretical concept just like reliability and validity. Because it relies on the difference between true and observed scores between groups of subjects, test bias cannot be measured directly, though it can be somehow estimated.

There are a number of methods to detect a construct test bias. For example, a method consists in calculating the discrimination measures for each item (see Section 2.8), separately for the different groups (e.g., females and males). When the discrimination indices calculated for every item within each group are about the same, then the items are unlikely biased; otherwise, we conclude that the behavior of certain items differs between groups.

Rank ordering of items based on their difficulty indices, separately for the different groups, is a further conventional procedure for test bias analysis. In the case of two groups, once item ranks have been obtained ordering the corresponding observed item scores, then the *Spearman rank-order correlation coefficient* is obtained as

$$\text{cor}_s = 1 - \frac{6 \sum_{j=1}^{J} d_j^2}{J(J^2 - 1)}, \tag{2.19}$$

where d_j is the difference between the ranks of item j in the first and second groups. If cor_s is less than 0.90, then the test may be suspected of being biased.

Another common method to study test bias is through factor analysis. Again, the procedure consists in performing a factor analysis separately for the groups of interest. If the analysis shows the existence of one factor within each group, then the test is unlikely biased because it implies an unidimensional structure and, therefore, a unique common construct.

Within the CTT framework, a predictive bias is usually assessed through regression analysis. Suppose that a test on reading abilities is administered to a large sample of pupils and that one is interested in checking if it is biased with respect to respondents' gender. Assume also that the two groups of males and females have a similar size and that the reading competencies are assessed through direct rating (outcome test scores). The procedure begins estimating the regression model for the ratings as a function of the reading observed test scores on the whole sample of pupils. This equation is called *common regression equation* and provides information on the capability of test scores (e.g., the scores observed on the reading test) to predict the outcome test scores (e.g., the ratings).

To assess if the common regression equation is equally applicable to males and females, the procedure goes further by comparing group-level regression equations, that is, regression equations estimated separately for males and females, with the common regression equation. If the group-level equations do not overlap with the common regression equations, then the test is likely biased. A test is said to show intercept bias when the size of the difference between the group-level equations and the common equations is constant across test scores, that is, one would get parallel regression lines for males and females. On the other hand, a test is said to show slope bias when the regression equations for the whole sample and the groups differ in their slopes. Most commonly, biased tests show both intercept and slope significant differences.

2.11 Generalizability Theory

In the context of CTT, the total variance of observed scores $V(Y_i.)$ is a function of two components, the true score variance and the error variance, as shown by Equation 2.2. Error variance is undifferentiated (Brennan, 2001), in the sense that it includes the effect of multiple sources of variation (the so-called facets), such as persons, raters, items, or time. *Generalizability theory* (Cronbach et al., 1963, 1972) is used to determine the reliability of a measurement, but unlike CTT, it accounts for more sources of variation and provides a decomposition of the observed score depending on the specific measurement situation (de Gruijter and van der Kamp, 2008). Its purpose is to quantify the amount of error caused by each facet and interaction between facets.

A *facet* is any different element involved in the measurement process. For instance, to measure a particular psychological construct, data might be collected in several ways, that is, involving an observer to rate respondents on a number of test items or entitling more than one observer, in order to get several independent measures on each item. In this example, items and observers represent different facets of the measurement strategy. A measurement method that involves only one observer would be defined as a one-facet design, the facet being the items of a test. On the other hand, a measurement method that involves more observers would be defined as a two-facet design.

Different facets have different effects on the quality of a measurement strategy. For example, it is possible that the items of a test are good measures of a psychological construct but different observers provide considerably different ratings or scores. The generalizability approach allows us to separate the effects of the multiple facets and to adjust measurement strategies accordingly.

From the perspective of generalizability theory, any facet is just a sample. For instance, items within a test are just a sample of the whole population of possible items that could be employed to measure a specific construct. Similarly, observers are only a sample of the universe of possible observers who might be recruited to conduct a study. Within the generalizability theory, the main issue is the degree to which scores obtained from a sample of items (and/or from a sample of observers) are representative of, and can be generalized to, the entire population of items. Thus, the main issue is the generalization from the observations to the appropriate universe of possible observations. This domain or universe is defined by all possible conditions of the facets of the study.

In generalizability theory, a distinction is made between a generalizability or G-study and a decision or D-study. A *G-study* is intended at identifying the degree to which the various facets affect the generalizability of the scores. A *D-study* aims at making decisions about future measurement strategies on the basis of the information provided by a G-study.

In practice, a G-study examines the factors affecting observed score variance, and the analysis of variance (ANOVA; de Gruijter and van der Kamp, 2008) is typically used to estimate variance components for each factor affecting the observed scores. In this respect, we remind that ANOVA can be used with a quantitative dependent variable. Therefore, it can be applied in the presence of quantitative items or, at least, polytomous ordinal items with enough modalities. A D-study estimates *coefficients of generalizability* that represent the degree to which differences among observed scores are consistent with the differences that would be obtained if the universe of all possible observations were used.

In multiple-facet designs, where more than one facet is involved, ANOVA is used to estimate variance components, as in one-facet design. However, multiple-facet designs add in the complexity of the components affecting variability. For instance, a two-facet design, where item and observer are the two facets, includes seven components: person, item, observer, person × item, person × observer, item × observer, and residual components. For a detailed and comprehensive description of such complex designs, see Cronbach et al. (1963), Brennan (2001), and de Gruijter and van der Kamp (2008).

2.12 Examples

In the following, we illustrate how to perform the CTT analyses described previously through the software Stata and R; the illustration is based on some of the datasets described in Chapter 1. First, we rely on the data collected by the grammar section of the INVALSI Italian test, characterized by a unidimensional set of binary items, and then we focus on the Russian Longitudinal Monitoring Survey (RLMS) data about job satisfaction, which was collected by a set of ordinal polytomous items. The first example is performed in `Stata`, whereas the second is based on the R software.

2.12.1 INVALSI Grammar Data: A Classical Analysis in `Stata`

After loading the INVALSI full dataset

```
. use INVALSI_full.dta, clear
```

we focus on the 10 items c1-c10 of the grammar section, here renamed Y1-Y10 to be coherent with the notation adopted along the book

```
. rename c* Y*
```

Then, with regard to the reliability of the questionnaire, we perform the split-half method and compute some internal consistency reliability measures.

As outlined in Section 2.6.3, the split-half method requires to divide the set of items into two halves, according to some possible criteria. A possible criterion consists in assigning even-numbered items to the first subset (form 1) and odd-numbered items to the second (form 2). Then the observed score is calculated separately for each test form and the reliability is estimated through the Spearman–Brown formula, defined in (2.11), as follows:

```
. * Standard correlation between scores
. gen Y_obs_even = Y2+Y4+Y6+Y8+Y10
. gen Y_obs_odd = Y1+Y3+Y5+Y7+Y9
. corr Y_obs_even Y_obs_odd
(obs=3774)
             | Y_obs_~n Y_obs_~d
-------------+------------------
  Y_obs_even |   1.0000
   Y_obs_odd |   0.4188   1.0000

. * Spearman-Brown correlation coefficient
. scalar rel = 2*r(rho)/(1+r(rho))
. display rel
.59037532
```

The estimated reliability is not completely satisfactory, being equal to 0.59.

Another possible criterion to split the overall set of items consists in (1) calculating the proportion of responses scored by 1 (i.e., correct answers) for each item, (2) reordering the items according to these proportions, and then (3) assigning even-numbered items to form 1 and odd-numbered items to form 2. For this, we first create a matrix with 10 rows (one for each item) and

2 columns indicating the item (column c1) and the corresponding proportion of responses 1 (column c2), respectively:

```
. matrix m1=J(10,2,0)
. forvalues num=1/10 {
  quietly summarize Y'num'
  scalar mean'num' = r(mean)
  matrix m1['num',1] = 'num'
  matrix m1['num',2] = mean'num'
  }
```

Then we use `Mata` to sort the matrix according to column 2

```
. mata
: st_matrix("m1", sort(st_matrix("m1"), 2))
: end
```

thus obtaining the new ranking of items:

```
. matrix list m1
m1[10,2]
           c1         c2
r1          4   .65315315
r2          6   .72760996
r3         10   .76603074
r4          2   .78431373
r5          7   .78881823
r6          3   .85744568
r7          5   .92077371
r8          1   .93561208
r9          8   .94859565
r10         9   .96396396
```

The estimation of the reliability coefficient continues in a similar way as previously described:

```
. * Standard correlation between scores
. gen Y_obs_reven =   Y6+Y2+Y3+Y1+Y9
. gen Y_obs_rodd = Y4+Y10+Y7+Y5+Y8
. corr Y_obs_reven Y_obs_rodd
(obs=3774)
             | Y_~reven Y_o~rodd
-------------+------------------
Y_obs_reven |   1.0000
 Y_obs_rodd |   0.4182   1.0000

. * Spearman-Brown correlation coefficient
. scalar rel2 = 2*r(rho)/(1+r(rho))
. display rel2
.58980838
```

We obtain a value of reliability very close to that previously obtained, confirming the not completely satisfactory level of reliability of the test.

In `Stata`, a quick alternative to the split-half method is represented by the internal consistency reliability method, consisting of calculating α-coefficients. More precisely, the Cronbach α-coefficient, see Equations 2.12 and 2.13, is obtained as

```
. * Cronbach's alpha
. alpha  Y1-Y10

Test scale = mean(unstandardized items)

Average interitem covariance:      .0149801
Number of items in the scale:            10
Scale reliability coefficient:       0.5702

. * Store Cronbach's alpha
. scalar alpha=r(alpha)
. display alpha
.57019487
```

and the standardized α-coefficient α^{st}, as defined in (2.14), is obtained by specifying the standardized item option, that is,

```
. alpha  Y1-Y10, std

Test scale = mean(standardized items)

Average interitem correlation:      0.1224
Number of items in the scale:           10
Scale reliability coefficient:      0.5825
```

In both cases, the reliability estimates are very close to those obtained with the split-half method.

The next step in the CTT analysis consists in determining the true score estimates for each individual. As outlined in Section 2.7, point estimates are given by the observed scores, which can be easily obtained by generating a new variable as follows:

```
. * Generate and summarize variable scores
. gen Y_obs = Y1+Y2+Y3+Y4+Y5+Y6+Y7+Y8+Y9+Y10
. summarize(Y_obs), detail

                              Y_obs
-------------------------------------------------------------
        Percentiles     Smallest
  1%         3               0
  5%         5               1
 10%         6               1        Obs              3774
 25%         7               1        Sum of Wgt.      3774

 50%         9                        Mean         8.346317
                           Largest    Std. Dev.    1.620859
 75%        10              10
 90%        10              10        Variance     2.627184
 95%        10              10        Skewness    -1.156607
 99%        10              10        Kurtosis     4.407204


* Mean score
. scalar mean_Y_obs = r(mean)
. display mean_Y_obs
8.3463169

* Standard deviation of scores
. scalar sd_Y_obs = r(sd)
. display sd_Y_obs
1.6208589
```

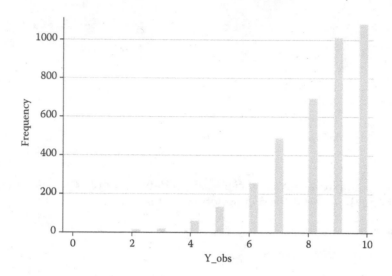

FIGURE 2.1
Histogram of observed scores.

As shown by the `Stata` output, the observed scores vary from 0 to 10 with a mean value of 8.346 and a standard deviation equal to 1.621. Moreover, through the histogram obtained as

```
. hist Y_obs, frequency
(bin=35, start=0, width=.28571429)
```

it can be further noted the negative skewness of the data (Figure 2.1), as also clearly outlined by the values of percentiles: note that the 25% of the best performers reach the maximum observed score.

Confidence interval estimates can be obtained by generating two new variables: a variable for the inferior limit (`l1`) and another for the superior limit (`l2`) of the interval. According to Equation 2.15, the limits of a 90% confidence interval may be calculated through the following commands, where `alpha` stands for the Cronbach's α previously obtained:

```
. * Limits of the 90% confidence interval
. gen l1 = Y_obs - 1.645*(sd_Y_obs*sqrt(1-alpha))
. gen l2 = Y_obs + 1.645*(sd_Y_obs*sqrt(1-alpha))
. * Display of the intervals for the first 5 subjects
. list Y_obs l1 l2 in 1/5
```

```
     +------------------------------+
     | Y_obs        l1          l2 |
     |------------------------------|
  1. |     4    2.251978    5.748022 |
  2. |     9    7.251978   10.74802 |
  3. |     8    6.251978    9.748022 |
  4. |     9    7.251978   10.74802 |
  5. |     8    6.251978    9.748022 |
     +------------------------------+
```

Other than the one-point estimate and the relative confidence intervals, we may also generate a further variable for the adjusted true score estimates as prescribed by Equation 2.16:

```
. gen Y_est = mean_Y_obs + alpha*(Y_obs - mean_Y_obs)
```

In our application, despite the low test reliability, the difference between the observed scores and adjusted true score estimates is slight. For instance, for the first five students, we have

```
. list Y_obs Y_est in 1/5

     +------------------+
     | Y_obs     Y_est |
     |------------------|
  1. |     4   5.868069 |
  2. |     9   8.719044 |
  3. |     8   8.148849 |
  4. |     9   8.719044 |
  5. |     8   8.148849 |
     |------------------|
```

The next step of our application is the item analysis (Section 2.8). As previously specified, in the case of binary items, item difficulties are represented by the proportion of individuals scored by 1 rather than 0. We may tabulate the difficulties of all 10 items by specifying option mean for the tabstat command (we also use option sd to calculate the standard deviation):

```
. tabstat Y1-Y10, column(statistics) stat(mean sd)

    variable |      mean        sd
-------------+--------------------
          Y1 |  .9356121  .2454752
          Y2 |  .7843137  .4113521
          Y3 |  .8574457  .3496641
          Y4 |  .6531532  .4760296
          Y5 |  .9207737  .2701274
          Y6 |    .72761  .4452485
          Y7 |  .7888182  .4082012
          Y8 |  .9485957  .2208503
          Y9 |   .963964  .1864045
         Y10 |  .7660307   .423409
----------------------------------
```

We may also visualize these quantities through a bar graph (Figure 2.2):

```
. graph hbar (mean) Y1-Y10, blabel(bar, format(%12.2f)) nolabel legend(off) showyvars
```

Overall, we observe that the difficulty distribution of the analyzed set of items is not completely satisfactory. In general, the highest discrimination levels are observed in correspondence with items having difficulty around 0.50, whereas much lower or higher levels of difficulty cannot differentiate individuals in a satisfactory way; in other words, when everyone passes (or fails) an item, it is equivalent to add a constant of 1 (or to add nothing)

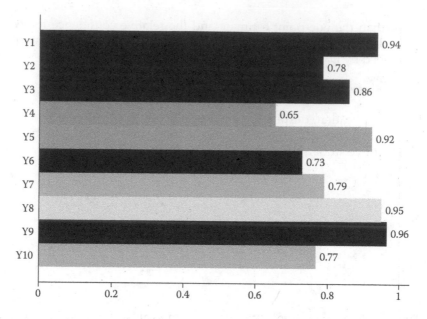

FIGURE 2.2
Horizontal bar plot of item difficulties.

to the observed scores to the other items. In light of this, we conclude that many items are too *easy*, as their difficulty is greater than 0.75. Also note that for items Y1, Y5, Y8, and Y9, the difficulty level is greater than 0.90.

As clarified at the end of Section 2.8, a useful instrument to evaluate difficulty and discrimination power of items is based on plotting item difficulties against the percentiles of the observed scores. As an example, in the following we provide the code to create this plot for item Y9, which presents the highest proportion of correct responses, in comparison to item Y4, having a proportion of correct responses much closer to 0.5. First, for item Y9, we calculate the difficulty levels corresponding to each decile of the observed score, so that the proportion of correct responses to item Y9 is provided for examinees who have observed scores less than each decile:

```
. quietly centile Y_obs, centile(10(10)100)
. forvalues cen=1/10 {
  scalar c`cen' = r(c_`cen')
  }
. forvalues cen=1/10 {
  quietly summarize Y9 if Y_obs <=c`cen'
  scalar mean9_`cen' = r(mean)
  display mean9_`cen'
  }
.86382114
.90316004
.92720764
.92720764
.94942358
.94942358
```

```
.94942358
.96396396
.96396396
.96396396
```

The same commands are repeated for item Y4 and may also be repeated for any other item. Thus, the results concerning the observed score percentiles can be saved (through copy-and-paste command) in a .dta file, named say, res1.dta, and then can be used to build the plot at issue for items Y4 and Y9 (Figure 2.3):

```
. use "res1.dta", clear
. label variable var1 "item Y9"
. label variable var2 "item Y4"
. label variable var3 "observed score percentiles"
. twoway connected var1 var2 var3, lpattern(solid dash) lcolor(black black)
mcolor(black black) msymbol(circle square) ylabel(#10) xlabel(#10)
```

As shown in Figure 2.3, the flat trend for item Y9 outlines that this item reaches immediately a difficulty level close to overall level of 0.96, and therefore, it does not discriminate appropriately between different ability levels. On the other hand, the difficulty level of item Y4 increases according to an approximately monotonic linear trend: as the overall performance of individuals on the test increases, the performance on item Y4 increases as well.

By specifying option item in function alpha, it is possible to obtain the point biserial correlations, as defined in (2.17), both on the total test

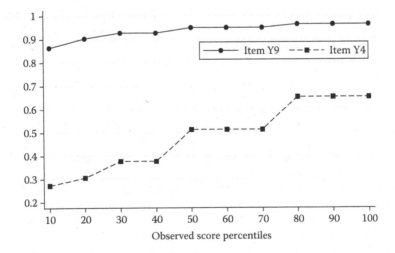

FIGURE 2.3
Plot of item difficulties (items Y4 and Y9) versus observed score percentiles.

score (column `item-test correlation` of the following output) and on the corrected total test score (column `item-rest correlation`) and Cronbach's α-if-item-omitted coefficients (column `alpha`):

```
. use INVALSI_full.dta, clear
. rename c* Y*
. gen Y_obs=Y1+Y2+Y3+Y4+Y5+Y6+Y7+Y8+Y9+Y10
. alpha  Y1-Y10, item

Test scale = mean(unstandardized items)
```

Item	Obs	Sign	item-test correlation	item-rest correlation	average interitem covariance	alpha
Y1	3774	+	0.3478	0.2050	.0165547	0.5563
Y2	3774	+	0.5469	0.3305	.0132956	0.5210
Y3	3774	+	0.4987	0.3103	.0142708	0.5292
Y4	3774	+	0.5298	0.2682	.0136646	0.5436
Y5	3774	+	0.4798	0.3361	.0149171	0.5299
Y6	3774	+	0.5604	0.3261	.012997	0.5221
Y7	3774	+	0.5464	0.3318	.0133114	0.5207
Y8	3774	+	0.3481	0.2205	.0166183	0.5547
Y9	3774	+	0.2650	0.1537	.0174661	0.5655
Y10	3774	+	0.3672	0.1132	.0167049	0.5877
Test scale					.0149801	0.5702

Note that the point biserial correlation coefficients here reported may also be obtained by function `pbis`, which in addition performs a *t*-test to evaluate the statistical significance of the estimated value. For instance, the point biserial correlation for item `Y1` is obtained as

```
. * Search and install function pbis; digit help(pbis) for the online help of function pbis
. findit pbis
. help(pbis)
. * Point biserial correlation coefficient
. pbis Y1 Y_obs

(obs= 3774)
Np= 3531  p= 0.94
Nq= 243  q= 0.06
-----------------+-------------------+------------------+------------------+
Coef.= 0.3478        t= 22.7814       P>|t| = 0.0001        df=   3772
```

where `Coef.` is the point biserial correlation, `t` is the test statistic, and `P>|t|` denotes the corresponding *p*-value.

The values of α-if-item-omitted coefficients obtained by command `alpha` are generally lower than Cronbach's α, with the exception of item `Y10`. Moreover, we observe positive medium–low values for point biserial correlation coefficients, likely due to the high item difficulties.

With regard to the biserial correlation coefficient defined in (2.18), `Stata` does not provide any function allowing for a direct calculation of this index. However, we may obtain it by using the point biserial correlations as follows:

```
* Biserial correlations
. forvalues num=1/10 {
    quietly pbis Y'num' Y_obs
    scalar z_p = invnormal($S_4/$S_3)
    scalar phi = normalden(z_p)
    scalar bis = $S_1*sqrt($S_5/$S_3)*sqrt($S_4/$S_3)/phi
    display "Item Y" 'num' ":  "  $S_1 "    " bis
  }
Item Y1:   .34777811     .67817908
Item Y2:   .54687015     .7683722
Item Y3:   .49859318     .77363662
Item Y4:   .52972867     .68297179
Item Y5:   .47969785     .87793013
Item Y6:   .56035908     .75117384
Item Y7:   .54633021     .7711625
Item Y8:   .34808204     .72900885
Item Y9:   .26497409     .62402818
Item Y10:  .3671458      .50702901
```

where the second column contains the point biserial correlation coefficient for each item, whereas the third column contains the biserial correlation coefficients. For instance, for item Y9, we obtain low values equal to 0.265 and 0.624 for the two indices, whereas for item Y4, we have more satisfactory values of 0.530 and 0.774.

Another useful method to investigate item discrimination is given by the extreme group method. First, we calculate the observed scores for the groups of the worst 25% and the best 25% performers:

```
. centile Y_obs, centile(25  75)

                                         -- Binom. Interp. --
                                         [95% Conf. Interval]
    Variable |    Obs  Percentile    Centile
-------------+------------------------------------------------------
       Y_obs |   3774          25          7            7          8
             |                 75         10           10         10
```

Then, we generate a new variable group to classify each individual into the worst, the best, and the intermediate performers (groups 0, 1, and ., respectively).

```
. gen group = .
(3774 missing values generated)
. replace group = 0 if Y_obs <= 7
(901 real changes made)
. replace group = 1 if Y_obs >= 10
(1085 real changes made)
```

Finally, we calculate the difficulty of each item separately for group 0 and group 1:

```
. forvalues num=1/10 {
    quietly  summarize Y'num' if group == 0
    scalar mean'num'0 = r(mean)
    quietly  summarize Y'num' if group == 1
    scalar mean'num'1 = r(mean)
    display "Y" 'num' " " mean'num'0 " " mean'num'1 " " mean'num'1-mean'num'0
  }
```

We obtain the following output:

```
Y1      .8216106      1      .1783894
Y2      .45158002     1      .54841998
Y3      .617737       1      .382263
Y4      .3058104      1      .6941896
Y5      .74617737     1      .25382263
Y6      .36289501     1      .63710499
Y7      .44240571     1      .55759429
Y8      .85219164     1      .14780836
Y9      .90316004     1      .09683996
Y10     .58409786     1      .41590214
```

where the difficulty level for the worst and the best performers and the difference between these two values are provided for each item, respectively. Note that the item difficulty of the best performers equals 1 for all items, since examinees belonging to this group positively endorsed all 10 items. We observe that only for items Y4 and Y6 the discrimination power is completely satisfactory, as the difference in the difficulties among the two groups is greater than 0.60. On the other hand, other items show a very weak discriminating power: this is specially true for items Y1, Y8, and Y9.

A further analysis concerns test validity (Section 2.9), in particular, the dimension of the test. For this, factor analysis is performed on the basis of the tetrachoric correlation matrix rather than on the Bravais–Pearson correlation matrix, as items are dichotomously scored:

```
. count
3774
. quietly tetrachoric  Y1-Y10
. factormat r(Rho), n(3774) pf
(obs=3774)
```

A crucial point is the choice of the number of factors, which may be based on several criteria. The most used rules of thumb are based on the number of eigenvalues greater than 1 or the presence of a clearly large difference between consecutive eigenvalues. As shown by the following output, both criteria coherently indicate the presence of only one factor as conjectured from the beginning of the application:

```
Factor analysis/correlation              Number of obs    =     3774
    Method: principal factors            Retained factors =        4
    Rotation: (unrotated)                Number of params =       34
```

Factor	Eigenvalue	Difference	Proportion	Cumulative
Factor1	2.85550	2.35726	1.0284	1.0284
Factor2	0.49824	0.37002	0.1794	1.2078
Factor3	0.12823	0.04415	0.0462	1.2540
Factor4	0.08407	0.12163	0.0303	1.2843
Factor5	-0.03756	0.05238	-0.0135	1.2708
Factor6	-0.08994	0.02757	-0.0324	1.2384
Factor7	-0.11751	0.03394	-0.0423	1.1960
Factor8	-0.15144	0.03463	-0.0545	1.1415
Factor9	-0.18607	0.02077	-0.0670	1.0745
Factor10	-0.20684	.	-0.0745	1.0000

```
[...] Output omitted
```

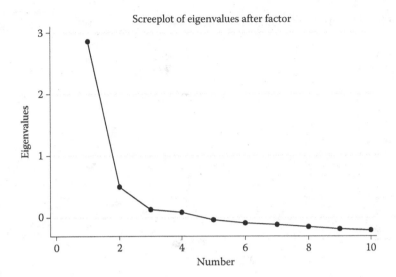

FIGURE 2.4
Screeplot of eigenvalues.

We may reach the same conclusion through the analysis of the corresponding screeplot that shows a large gap between the first and the second eigenvalues (Figure 2.4) and, therefore, we consider the assumption of unidimensionality as valid:

```
. screeplot
```

Finally, we study the possible presence of test bias with regard to respondents' gender. As outlined in Section 2.10, we may base this analysis on several criteria. First, we propose to calculate the item discrimination measures separately for males and females: marked differences between the two groups raise doubts about the correct functioning of the related item. The commands and the corresponding output showing the differences between genders of the α-if-item-omitted coefficients are as follows:

```
* Differences between genders for alpha-if-item-omitted
. quietly alpha  Y1-Y10 if gender == 0, item
. matrix Alpha0 = r(Alpha)
. quietly alpha  Y1-Y10 if gender == 1, item
. matrix Alpha1 = r(Alpha)
. matrix diff = Alpha1 - Alpha0
. matrix list diff

diff[1,10]
          Y1           Y2           Y3           Y4           Y5
r1  -.00654292   -.01774576   -.01665062   -.03981482   -.01560844

          Y6           Y7           Y8           Y9          Y10
   -.04989398   -.02220293   -.02279402   -.02517821   -.02390587
```

while for the point biserial correlation coefficients, we have

```
* Differences between genders for point biserial correlation coefficients
. forvalues num=1/10 {
    quietly  pbis Y'num' Y_obs if gender == 0
    scalar pbis'num'0 = $S_1
    quietly  pbis Y'num' Y_obs if gender == 1
    scalar pbis'num'1 = $S_1
    scalar diff'num' = pbis'num'1-pbis'num'0
    display "difference between groups in pbis for Y"'num' " = " diff'num'
    }
difference between groups in pbis for Y1 = -.10759786
difference between groups in pbis for Y2 = -.02877808
difference between groups in pbis for Y3 = -.03580114
difference between groups in pbis for Y4 = .04488403
difference between groups in pbis for Y5 = -.05073961
difference between groups in pbis for Y6 = .05781502
difference between groups in pbis for Y7 = -.0202989
difference between groups in pbis for Y8 = -.02147398
difference between groups in pbis for Y9 = .0009627
difference between groups in pbis for Y10 = .01733653
```

Both indices show very slight differences between males and females. When looking at the α-if-item-omitted coefficients, all differences are close to zero, except for items Y4 and Y6, where differences are equal to 0.04 and −0.05, respectively. Besides, in the half of items, observed gender differences through point biserial correlation are between 0.03 and 0.06, being greater than 0.10 only for item Y1.

While item discrimination measures calculated separately for males and females allow us to assess item bias, the Spearman's rank-order correlation coefficient, see Equation 2.19, allows us to globally evaluate the test bias. To determine this coefficient, first we need to compute the item difficulty estimates for each group:

```
. forvalues num=1/10 {
    quietly  summarize Y'num' if gender == 0
    scalar mean'num'0 = r(mean)
    quietly  summarize Y'num' if gender == 1
    scalar mean'num'1 = r(mean)
    display mean'num'0 "     " mean'num'1
    }
.92636073    .94473684
.76467449    .80368421
.84631804    .86842105
.6488794     .65736842
.90554963    .93578947
.72091782    .73421053
.76520811    .81210526
.93916756    .95789474
.95997866    .96789474
.75987193    .77210526
```

Then, we save these difficulties in a .dta file (say res2.dta) and we apply function spearman to variables diff0 and diff1, representing the difficulties for group 0 and for group 1, respectively:

```
. use "res2.dta", clear
. spearman diff0 diff1

 Number of obs =      10
Spearman's rho =    1.0000

Test of Ho: diff0 and diff1 are independent
     Prob > |t| =     0.0000
```

The Spearman coefficient equals 1, so the test may be considered unbiased with respect to gender.

To conclude, through factor analyses performed separately on males and females, we may further validate the absence of test bias. Indeed, results shown in the following outputs and the related screeplots (Figure 2.5) agree in detecting only one latent trait, both for males and for females:

```
. use INVALSI_full.dta, clear
. rename c* Y*
. count if gender == 0
  1874
. quietly tetrachoric  Y1-Y10 if gender == 0
. factormat r(Rho), n(1874) pf
(obs=1874)

Factor analysis/correlation                Number of obs    =      1874
     Method: principal factors             Retained factors =         4
     Rotation: (unrotated)                 Number of params =        34

    --------------------------------------------------------------------
        Factor |  Eigenvalue  Difference       Proportion   Cumulative
    -------------+------------------------------------------------------
       Factor1 |     2.87006     2.32791           0.9916       0.9916
       Factor2 |     0.54216     0.34502           0.1873       1.1790
       Factor3 |     0.19714     0.09304           0.0681       1.2471
       Factor4 |     0.10410     0.15325           0.0360       1.2831
       Factor5 |    -0.04915     0.01862          -0.0170       1.2661

[...] Output omitted

. screeplot, title("Screeplot of eigenvalues (gender = 0)")

. count if gender == 1
  1900
. quietly tetrachoric  Y1-Y10 if gender == 1
. factormat r(Rho), n(1900) pf
(obs=1900)

Factor analysis/correlation                Number of obs    =      1900
     Method: principal factors             Retained factors =         5
     Rotation: (unrotated)                 Number of params =        40

    --------------------------------------------------------------------
        Factor |  Eigenvalue  Difference       Proportion   Cumulative
    -------------+------------------------------------------------------
       Factor1 |     2.83447     2.25171           0.9783       0.9783
       Factor2 |     0.58276     0.42741           0.2011       1.1794
       Factor3 |     0.15536     0.10409           0.0536       1.2330
       Factor4 |     0.05126     0.01640           0.0177       1.2507
       Factor5 |     0.03487     0.11804           0.0120       1.2627

[...] Output omitted

. screeplot, title("Screeplot of eigenvalues (gender = 1)")
```

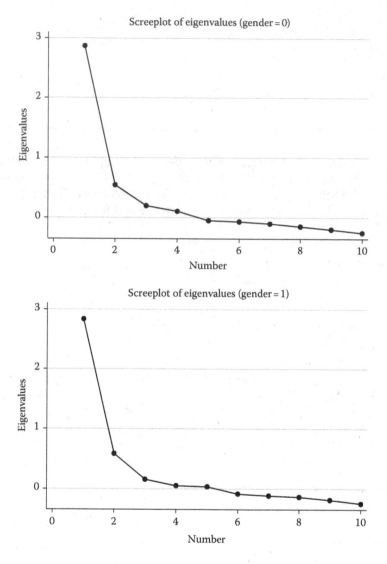

FIGURE 2.5
Screeplot of eigenvalues by gender.

2.12.2 RLMS Data: A Classical Analysis in R

First, we load the RLMS data concerning the job satisfaction and drop the covariates, because they are not of interest here. Then, we perform a summary to check for the presence of missing item responses:

```
> load("RLMS.RData")
> # Keep item responses
> data = data[,6:9]
```

```
> summary(data)
      Y1                Y2                Y3                Y4
 Min.   :0.000    Min.   :0.000    Min.   :0.000    Min.   :0.000
 1st Qu.:1.000    1st Qu.:1.000    1st Qu.:1.000    1st Qu.:1.000
 Median :1.000    Median :1.000    Median :2.000    Median :2.000
 Mean   :1.404    Mean   :1.449    Mean   :2.179    Mean   :1.933
 3rd Qu.:2.000    3rd Qu.:2.000    3rd Qu.:3.000    3rd Qu.:3.000
 Max.   :4.000    Max.   :4.000    Max.   :4.000    Max.   :4.000
 NA's   :16       NA's   :16       NA's   :24       NA's   :61
```

Consequently, we drop from the dataset the sample units with at least one missing response. Besides, we arrange the response categories in increasing order and then from absolutely unsatisfied (0) to absolutely satisfied (4), so to have a clearer interpretation of the results:

```
> ind = which(apply(is.na(data),1,any))
> # Drop records with missing observations
> data = data[-ind,]
> # Reverse  response categories
> data = 4-data
> (n = nrow(data))
[1] 1418
```

Note that vector `ind` created through the previous commands contains the labels of the subjects with at least one missing response. Also note that, after removing these subjects, the sample size reduces to $n = 1418$ as indicated in Section 1.8.2. As a general rule, we remind that the definition of an R object is immediately displayed on the console if it is put in brackets.

Based on the resulting dataset, the distribution of the response categories may be obtained for each item as follows:

```
> table(data[,1])

  0   1   2   3   4
 54 180 277 668 239
> table(data[,2])

  0   1   2   3   4
 64 183 294 645 232
> table(data[,3])

  0   1   2   3   4
228 380 309 399 102
> table(data[,4])

  0   1   2   3   4
162 319 337 459 141
```

We note that the responses to the third item, concerning satisfaction with respect to earnings, tend to be in lower categories with respect to the other items, in particular with respect to the first item concerning general satisfaction about work.

Given that the dataset contains only four items, we consider the study of reliability using the split-half method (Section 2.6.3) and some internal consistency reliability measures (Section 2.6.4). To apply the first method, we separately consider two subsets of items and apply the Spearman–Brown formula, see Equation 2.11, as follows:

```
> Y_obs_even = data$Y2+data$Y4
> Y_obs_odd = data$Y1+data$Y3
> rho = cor(Y_obs_even,Y_obs_odd)
> rel = 2*rho/(1+rho)
> rel
[1] 0.843689
```

The estimated reliability is completely satisfactory, being equal to 0.844.

In order to study reliability, we can compute Cronbach's α in the formulation for quantitative variables, see Equation 2.12, and its standardized version, see Equation 2.14, as follows:

```
> require(ltm)
Loading required package: ltm
Loading required package: MASS
Loading required package: msm
Loading required package: mvtnorm
Loading required package: polycor
Loading required package: sfsmisc
> (cron = cronbach.alpha(data))

Cronbach's alpha for the 'data' data-set

Items: 4
Sample units: 1418
alpha: 0.826

> cronbach.alpha(data, standardized = TRUE)

Standardized Cronbach's alpha for the 'data' data-set

Items: 4
Sample units: 1418
alpha: 0.832
```

Note that these commands require the package ltm that must be previously installed in R. In both cases, the reliability estimates are very high.

The point estimates of the individual scores are obtained by the command rowSums(), which computes the sum of each row of a matrix, as follows:

```
> # individual scores
> Y_obs = rowSums(data)
> # summary statistics and store standard deviation
> summary(Y_obs)
   Min. 1st Qu.  Median    Mean 3rd Qu.    Max.
  0.000   7.000  10.000   9.073  12.000  16.000
> (sd_Y_obs = sd(Y_obs))
[1] 3.630957
```

We observe that the average score is 9.073 with a standard deviation of 3.63. Moreover, around one-half of the sample has a score between 4 (1st quartile) and 9 (3rd quartile). These results are confirmed by the histogram of the scores, reported in Figure 2.6, which is simply obtained as follows:

```
> hist(Y_obs)
```

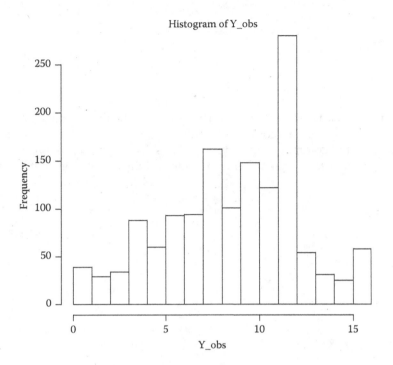

FIGURE 2.6
Histogram of observed scores.

We can also create a matrix with estimates of the score, subject by subject, and the 95% confidence interval, computed according to (2.15), as follows:

```
> Conf = cbind(Y_obs,11=Y_obs-1.96*(sd_Y_obs*sqrt(1-cron$alpha)),
               12=Y_obs+1.96*(sd_Y_obs*sqrt(1-cron$alpha)))
> head(Conf)
  Y_obs        11         12
2    10  7.031776  12.968224
3    12  9.031776  14.968224
4     9  6.031776  11.968224
5     8  5.0317756 10.968224
6    14 11.031776  16.968224
7     4  1.031776   6.968224
```

Note that in the previous output, obtained by command head(), subject 1 is missing, so that the first row of the matrix Conf is indicated by 2, as he or she has at least one missing response. In a similar way, we may obtain the adjusted true score estimates according to Equation 2.16.

Regarding the item analysis (Section 2.8), we limit the illustration to the computation of the mean and standard deviation for each test item and to that of some discrimination indices. In particular, the mean of each column of the data matrix is obtained by the command colMeans(), whereas its standard deviation is obtained by the command apply() in combination with sd():

```
> cbind(mean=colMeans(data),sd=apply(data,2,sd))
         mean        sd
Y1 2.605078 1.029160
Y2 2.562764 1.050226
Y3 1.835684 1.205935
Y4 2.069111 1.183045
```

This output leads us to conclude that the item to which subjects tend to respond with lower categories is the third.

By using the command `cronbach.alpha()`, it is then possible to obtain Cronbach's α-if-item-omitted and the item-to-total correlation coefficients as follows:

```
> Disc = NULL
> for(j in 1:4){
+   Disc = rbind(Disc,
+              c(cronbach.alpha(data[,-j], standardized = TRUE)$alpha,
+                cor(data[,j],Y_obs)))
+ }
> colnames(Disc) = c("alpha","cor")
> Disc
          alpha       cor
[1,]  0.7485196 0.8528037
[2,]  0.7615985 0.8366894
[3,]  0.8331336 0.7642574
[4,]  0.8009881 0.8054880
```

On the basis of these results, we conclude that the third item has the smallest discriminating power with respect to the other items. In fact, Cronbach's α increases with respect to the global level (0.832) when this item is removed. Accordingly, for this item, we have the lowest correlation level with the observed score.

Exercises

1. Using the dataset `aggression.dat` (downloadable from `http://www.gllamm.org/aggression.dat`) with categories 1 and 2 collapsed, perform the following analyses through `Stata`:

   (a) Apply the split-half method (according to one of the criteria illustrated in Section 2.6.3) and calculate the Cronbach's α.

   (b) On the basis of the results previously obtained, discuss the reliability of the test.

   (c) Compute the item difficulties and analyze the corresponding empirical distribution, by calculating the mean and standard deviation and by plotting the values obtained.

(d) For each item, compute the α-if-item-omitted and the point biserial correlation coefficients.

(e) On the basis of the results obtained at point (d), discuss the contribution of each item to the reliability of the test.

2. Using the dataset `delinq.txt` (downloadable from `http://www.gllamm.org/delinq.txt`), perform the following analyses through Stata:

(a) Compute the observed score of each individual and analyze the corresponding empirical distribution, by calculating the mean and standard deviation and by plotting the values obtained.

(b) Compute the individuals' adjusted true scores and compare them to the observed scores.

(c) Apply the extreme group method and discuss the discrimination power of each item.

(d) For each item and separately for males and females, compute the α-if-item-omitted and the point biserial correlation coefficients and discuss the item bias with respect to variable `sex`.

(e) Separately for males and females, compute the Spearman rank-order correlation coefficient and perform a factor analysis and discuss the test bias with respect to variable `sex`.

3. Consider the dataset `naep` available in the R package `MultiLCIRT`. Using the appropriate R commands, perform the following analyses:

(a) Estimate reliability using the split-half method.

(b) Estimate the KR20 coefficient and the standardized α-coefficient and comment the results obtained.

(c) For each respondent, estimate the true score and compute the corresponding 95% level confidence intervals.

(d) Represent and analyze the distribution of the scores obtained at point (c).

(e) Apply the extreme group method and estimate the point biserial correlation coefficient for each item; comment the results from a perspective of item analysis.

(f) Split the overall sample of respondents in two hypothetical subsamples of equal size, including the first 755 subjects in the first subsample and the other 755 subjects in the second; perform a test bias analysis based on the Spearman rank-order correlation coefficient between item difficulties in the two groups.

4. On the basis of the dataset `hads` with ordinal responses available in the R package `MultiLCIRT`:

(a) Perform the analysis of dimensionality, based on factor analysis, of the whole set of items and only for the items measuring depression (use the R packages `polycor` and `psych` as illustrated in `http://www.uni-kiel.de/psychologie/rexrepos/posts/multFApoly.html`).

(b) Limited to the items referring to depression, perform the same analyses illustrated in Section 2.12.2.

3

Item Response Theory Models for Dichotomous Items

3.1 Introduction

Item response theory (IRT) is a model-based theory founded on the idea that the responses to each test item depend on some person and item characteristics, according to specific probabilistic relations.

The first factor affecting the item responses is the respondent's level on the *latent trait*. Obviously, a student with a high level of mathematical ability has a higher probability to respond correctly to an item in mathematics than a student with a low level of this ability. Similarly, an individual suffering from deep depression will be more likely to provide a discouraging response to an item measuring this illness than an individual with a low level of depression. Therefore, all IRT models for dichotomous items assume that the probability of responding correctly to an item is a monotonic nondecreasing function of the latent trait they measure.

However, other factors affect the probability to choose a certain item response, and they are referred to some item characteristics as the difficulty level, the discrimination power, and the easiness of guessing. These names are particularly meaningful in the educational field, where IRT models have had their first developments (see Section 1.6). These item characteristics are represented by suitable model parameters. It is less likely that a difficult item will be answered correctly than an easy item. On the other hand, the higher the discrimination power, the neater is, in general, the difference between respondents with different ability levels. Depending on the complexity of the adopted item parameterization, different types of IRT model are defined. The simplest parameterization corresponds to the well-known Rasch model, or one-parameter logistic model. According to this model, the probability to answer correctly an item depends on the respondent's ability level and the item difficulty. The difficulty level is thus the only parameter describing the item, whereas the discrimination power is assumed to be constant across items (and, more specifically, to be equal to any constant value in the one-parameter logistic model and to 1 in the Rasch model). On the other hand, in the two-parameter logistic (2PL) model, two parameters are used to describe each item, corresponding to the difficulty level and the discrimination power. Finally, the three-parameter logistic (3PL) model adds a pseudo-guessing parameter for each item, which defines the lowest horizontal asymptote at which the probability to endorse an item tends, even in the case of respondents with a very low ability level, as a result of guessing.

Further relevant differences between IRT models, other than the complexity of the parameterization of the conditional probability of endorsing an item and the number of latent traits, are related to the formulation of the latent structure. First, we have to disentangle the so-called fixed-effects approach from the random-effects approach. In the first case, every subject's latent trait level is included in the model as a fixed parameter that is estimated together with the item parameters or is somehow eliminated. In the second case, the latent trait level is seen as a realization of a random variable having a certain distribution in the population from which the observed sample has been drawn. This distribution may be continuous, typically normal, or discrete, giving rise to latent classes in the population.

The present chapter is organized as follows. The next section illustrates the main assumptions characterizing IRT models for dichotomously scored items together with the concept of an *item characteristic curve* (ICC), which is used to formalize the conditional probability of answering to an item given the latent trait. Then, the three main models used in the presence of dichotomous items are described in Sections 3.3 through 3.5. Section 3.6 is devoted to entangle these models in the framework of random-effects models under the normality assumption for the latent trait and under that of discreteness of the latent trait. Some hints about the approaches for estimating models at issue are provided in Section 3.7. To conclude, some illustrative examples are provided in Section 3.8.

3.2 Model Assumptions

In the binary case, where the response variables Y_{ij}, $i = 1, \ldots, n$, $j = 1, \ldots, J$, are equal to 0 or 1 and typically correspond to a wrong or a correct response, respectively, the main assumptions of IRT models may be summarized as follows:

1. *Unidimensionality*: For every subject i, the responses to the J items depend on the same latent trait level θ_i, which is unidimensional and belongs to $\mathbb{R}$.

2. *Local independence*: For each subject i, the responses to the J items are independent given θ_i.

3. *Monotonicity*: The conditional probability of responding correctly to item j, denoted by

$$p_j(\theta_i) = p(Y_{ij} = 1 | \theta_i)$$

and known as ICC or *item response function*, is a monotonic nondecreasing function of θ_i.

A first comment concerns the number of the latent traits and, then, the dimension of the model. Typically, an IRT model assumes that responses to the test items can be explained by latent traits that are much fewer in number than the items. Most IRT models assume that a single latent trait is involved in the response process (Crocker and Algina, 1986). Such models are referred to as *unidimensional*, as opposed to *multidimensional* IRT models, which assume more latent traits underlying the item responses. In this chapter, the attention is focused on models that assume unidimensionality, whereas the extension to the multidimensional case is considered in Chapter 6.

A further basic assumption characterizing IRT models is that of local independence. With reference to a unidimensional IRT model, local independence means that the responses to the test items are conditionally independent given the latent trait of interest. Thus, if the individual's ability level was known exactly, the response to an item would not add any relevant information in predicting the response of the same subject to any other item. In other words, local independence means that the latent trait is the only factor that explains the difference between the response patterns provided by two subjects. Moreover, if a subject responds better than another subject to an item, he or she will tend to respond better to any other item due to a higher level of ability. It is important to note that the assumption of local independence characterizes many statistical models based on latent variables. An important example is the latent class (LC) model (Lazarsfeld and Henry, 1968; Goodman, 1974), which assumes a discrete latent variable to explain the association between categorical responses and is used to cluster sample units on the basis of these responses. It should also be noted that unidimensionality and local independence are different notions. In fact, the first is ensured when there is only a single latent trait, whereas the notion of local independence also has relevance for multidimensional tests (Raykov and Marcoulides, 2011).

Unidimensionality and local independence imply that, for any two items j_1 and j_2, the joint distribution of the corresponding pair of responses y_{ij_1} and y_{ij_2} is

$$p(y_{ij_1}, y_{ij_2} | \theta_i) = p(y_{ij_1} | \theta_i) p(y_{ij_2} | \theta_i), \quad y_{ij_1}, y_{ij_2} = 0, 1,$$

meaning that if one knew the true ability level θ_i, the response to an item would not add relevant information to predict the response to any other item, as already explained. Considering that

$$p(y_{ij}|\theta_i) = p_j(\theta_i)^{y_{ij}}[1 - p_j(\theta_i)]^{1-y_{ij}}, \quad y_{ij} = 0, 1,$$

for the overall sequence of J items $y_i = (y_{i1}, \ldots, y_{iJ})'$ given θ_i, we have the joint distribution

$$p(y_i|\theta_i) = \prod_{j=1}^{J} p_j(\theta_i)^{y_{ij}}[1 - p_j(\theta_i)]^{1-y_{ij}}. \tag{3.1}$$

To clarify the assumption of local independence, it is useful to consider explicitly the *random-effects approach*, in which θ_i is considered as a realization of the random variable Θ_i. In contrast, under the *fixed-effects approach*, each θ_i is considered as a parameter to be estimated together with the item parameters. Under the random-effects approach, the expression in (3.1) corresponds to the conditional probability of y_i given θ_i. Moreover, the marginal distribution of y_i, which is called *manifest distribution*, is obtained by integrating out the latent trait as follows:

$$p(y_i) = \int_{\mathbb{R}} p(y_i|\theta_i)f(\theta_i)\,d\theta_i, \tag{3.2}$$

where $f(\theta_i)$ is the density function of Θ_i, which is common to all subjects in the sample. Note that, as will be clarified in Section 3.6.2, the integral in (3.2) becomes a sum for a latent trait having a discrete distribution. However, in general, we have that

$$p(y_i) \neq \prod_{j=1}^{J} p(y_{ij}),$$

where

$$p(y_{ij}) = \int_{\mathbb{R}} p(y_{ij}|\theta_i)f(\theta_i)\,d\theta_i$$

corresponds to the marginal distribution of the response to item j. This means that the assumption of local independence does not imply that the item responses are marginally independent. The main difference between the conditional probability $p(y_{ij}|\theta_i)$ and the marginal probability $p(y_{ij})$ is that the first one is referred to the subpopulation of subjects having the same level of latent trait θ_i, whereas the second probability is referred to the whole population of subjects and, then, it is computed for a random subject in this population. In this second case, the knowledge of the response to an item is a relevant information for predicting the response to another item. The topic at issue is highly debated in the IRT literature; see Bartolucci and Forcina (2000) and the references therein.

A further comment regarding ICC is that $p_j(\theta_i)$ is a function of θ_i and of a set of parameters that describe the item characteristics. Different models arise according to the assumed parameterization of the ICC and the number of item parameters. The most typical IRT models, which will be described in the following sections, rely on the logistic function. However, note that any other link function used for generalized linear models for binary response variables (McCullagh and Nelder, 1989) may be adopted, provided that monotonicity of $p_j(\theta_i)$ holds. For instance, the probit link function is used in the IRT normal ogive model (Lord and Novick, 1968).

The Rasch model (Rasch, 1960, 1961) is based on an ICC of logistic type, which relies on a very simple parameterization, whereas 2PL and 3PL models (Birnbaum, 1968; Lord, 1980) are based on logistic-shaped ICCs having a more complex characterization in terms of item parameters. We recall that the *logistic function* is defined, in general, as

$$p = \frac{e^\psi}{1 + e^\psi},$$

for any real argument ψ, so that the function value p is always between 0 and 1.

When a logit or a probit link function is adopted, then the resulting ICC has an increasing S-shape, which approaches 0 for θ_i approaching $-\infty$ and 1 for θ_i approaching $+\infty$. An example of an S-shaped ICC is shown in Figure 3.1. In principle, other monotonic nondecreasing shapes may also be admitted, such as the step function shown in Figure 3.2, meaning that the probability to answer the item correctly increases for specific values of θ_i (the so-called steps), whereas it is constant for the remaining ones.

To conclude this section, it is important to recall that an implicit assumption of IRT models is that the response vectors corresponding to different subjects in the sample are independent to each other. Consequently, the conditional probability of observing the response matrix

$$Y = \begin{pmatrix} y_{11} & \cdots & y_{1J} \\ \vdots & \ddots & \vdots \\ y_{n1} & \cdots & y_{nJ} \end{pmatrix} = \begin{pmatrix} y_1' \\ \vdots \\ y_n' \end{pmatrix},$$

given the ability vector $\theta = (\theta_1, \ldots, \theta_n)'$, is equal to

$$p(Y|\theta) = \prod_{i=1}^{n} p(y_i|\theta_i) = \prod_{i=1}^{n} \prod_{j=1}^{J} p_j(\theta_i)^{y_{ij}} [1 - p_j(\theta_i)]^{1 - y_{ij}}. \tag{3.3}$$

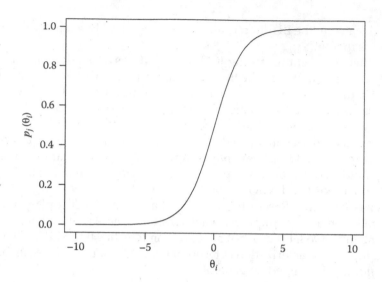

FIGURE 3.1
Example of S-shaped ICC.

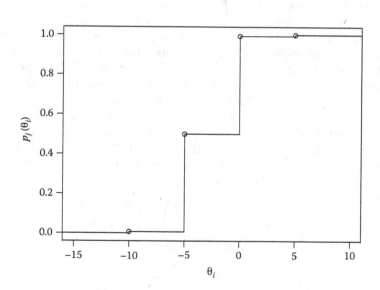

FIGURE 3.2
Example of step ICC.

Under the random-effects approach, we can also express the manifest probability of Y as

$$p(Y) = \prod_{i=1}^{n} p(y_i).$$

The previous probabilities are of crucial importance for a maximum likelihood estimation of the specific IRT model of interest, as will be clarified in Sections 5.3 and 5.4.

3.3 Rasch Model

The *Rasch model*, introduced by Rasch (1960, 1961), is the most well-known IRT model for binary responses. It uses only one parameter to describe each item of the questionnaire. This parameter is commonly named *item difficulty* as it is negatively related with the probability to endorse an item.

The Rasch model is based on the assumptions illustrated in Section 3.2 and, further, on the following characterization of ICC:

$$p_j(\theta_i) = \frac{e^{\theta_i - \beta_j}}{1 + e^{\theta_i - \beta_j}}, \quad i = 1, \ldots, n, \, j = 1, \ldots, J, \tag{3.4}$$

where β_j is the *difficulty parameter* of item j. It is important to outline that both θ_i and β_j are measured on the same scale and lie on $\mathbb{R}$. This allows for a direct comparison between values of ability and difficulty. In particular, we have that

- if $\theta_i = \beta_j$, then $p_j(\theta_i) = 0.50$;
- if $\theta_i > \beta_j$, then $p_j(\theta_i) > 0.50$;
- if $\theta_i < \beta_j$, then $p_j(\theta_i) < 0.50$.

In other words, the item difficulty represents the level of ability required to have a 50% probability of answering correctly (or wrongly) that item. Obviously, a difficult item requires a relatively high latent trait level to be endorsed, whereas an easy item requires a low latent trait level.

As already stated in Section 3.2, the adopted parameterization is such that $p_j(\theta)$ is monotonic nondecreasing in θ and it goes from 0, for θ_i tending to $-\infty$, to 1, for θ_i tending to $+\infty$. Moreover, the probability of success decreases with the parameter β_j, which is common to all subjects. This enforces the interpretation of this parameter as the difficulty level of the item. These aspects are clarified by Figure 3.3, which represents the function $p_j(\theta_i)$ with respect to

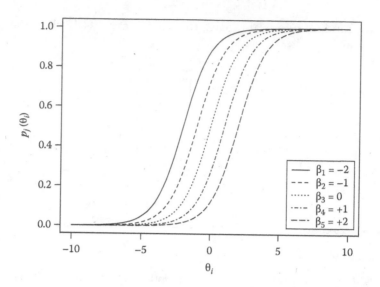

FIGURE 3.3
ICCs for items $j = 1, \ldots, 5$ under the Rasch model (different levels of difficulty are assumed).

θ_i for five hypothetical items having different difficulty levels ($\beta_1, \ldots, \beta_5$). It is worth noting that since Equation 3.4 is based on an additive form in the parameters, subject–item interactions are ruled out. Moreover, $p_j(\theta_i)$ is invariant with respect to a transformation of the parameters consisting of adding a constant to every ability parameter θ_i and to every difficulty parameter β_j, that is, if $\theta_i^* = \theta_i + b$ and $\beta_j^* = \beta_j + b$, then $p_j^*(\theta_i^*) = p_j(\theta_i)$ for any value of b, where

$$p_j^*(\theta_i^*) = \frac{e^{\theta_i^* - \beta_j^*}}{1 + e^{\theta_i^* - \beta_j^*}}.$$

This makes the model nonidentifiable, and as will be discussed in more detail in Chapter 5, an estimation requires *identifiability* constraints in order to be performed properly. A typical constraint is that the average ability level is equal to 0 or, alternatively, that the difficulty of a certain item, called the *reference item*, is equal to 0.

　　The plots in Figure 3.3 confirm that $p_j(\theta_i)$ is increasing in θ_i and is decreasing in β_j. Moreover, the value assumed by the latent trait for a given subject i indicates the point on the latent continuum where this subject is located with respect to a specific ICC, whereas the value assumed by β_j corresponds to the position of the ICC with respect to the x-axis. When the difficulty parameter β_j increases (decreases), the ICC shifts toward the right (left).

　　It should be clear that the item difficulty in the IRT setting has a different meaning with respect to the item difficulty in the classical test theory (CTT) setting. In the first case, it is a location parameter, as it identifies the point on the latent continuum at which the individual's latent trait is located. On the contrary, in CTT, the item difficulty corresponds to the relative frequency of individuals that endorsed a certain binary item (see Section 2.8).

An important characteristic of the Rasch model is that, as items have the same discriminating capacity, all ICCs have the same slope, that is, they are "parallel." From a practical point of view, this means that the ranking of item difficulty does not depend on the value of θ_i. For instance, consider two of the items illustrated in Figure 3.3, say, the first and the fifth. We have that $\beta_1 < \beta_5$ and then, for any level of ability θ_i, the probability to endorse item 1 is greater than the probability to endorse item 5, that is, $p_1(\theta_i) > p_5(\theta_i)$; moreover, a higher level of θ_i is needed for item 5 in order to have the same probability of success of item 1.

According to a typical interpretation of the econometric literature on discrete choice models (see, for instance, Wooldridge, 2002), assumption (3.4) may be motivated by considering $p_j(\theta_i)$ as the probability that the variable Y_{ij} is equal to 1 when this variable corresponds to the dichotomized version of an underlying continuous variable Y_{ij}^*. More precisely, we have

$$Y_{ij} = 1\{Y_{ij}^* \geq 0\}, \tag{3.5}$$

where $1\{\cdot\}$ denotes the indicator function, which is equal to 1 if its argument is true and to 0 otherwise, and Y_{ij}^* is equal to $\theta_i - \beta_j$ plus an error term with standard logistic distribution, which accounts for accidental factors. If Y_{ij}^*, which may be seen as a noisy version of the propensity of a correct response, is high enough, then the subject responds correctly to the item.

Assumption (3.4), together with assumption (3.1), implies that the joint probability of the response vector y_i given θ_i is equal to

$$p(y_i|\theta_i) = \prod_{j=1}^{J} \frac{e^{y_{ij}(\theta_i - \beta_j)}}{1 + e^{\theta_i - \beta_j}} = \frac{e^{y_{i\cdot}\theta_i - \sum_{j=1}^{J} y_{ij}\beta_j}}{\prod_{j=1}^{J}(1 + e^{\theta_i - \beta_j})}, \tag{3.6}$$

where $y_{i\cdot}$ may be considered as a raw measure of the subject's ability, as it is the sum of the responses provided by the same subject that, in turn, is equal to the number of correct responses (see Section 1.7). Finally, for the overall set of individuals, from Equation 3.3 we have

$$p(Y|\theta) = \prod_{i=1}^{n} \frac{e^{y_{i\cdot}\theta_i - \sum_{j=1}^{J} y_{ij}\beta_j}}{\prod_{j=1}^{J}(1 + e^{\theta_i - \beta_j})} = \frac{e^{\sum_{i=1}^{n} y_{i\cdot}\theta_i - \sum_{j=1}^{J} y_{\cdot j}\beta_j}}{\prod_{i=1}^{n}\prod_{j=1}^{J}(1 + e^{\theta_i - \beta_j})}, \tag{3.7}$$

where $y_{\cdot j}$ may be considered as a raw measure of the difficulty of the item.

Under the fixed-parameter approach, in which the ability levels $\theta_1, \ldots, \theta_n$ are seen as fixed parameters, Equation 3.7 implies that $(y_{1.}, \ldots, y_{n.})'$ is a vector of minimal sufficient statistics for these ability parameters, which are collected in the vector θ. Similarly, $(y_{.1}, \ldots, y_{.J})'$ is a set of minimal sufficient statistics for the parameter vector $\beta = (\beta_1, \ldots, \beta_J)'$. In fact, the distribution with probability function (3.7) belongs to the exponential family (Barndorff-Nielsen, 1978) with canonical parameters being a linear function of θ and β.

The sufficiency of raw scores implies that individuals with the same score obtain the same estimate of the ability, independently of the specific pattern of responses. As will be clear in Chapter 5, the sufficiency of vectors $(y_{1.}, \ldots, y_{n.})'$ and $(y_{.1}, \ldots, y_{.J})'$ strongly simplifies the estimation of the item and person parameters, and it allows us to compare the ability of different individuals in an objective way, that is, without the influence of the set of items (*item-free persons calibration*) and to compare item difficulties without the influence of the set of individuals (*person-free item calibration*). This characteristic corresponds to the fact that the ICCs are *parallel* and is known as *parameter separability*.

Parameter separability concurs to define the *specific objectivity* of a measurement (Rasch, 1967, 1977; Hambleton and Swaminathan, 1985; Fischer, 1995), which asserts that the comparison of two objects should be independent of everything but the two objects and their observed reactions. This property is certainly true in the physical sciences. If a person is taller than another under a given measurement system (e.g., meters), such a relationship holds true also under any other measurement system (e.g., feet). Obviously, this property is also desirable in the social and psychological sciences. If the ability in mathematics of a student is higher than that of another student under a given test, this relationship should remain true even if the test items are modified. The specific objectivity property implies that comparisons between respondents are independent of the particular items used and, symmetrically, comparisons of items are independent of the subjects who answer them.

In the class of IRT models, it can be shown that both sufficiency of raw scores and specific objectivity imply a family of models characterized by the same discrimination parameter and, therefore, by parallel ICCs, as it results from assumption (3.4); see Fischer (1995) for details. Therefore, the Rasch model plays a crucial role as it is the only IRT model for dichotomous items that guarantees specific objectivity. For this reason, a certain stream of literature considers the Rasch model as the only acceptable measurement model in the presence of binary item responses. This in turn implies that the objective measurement of a latent trait is possible only when the data agree with the Rasch model: data for which this model has a bad fit have to be excluded by the analysis. As opposed to this approach, the statistical approach to measurement aims at developing models with a good fit to the observed data. From a statistical point of view, the item responses are modeled in order to explain at the best the observed data, that is, the model has to fit the data rather than vice versa. Within the statistical approach, the specific objectivity is a secondary aspect with respect to the goodness-of-fit of the model.

3.4 2PL Model

The 2PL model (Birnbaum, 1968) is an extension of the Rasch model in which

$$p_j(\theta_i) = \frac{e^{\lambda_j(\theta_i - \beta_j)}}{1 + e^{\lambda_j(\theta_i - \beta_j)}}, \quad i = 1, \ldots, n, \quad j = 1, \ldots, J, \tag{3.8}$$

where λ_j is the *discriminating parameter* for item j, which measures the capacity of that item to distinguish between individuals with different ability levels. Indeed, λ_j defines the slope of the ICC for item j: the higher the slope of the ICC, the better the discriminating power between individuals with different levels of ability, as illustrated in Figure 3.4. Note that, because of the presence of horizontal asymptotes in the ICCs, the discrimination between individuals' abilities is mainly possible for intermediate values of θ_i (i.e., close to the difficulty level).

The parameter λ_j is typically assumed to be positive. Besides, its inclusion is particularly useful when one wants to spot *anomalous* items, that is, items with a low ($\lambda_j \approx 0$) or even a negative ($\lambda_j < 0$) dependence on the ability.

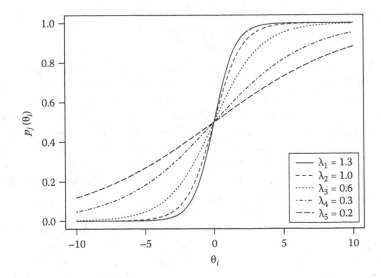

FIGURE 3.4
ICCs for items $j = 1, \ldots, 5$ under the 2PL model (different levels of discriminating power and the same levels of difficulty, $\beta_1 = \cdots \beta_5 = 0$, are assumed).

These items need to be eliminated since they are not effective in measuring the ability of an individual.

Under the 2PL model, the probability of endorsing an item is invariant with respect to a linear transformation. Actually, $p_j(\theta_i)$ does not change when multiplying both θ_i and β_j (and possibly summing a constant to these parameters) and, at the same time, dividing λ_j by the same constant. More precisely, if $\theta_i^* = a\theta_i + b$, $\beta_j^* = a\beta_j + b$, and $\lambda_j^* = \lambda_j/a$, then $p_j^*(\theta_i^*) = p_j(\theta_i)$ for any value of a different from 0 and any value of b, where

$$p_j^*(\theta_i^*) = \frac{e^{\lambda_j^*(\theta_i^* - \beta_j^*)}}{1 + e^{\lambda_j^*(\theta_i^* - \beta_j^*)}}.$$

Assumption (3.8) implies that the conditional probability of the response vector y_i is given by

$$p(y_i|\theta_i) = \frac{e^{y_{i\cdot}^*\theta_i - \sum_{j=1}^{J} y_{ij}\lambda_j\beta_j}}{\prod_{j=1}^{J}[1 + e^{\lambda_j(\theta_i - \beta_j)}]}, \tag{3.9}$$

where $y_{i\cdot}^* = \sum_{j=1}^{J} \lambda_j y_{ij}$ is equal to a *weighted score*, obtained by weighting each item score by its relative importance, which is measured by the discrimination parameter. Finally, for the overall set of items, we have

$$p(Y|\theta) = \frac{e^{\sum_{i=1}^{n} y_{i\cdot}^*\theta_i - \sum_{j=1}^{J} y_{\cdot j}\lambda_j\beta_j}}{\prod_{i=1}^{n}\prod_{j=1}^{J}[1 + e^{\lambda_j(\theta_i - \beta_j)}]}. \tag{3.10}$$

Due to the inclusion of the discriminant parameters, the property of sufficiency of the scores and that of specific objectivity are lost. The violation of the specific objectivity assumption implies that the comparison between the abilities of two or more individuals depends on the specific set of items used to measure the ability, and similarly, the comparison between the difficulties of two or more items depends on the specific set of individuals. Such a violation is clearly illustrated by Figure 3.4. For instance, an item with $\lambda_j = 0.2$ is more difficult than an item with $\lambda_j = 1.3$ for positive values of the ability and vice versa for negative values of the ability.

The violation of the property of sufficiency means that individuals with the same score, but different response patterns, obtain different ability estimates. As an example, Table 3.1 references four dichotomous items, ordered from the easiest to the most difficult, the respondents' scores and ability estimates under both the 2PL model and the Rasch model. These estimates have been obtained by one of the estimation methods outlined in Section 3.7; the method and the data on the basis of which they have been obtained are relatively unimportant at this stage. The main point is that, as can be noted from

TABLE 3.1

Comparison between the 2PL and the Rasch Models

Individual	Response Pattern	Score	Estimated Ability	
			2PL	**Rasch**
1	0000	0	−0.82	−0.84
2	1000	1	−0.27	−0.22
3	0100	1	−0.21	−0.22
4	0010	1	−0.19	−0.22
5	0001	1	−0.01	−0.22
6	1100	2	0.14	0.22
7	1010	2	0.15	0.22
8	0110	2	0.19	0.22
9	1001	2	0.31	0.22
10	0101	2	0.36	0.22
11	0011	2	0.37	0.22
12	1110	3	0.52	0.71
13	1101	3	0.72	0.71
14	1011	3	0.74	0.71
15	0111	3	0.80	0.71
16	1111	4	1.35	1.36

the table, individuals with the same score (e.g., with a score of 2) obtain the same ability estimates (e.g., 0.22) under the Rasch model but different ability estimates under the 2PL model, as their response patterns are different. For instance, subject 9 has a higher ability than subject 6 (0.31 versus 0.14), although he or she answered to a more difficult item (item 4 rather than item 2); we recall that items are ordered according to the difficulty level. It is also worth noting that, according to the Rasch approach, the response pattern of individual 9 (1001) is highly improbable (see also the examples at the end of this chapter) and then this response configuration should appear with a very low chance in the dataset. In this regard, Chapter 5 proposes some statistics that can be used to test the goodness-of-fit at individual and item levels.

3.5 3PL Model

A specific generalization (Lord, 1980) of the 2PL model is obtained as follows:

$$p_j(\theta_i) = \delta_j + (1 - \delta_j)\frac{e^{\lambda_j(\theta_i - \beta_j)}}{1 + e^{\lambda_j(\theta_i - \beta_j)}}, \quad i = 1, \ldots, n, \ j = 1, \ldots, J, \qquad (3.11)$$

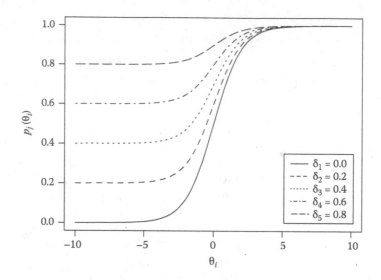

FIGURE 3.5
ICCs for items $j = 1, \ldots, 5$ under the 3PL model (different levels of pseudo-guessing and same levels of difficulty, $\beta_j = 0$, and discriminating power, $\lambda_j = 1$, are assumed).

where δ_j is the *pseudo-guessing parameter*, which corresponds to the probability that a subject, whose ability approaches $-\infty$, endorses the item. Each parameter δ_j lies on the interval $(0, 1)$. As shown in Figure 3.5, the pseudo-guessing parameter defines a shift of the lower horizontal asymptote of $p_j(\theta_i)$ toward positive values.

The 3PL model has a more limited use with respect to Rasch and 2PL models: it is mainly used in the educational setting, where it is reasonable to assume that students try to guess the correct answer, especially when their ability level is low. Note that, for the 3PL model, expressions more complex than (3.10) exist for the probability $p(Y|\theta)$ of the item responses provided by all subjects.

3.6 Random-Effects Approach

Until now, we indifferently considered θ_i as a fixed-person parameter or as a specific value assumed by the latent variable Θ_i for person i. In principle, the fixed-effects approach has to be adopted when we are referring to a fixed group of subjects, so that the variability between hypothetical repeated responses provided by the same subject to the same item is only due to accidental factors (Holland, 1990). The random-effects approach has to be adopted when the group of subjects who respond to the questionnaire is a sample drawn from a population in which the ability has a certain continuous or discrete distribution. As will be made clear in Chapter 5, the two approaches have important implications on the estimation methods that can be used, and this mainly drives the choice of the approach that is adopted.

As will be illustrated in the following sections, under the random-effects approach, many IRT models can be cast in the class of nonlinear mixed models. Moreover, a type of IRT model characterized by homogeneous classes of individuals is defined when a discrete distribution is assumed for Θ_i.

3.6.1 Nonlinear Mixed Framework

Under the normality assumption for Θ_i, certain IRT models can be conceptualized as *nonlinear mixed models* (Rijmen et al., 2003; De Boeck and Wilson, 2004b). For an overview on mixed models, which are closely related to multilevel or hierarchical models, see McCulloch and Searle (2001), Goldstein (2003), Skrondal and Rabe-Hesketh (2004), and Snijders and Bosker (2012), among others. See also Skrondal and Rabe-Hesketh (2004) and Raykov and Marcoulides (2011) for the connections between IRT models and nonlinear factor analysis models. The unifying framework of nonlinear mixed models is especially useful as it allows traditional IRT models to be readily extended and adapted to complex and realistic settings involving longitudinal data, hierarchical data, and individual covariates (see Chapter 6 for details). Moreover, both traditional and newly formulated models can be easily estimated using general statistical software for generalized linear and nonlinear mixed models, such as Stata and R, rather than dedicated software that usually are not very flexible.

In the following, we illustrate how the Rasch model and the 2PL model can be formulated as nonlinear mixed models. Note that some IRT models do not fit within this framework (Rijmen et al., 2003), such as IRT models that include a pseudo-guessing parameter (i.e., the 3PL model illustrated in Section 3.5).

The typical setting of IRT models relies on hierarchical data structures with item responses nested within individuals. Therefore, items are considered as first-level units and individuals as second-level units. Moreover, every random variable Θ_i is assumed to follow a normal distribution with mean 0 and variance common to all individuals.

The first-level model (or item model) is defined for the item responses as follows:

$$\text{logit}[p_j(\theta_i)] = \gamma_{i0} + \gamma_{i1}1\{j = 1\} + \cdots + \gamma_{iJ}1\{j = J\}, \quad i = 1, \ldots, n, \, j = 1, \ldots, J. \tag{3.12}$$

In addition, being the intercept γ_{i0} and the slopes $\gamma_{i1}, \ldots, \gamma_{iJ}$ subject specific, it is necessary to formulate a second-level model (or person model) for these parameters. By assuming a random intercept and fixed slopes, the following model is obtained:

$$\begin{cases} \gamma_{i0} = \beta_0 + \Theta_i, & \Theta_i \sim N(0, \sigma_\theta^2), \\ \gamma_{i1} = \quad -\beta_1, \\ \vdots \qquad \vdots \\ \gamma_{iJ} = \quad -\beta_J. \end{cases} \qquad (3.13)$$

Note that, according to the terminology of mixed models, Θ_i is conceived as a random-effect or second-level residual and its mean is conventionally constrained to zero, so that the ability of individual i may be interpreted as deviation from the population mean ability. Finally, by substituting Equation 3.13 in Equation 3.12, the complete model is obtained as

$$\text{logit}[p_j(\theta_i)] = \beta_0 - \beta_1 1\{j = 1\} \cdots - \beta_J 1\{j = J\} + \theta_i. \qquad (3.14)$$

Constraining $\beta_0 = 0$ to ensure the model identifiability, we obtain

$$p_j(\theta_i) = \frac{e^{\theta_i - \beta_1 1\{j=1\} \cdots - \beta_J 1\{j=J\}}}{1 + e^{\theta_i - \beta_1 1\{j=1\} \cdots - \beta_J 1\{j=J\}}} = \frac{e^{\theta_i - \beta_j}}{1 + e^{\theta_i - \beta_j}},$$

which is the same as Equation 3.4. Alternatively, it is possible to impose $\beta_j = 0$ with j belonging to $\{1, \ldots, J\}$ (usually, the first or the last one) or $\sum_{j=1}^{J} \beta_j = 0$. Therefore, the Rasch model is formally equivalent to a nonlinear mixed model and, more precisely, to a two-level random intercept logistic model (Agresti et al., 2000; Kamata, 2001) based on a suitable parameterization of the covariates.

The mixed logistic model formulated previously may be extended by including loading parameters $\lambda_1, \ldots, \lambda_J$ in Equations 3.12 and 3.13 as follows:

$$\text{logit}[p_j(\theta_i)] = \gamma_{i0}(\lambda_1 1\{j = 1\} + \cdots + \lambda_J 1\{j = J\}) \\ + \gamma_{i1} 1\{j = 1\} + \cdots + \gamma_{iJ} 1\{j = J\},$$

and

$$\begin{cases} \gamma_{i0} = \quad \Theta_i, & \Theta_i \sim N(0, \sigma_\theta^2), \\ \gamma_{i1} = -\lambda_1 \beta_1, \\ \vdots \qquad \vdots \\ \gamma_{iJ} = -\lambda_J \beta_J. \end{cases}$$

Assuming item-specific and person-fixed loading parameters, the 2PL model described in Section 3.4 is obtained, where the new parameters are interpreted as discriminant indices.

3.6.2 Latent Class Approach

The IRT models based on assumptions (3.4), (3.8), and (3.11) are also compatible with the assumption that the population under study is composed of homogeneous classes (or subpopulations) of individuals who have very similar unobservable characteristics. This is the formulation adopted in the LC model (Lazarsfeld and Henry, 1968; Goodman, 1974). In some situations, this assumption is particularly convenient, as it allows us to cluster individuals. In health care, for instance, by introducing the discreteness of the latent trait it is possible to single out a certain number of clusters of patients that will receive the same clinical treatment. Moreover, this assumption implies several advantages for the estimation process, as will be discussed in more detail in Chapter 5. See, among others, Langheine and Rost (1988), Heinen (1996), Fieuws et al. (2004), and Formann (2007a) for a critical discussion about discretized variants of IRT models; and Rost (1990), Lindsay et al. (1991), Formann (1995), Rost and von Davier (1995), and Bartolucci (2007) for some examples of this class of models for dichotomously scored items.

In the LC-IRT approach here discussed, every random variable Θ_i, $i = 1, \ldots, n$, is assumed to have a discrete distribution with support points $\xi_1, \ldots, \xi_k$ and corresponding weights $\pi_1, \ldots, \pi_k$. Each weight π_v ($v = 1, \ldots, k$) represents the probability that a subject belongs to class v, that is,

$$\pi_v = p(\Theta_i = \xi_v),$$

with $\sum_{v=1}^{k} \pi_v = 1$ and $\pi_v \geq 0$, $v = 1, \ldots, k$. The number of latent classes can either be assumed a priori, on the basis of theoretical knowledges or substantial reasons (e.g., it is necessary to cluster students in five classes because it is required by certain academic rules), or selected by comparing the fit of the model under different values of k. Procedures to select this number of components will be discussed in Section 5.7.1.

Another relevant element is represented by the ICC $p_j(\xi_v) = p(y_{ij} = 1 | \Theta_i = \xi_v)$, which can be specified as in Equations 3.4, 3.8, and 3.11 according to parameterizations of type Rasch, 2PL, and 3PL, respectively, letting θ_i equal to ξ_v. Note that, as usual in the LC model, individuals do not differ within latent classes, as the same ability level ξ_v is assumed for all individuals in class v. Moreover, the item parameters β_j, λ_j, and δ_j are supposed to be constant across classes. Both these hypotheses may be relaxed so as to obtain a more general model, even if it is less parsimonious and with a more complex interpretation. An example of such a generalized LC-IRT model is provided by the mixed Rasch model (Rost, 1990). This model allows for class-specific item parameters β_{jv} describing the difficulty of item j when it is answered by individuals in latent class v. The resulting ICC is given by

$$p_j(\xi_v) = \frac{e^{\xi_v - \beta_{jv}}}{1 + e^{\xi_v - \beta_{jv}}}, \quad v = 1, \ldots, k. \tag{3.15}$$

In an LC-IRT model for binary items, the basic assumptions of unidimensionality, local independence, and monotonicity of the probability of endorsing an item (see Section 3.2) are still valid. Therefore, according to Equation 3.1, the joint distribution of the response vector y_i given the latent class to which subject i belongs is defined as

$$p(y_i|\xi_v) = \prod_{j=1}^{J} p_j(\xi_v)^{y_{ij}} [1 - p_j(\xi_v)]^{1-y_{ij}}, \tag{3.16}$$

whereas the manifest distribution of y_i is obtained by substituting the integral in Equation 3.2 with the weighted sum:

$$p(y_i) = \sum_{v=1}^{k} \pi_v p(y_i|\xi_v). \tag{3.17}$$

3.7 Summary about Model Estimation

The choice between the fixed-effects and the random-effects approach has relevant consequences on the method adopted for the parameter estimation. Here, we provide a general summary and refer the reader to Chapter 5 for details.

The most commonly used estimation methods for IRT models are

- *Joint (or unconditional) maximum likelihood* (JML) method
- *Conditional maximum likelihood* (CML) method
- *Marginal maximum likelihood* (MML) method

The basic method to estimate the parameters of an IRT model is the JML method (Birnbaum, 1968; Wright and Masters, 1982), which is specific of the fixed-parameters approach. It consists of maximizing the likelihood of the model, corresponding to the probability of the observed data matrix Y given in (3.3), with respect to the ability and item parameters jointly. Though the method is very simple to implement, it has some drawbacks, the main of which is the lack of consistency of the resulting estimator for J fixed as n grows to infinity. The reason is that the number of the ability parameters increases with the sample size.

An alternative method of estimation, which may be used only for Rasch-type models (i.e., the Rasch model in the binary case), is the CML method (Andersen, 1970, 1972; Chamberlain, 1980). It is typically used to estimate only the vector β of the item parameters and is based on the maximization of the conditional likelihood of these parameters given a set of minimal sufficient statistics for every θ_i, which, as outlined in Section 3.3, corresponds to the vector of raw scores $(y_1., \ldots, y_n.)'$. The vector θ is eliminated by conditioning on the sufficient statistics and need not be estimated; therefore, the resulting estimator of β is consistent. To estimate the vector θ, another procedure has to be used: usually, the item parameter estimates are treated as known, and then, ability estimates are obtained by a simplified version of the JML method.

Finally, the MML method, suitable for IRT models formulated under a random-effects approach, is of great interest. It consists in maximizing the marginal likelihood corresponding to the manifest probability $p(y_i)$. The distribution that is assumed on the latent trait may be treated in different ways, as outlined in the previous section. A common practice consists in assuming that Θ_i is normally distributed, and then, the marginal likelihood is obtained by integrating the probability of the item response patterns over the ability distribution, as in Equation 3.2; see Bock and Lieberman (1970) and Bock and Aitkin (1981). In such a way, the random ability parameters are removed, and the item parameter estimates (other than the variance of the latent trait) are obtained by maximizing the resulting marginal log-likelihood. The main advantage of the MML method is that the parameter estimates are consistent under the aforementioned assumption of normality. However, if this assumption does not hold, parameter estimates are typically biased; moreover, the method is more complex to implement than JML, due to the presence of the integral involved in the expression of the marginal likelihood. There are several approaches that exist to deal with this integral, such as Gaussian quadrature–based methods (Abramowitz and Stegun, 1965), which include the adaptive Gaussian quadrature (Pinheiro and Bates, 1995) and Monte Carlo–based integration methods. Each of these methods replaces the integral in Equation 3.2 by a finite sum, and the resulting expression is then maximized through direct algorithms, such as the Newton–Raphson or the Fisher-scoring, or indirect algorithms, such as the expectation–maximization (EM, Dempster et al., 1977).

An alternative to the MML method is based on assuming that the latent trait has a discrete distribution with support points and weights which need to be estimated together with the item parameters. In such an approach, the marginal likelihood is obtained by summing the probability of the observed item response patterns over the ability distribution, as in Equation 3.17. The approach will be denoted in the following as MML-LC, to distinguish it from that based on the normality of the latent trait. For the Rasch model, this method was studied in detail by Lindsay et al. (1991), who showed that, under certain conditions, the estimates of the item parameters obtained from the MML-LC method are equal to those obtained from the CML method and are then consistent (see also Formann, 1995, 2007a). Moreover, the maximization of the log-likelihood based on Equation 3.17 is easier to perform than the maximization of that based on Equation 3.2 as the problem of solving the integral involved in the MML approach is skipped. Typically, the MML-LC estimates are obtained through an EM algorithm (Bartolucci, 2007).

Once the item parameters have been estimated, person parameters can be estimated by treating item parameters as known and maximizing the log-likelihood with respect to the latent trait or, alternatively, using the expected value under the corresponding posterior distribution or the maximum of the posterior probabilities.

About the MML-LC method, a final point concerns the choice of the number of support points of the latent distribution or, equivalently, latent classes (k). This choice is crucial in applications where k cannot be *a priori* fixed. Among the possible selection criteria, one of the most relevant in this literature is the Bayesian information criterion (BIC, Schwarz, 1978), which is based on the minimization of an index that may be seen as a penalized version of the log-likelihood for the number of free parameters. This criterion aims at selecting the model, in our case the value of k, corresponding to the best compromise between goodness-of-fit and model complexity.

3.8 Examples

In the following, we illustrate, through data analysis, how to estimate in Stata and R the IRT models described in the previous sections. First, we illustrate the Stata commands through the analysis of a dichotomized version of the RLMS dataset (see Section 1.8.2 for the dataset description). Then, we provide an analysis in R of the INVALSI mathematical dataset illustrated in Section 1.8.1.

3.8.1 RLMS Binary Data: Rasch and 2PL Models in Stata

For Stata users, there are several possibilities to estimate IRT models. The most flexible routine is represented by gllamm (Rabe-Hesketh et al., 2004), which can be freely downloaded by typing

```
. ssc install gllamm, replace
. help gllamm
```

in the Stata console. The acronym gllamm stands for generalized linear latent and mixed models, which refers to a wide class of multilevel latent variable models for multivariate responses of mixed types including continuous, dichotomous, and ordered and unordered polytomous responses. Moreover, the latent variables can be assumed to be discrete or multivariate normally distributed. In such a class, IRT models are also included, as described in Section 3.6.1. Under the assumption of normality, gllamm uses the MML estimation method with adaptive Gaussian quadrature (see Chapter 5 for details); this is the main estimation method used to obtain the results in this section.

3.8.1.1 Data Organization

We begin loading the dichotomized version of the RLMS dataset, denoted as RLMS_bin; we only keep the four items and drop the covariates:

```
. use "RLMS_bin.dta", clear
. keep Y1-Y4
```

Then, we need to properly prepare the data to be analyzed in the context of mixed models. First, it is useful to collapse the data to obtain a record for each distinct response pattern instead of a record for each individual. This data collapsing typically results in a quicker estimation process, in the presence of binary data and relatively few items, as the frequency of repeated patterns among different individuals may be substantially high. To collapse the data, we generate a new variable corresponding the frequency (or weight, wt2) of each response pattern, and then, we use function collapse as follows:

```
. gen cons=1
. collapse (sum) wt2=cons, by (Y1-Y4)
```

In this way, the original 1418 records are reduced to only 16 records.

Second, we need to convert our dataset from the wide shape to the long shape (unless data are already in the long format), coherently with hierarchical data structures treated by mixed models. In other words, for an IRT model formulated according to Equations 3.12 and 3.13, it is necessary to have a record for each individual–item combination (or response pattern–item combination, when the data are collapsed).

```
. gen ind=_n
. reshape long Y, i(ind) j(item)
(note: j = 1 2 3 4)

Data                            wide    ->   long
-----------------------------------------------------------------
Number of obs.                    16    ->     64
Number of variables                6    ->      4
j variable (4 values)                   ->   item
xij variables:
                        Y1 Y2 ... Y4    ->   Y
-----------------------------------------------------------------
```

The new variable Y contains the observed item responses, whereas ind corresponds to the individuals or response patterns, and item indicates the item corresponding to each value reported in Y. Finally, $J = 4$ dummies are generated to identify the four items, as shown in Equation 3.12.

```
. qui tab item, gen(d)
```

In this way, the original dataset is reshaped and appears as follows, where ind denotes the response pattern, Y is the response, wt2 is the absolute frequency of the corresponding response pattern, and d1–d4 are the item indicators:

```
. list in 1/10
```

```
     +-------------------------------------------------------+
     |  ind    item        Y     wt2    d1    d2    d3    d4  |
     |-------------------------------------------------------|
 1.  |   1       1         0     361    1     0     0     0   |
 2.  |   1       2         0     361    0     1     0     0   |
 3.  |   1       3         0     361    0     0     1     0   |
 4.  |   1       4         0     361    0     0     0     1   |
 5.  |   2       1         0      35    1     0     0     0   |
     |-------------------------------------------------------|
 6.  |   2       2         0      35    0     1     0     0   |
 7.  |   2       3         0      35    0     0     1     0   |
 8.  |   2       4         1      35    0     0     0     1   |
 9.  |   3       1         0      18    1     0     0     0   |
10.  |   3       2         0      18    0     1     0     0   |
     +-------------------------------------------------------+
```

To avoid estimates of the difficulty parameters with opposite sign with respect to the traditional notation used in Equation 3.4, we create new indicators nd_j ($j = 1, \ldots, 4$), which assume values -1 and 0.

```
forvalues j=1/4 {
    generate nd`j'=-d`j'
    }
```

The reshaped dataset may be visualized as usual through a descriptive tabulate, which provides the absolute frequencies of unsatisfied (Y = 0) and satisfied (Y =1) individuals out of the 1418 complete questionnaires.

```
. tab Y item  [fweight=wt2]
```

Y	item 1	2	3	4	Total
0	511	541	917	818	2,787
1	907	877	501	600	2,885
Total	1,418	1,418	1,418	1,418	5,672

3.8.1.2 Analysis in gllamm under the Assumption of Normality of the Latent Trait

We start the analysis of the data by estimating the Rasch model. As shown in Section 3.6.1, this model may be formulated as a two-level logistic model with random intercept corresponding to the latent trait (i.e., job satisfaction). In the used package, this type of model is easily estimated by invoking the gllamm function and by specifying the item response variable (Y), the item indicators (nd1, ..., nd4), the link function (logit, but also link probit is admitted), the second-level units (ind), and the frequencies of the response patterns (wt). Further options are also admitted (see the gllamm manual by Rabe-Hesketh et al., 2004, for further explanations). Here, we just point out that a model with discrete random effects is estimated by adding options ip(f) and nip(#), where # denotes the number of latent classes of individuals.

```
. * Rasch model
. gllamm Y nd1-nd4, nocons link(logit) fam(bin) i(ind) w(wt) adapt dots
```

The output of the gllamm function appears as follows:

```
Running adaptive quadrature
....Iteration 0:    log likelihood = -3569.644
....Iteration 1:    log likelihood = -3084.3225
....Iteration 2:    log likelihood = -3019.6748
....Iteration 3:    log likelihood = -3018.9945
....Iteration 4:    log likelihood = -3018.9919

Adaptive quadrature has converged,
running Newton-Raphson
.........Iteration 0:log likelihood = -3018.9919
.........Iteration 1:log likelihood = -3018.9919
.........Iteration 2:log likelihood = -3018.9918

number of level 1 units = 5672
number of level 2 units = 1418

Condition Number = 2.3868812

gllamm model

log likelihood = -3018.9918

------------------------------------------------------------------------------
       Y |      Coef.   Std. Err.      z    P>|z|     [95% Conf. Interval]
---------+--------------------------------------------------------------------
     nd1 |  -1.191568     .11953    -9.97   0.000    -1.425843   -.9572939
     nd2 |  -.9976521     .11812    -8.45   0.000    -1.229163   -.7661411
     nd3 |   1.305122    .1197148   10.90   0.000     1.070486    1.539759
     nd4 |   .6919077    .1160341    5.96   0.000      .464485    .9193303
------------------------------------------------------------------------------

Variances and covariances of random effects
------------------------------------------------------------------------------
***level 2 (ind)
    var(1): 8.4114186 (.68772446)
------------------------------------------------------------------------------
```

The output displayed earlier contains some estimation process details and certain information about the multilevel data structure. In particular, we note that the RLMS dataset is composed by 5672 first-level units (i.e., four item responses for each individual) and 1418 second-level units (i.e., individuals). Then, the condition number and the log-likelihood value are displayed. The condition number is defined as the square root of the ratio of the largest to smallest eigenvalues of the Hessian matrix: it is large when the Hessian matrix is nearly singular indicating that the model is not well identified. In our case, we obtain a satisfactory value equal to 2.387.

Finally, the parameter estimates are shown. First, the estimated difficulties $\hat{\beta}_j$ are displayed for each item j (column named `Coef.` in correspondence of rows `nd1` to `nd4` for items 1–4, respectively), with the corresponding standard errors (column `Std. Err.`), z-values, p-values, and inferior and superior limits of the confidence intervals at the 95% level. According to this output, the most difficult item is the third ($\hat{\beta}_3 = 1.305$), followed by item 4 ($\hat{\beta}_4 = 0.692$), and then by items 2 ($\hat{\beta}_2 = -0.998$) and 1 ($\hat{\beta}_1 = -1.192$). We recall that the random intercept is assumed to be normally distributed with mean equal to 0. Its estimated variance is equal to 8.412 with a standard error of 0.688.

Note that the same model may be estimated by constraining one difficulty parameter, say, the last one, to 0 and by including the constant term, which represents the mean value of the latent variable, as follows:

```
. * Rasch model
. gllamm Y nd1-nd3, link(logit) fam(bin) i(ind) w(wt) adapt dots


[…] Output omitted

--------------------------------------------------------------------------
         Y |      Coef.    Std. Err.      z     P>|z|     [95% Conf. Interval]
-----------+--------------------------------------------------------------
       nd1 |  -1.883501    .1225961    -15.36   0.000    -2.123785    -1.643217
       nd2 |  -1.689583    .1200026    -14.08   0.000    -1.924784    -1.454382
       nd3 |   .6132173    .1126315      5.44   0.000     .3924635     .833971
     _cons |  -.6919448    .1160325     -5.96   0.000    -.9193644    -.4645251
--------------------------------------------------------------------------

Variances and covariances of random effects
--------------------------------------------------------------------------
***level 2 (ind)

  var(1): 8.4118768 (.68777541)
--------------------------------------------------------------------------
```

This output is perfectly equivalent to the previous one. In fact, we may shift from one output to the other by adding $\_cons = 0.692 = \hat{\beta}_4$ to each difficulty parameter.

Typically, in the analysis of a questionnaire, we are interested not only in the item parameter estimation but also in the ranking of individuals on the latent trait. Thus, one may estimate the job satisfaction level θ_i for each response pattern through function `gllapred`:

```
. * Prediction of ability levels
. gllapred thetaRM, u
(means and standard deviations will be stored in
 thetaRMm1 thetaRMs1)
Non-adaptive log-likelihood: -3027.8949
-3018.7229 -3018.9897 -3018.9911 -3018.9915 -3018.9914
log-likelihood:-3018.9914
```

For each record of the long-shaped dataset, two new variables are generated containing the empirical Bayes estimates (Rabe-Hesketh et al., 2004) of the latent variable (`thetaRMm1`) and the corresponding standard errors (`thetaRMs1`).

```
. list ind wt2 thetaRMm1 thetaRMs1 if d1==1
. * Alternatively:
. * collapse theta*, by(ind)
. * list

     +------------------------------------+
     | ind    wt2    thetaRMm1   thetaRMs1 |
     |------------------------------------|
  1. |   1    361   -3.4198103   1.7241539 |
  5. |   2     35   -1.3743963   1.2217322 |
  9. |   3     18   -1.3743963   1.2217322 |
 13. |   4      4    -.04588027  1.1209682 |
 17. |   5     62   -1.3743963   1.2217322 |
     |------------------------------------|
 21. |   6     12    -.04588027  1.1209682 |
 25. |   7      8    -.04588027  1.1209682 |
 29. |   8     11    1.2912304   1.2287342 |
 33. |   9     75   -1.3743963   1.2217322 |
 37. |  10     18    -.04588027  1.1209682 |
     |------------------------------------|
 41. |  11     16    -.04588027  1.1209682 |
 45. |  12     14    1.2912304   1.2287342 |
 49. |  13    186    -.04588027  1.1209682 |
 53. |  14    168    1.2912304   1.2287342 |
 57. |  15     92    1.2912304   1.2287342 |
     |------------------------------------|
 61. |  16    338    3.3642703   1.7365411 |
     +------------------------------------+
```

These estimates may be stored to be used in the following:

```
. estimates store RM
```

After estimating the item parameters and predicting the latent trait levels, it is possible to calculate the probability to endorse a specific item by a given subject, according to Equation 3.4. In order to understand how the probability of endorsing an item varies with the job satisfaction, some values of this probability are shown in Table 3.2 for low, intermediate, and high levels of the latent trait (as it results from variable `thetaRMm1`) and for the four items of

TABLE 3.2

Estimated Probability to Endorse an Item for Some Values of
Job Satisfaction and Item Difficulty under the Rasch Model

j	$\hat{\theta}_i$	$\hat{\beta}_j$	$\hat{p}_j(\hat{\theta}_i)$
1	−1.374	−1.192	0.455
2	−1.374	−0.998	0.407
3	−1.374	0.692	0.112
4	−1.374	1.305	0.064
1	−0.046	−1.192	0.759
2	−0.046	−0.998	0.722
3	−0.046	0.692	0.323
4	−0.046	1.305	0.206
1	1.291	−1.192	0.923
2	1.291	−0.998	0.908
3	1.291	0.692	0.645
4	1.291	1.305	0.497

the Russian dataset. It is confirmed that the probability to positively answer an item decreases as its difficulty increases (job satisfaction level being constant) and increases as the job satisfaction level increases (the item difficulty being constant).

We now consider the 2PL model that may be estimated by `gllamm` similar to the Rasch model, the main difference being represented by the presence of the discriminating parameters. To introduce these multiplicative factors, it is necessary to define an equation containing the names of the items with different slopes and then to retrieve them in the `gllamm` command by the `eqs` option.

```
. * 2PL model
. local ll=e(ll)
. eq load: nd1-nd4
. gllamm Y nd1-nd4, nocons link(logit) fam(bin) i(ind) eqs(load) w(wt) adapt dots

[...] Output omitted

---------------------------------------------------------------------
       Y |      Coef.   Std. Err.      z    P>|z|     [95% Conf. Interval]
---------+-----------------------------------------------------------
     nd1 |  -1.782488   .2533724    -7.04   0.000    -2.279089   -1.285888
     nd2 |  -1.395524   .2044757    -6.82   0.000    -1.796289    -.9947592
     nd3 |   1.009236   .1105253     9.13   0.000     .7926103    1.225862
     nd4 |   .5111746   .1009975     5.06   0.000     .3132232     .709126
---------------------------------------------------------------------

Variances and covariances of random effects
---------------------------------------------------------------------

***level 2 (ind)

   var(1): 19.464344 (4.1315634)
```

```
loadings for random effect 1
nd1: 1 (fixed)
nd2: .91289987 (.14573699)
nd3: .48128091 (.06563445)
nd4: .48891847 (.06562459)
```

The output for the 2PL model is similar to that described previously for the Rasch model. In addition, it contains information on the discriminating parameter estimates, denoted by `loadings for random effect 1`. Note that $\lambda_1 = 1$ by default and that `gllamm` includes the multiplicative factor only on the latent trait (i.e., the exponent in the right hand side of Equation 3.8 is specified as $\lambda_j \theta_i - \beta_j^*$ rather than $\lambda_j(\theta_i - \beta_j)$); therefore, in order to have a parameterization comparable to that adopted in this book, the estimates of the item difficulties must be divided by the estimates of the discriminating parameters.

Concerning the discriminating parameters, we observe that items 1 and 2 have very similar slopes; the same holds for items 3 and 4. Regarding difficulties, the ranking of items is the same as in the Rasch model. Moreover, for each response pattern, we may obtain the empirical Bayes estimates of job satisfaction and store them for a subsequent analysis.

```
. * Prediction of ability levels
. gllapred theta2PLM, u
(means and standard deviations will be stored in theta2PLMm1 theta2PLMs1)
Non-adaptive log-likelihood: -3007.9628
-2987.9935 -2992.1499 -2992.2549 -2992.2637 -2992.2634 -2992.2635
-2992.2634
log-likelihood:-2992.2634

. list ind theta2PLMm1 theta2PLMs1 if d1==1

      +-------------------------------+
      | ind    theta2P~m1    theta2~s1 |
      |-------------------------------|
  1.  |   1    5.3009924    2.4898138  |
  5.  |   2     3.142685     1.75689   |
  9.  |   3    3.1666255    1.7634459  |
 13.  |   4    1.8827706    1.5232518  |
 17.  |   5    2.0165103    1.5362076  |
      |-------------------------------|
 21.  |   6     .90321498   1.5180752  |
 25.  |   7     .92080303   1.5168771  |
 29.  |   8    -.31292648   1.684569   |
 33.  |   9    1.8140526    1.5177751  |
 37.  |  10     .70023846   1.5348686  |
      |-------------------------------|
 41.  |  11     .71823475   1.533166   |
 45.  |  12    -.56886429   1.7350812  |
 49.  |  13    -.39880887   1.7010878  |
 53.  |  14    -2.1343595   2.0990849  |
 57.  |  15    -2.1009547   2.0907405  |
      |-------------------------------|
 61.  |  16    -4.9574602   2.8115066  |
      +-------------------------------+

. estimates store TWOPLM
```

TABLE 3.3

Estimated Probability to Endorse an Item for Some Values of Job Satisfaction, Item Difficulty, and Item Discriminating Power under the 2PL Model

j	$\hat{\theta}_i$	$\hat{\beta}_j$	$\hat{\lambda}_j$	$\hat{p}_j(\hat{\theta}_i)$
2	−3.160	−1.392	0.913	0.166
3	−3.160	1.011	0.482	0.118
2	0.401	−1.392	0.913	0.837
3	0.401	1.011	0.482	0.427
2	2.135	−1.392	0.913	0.962
3	2.135	1.011	0.482	0.632

As for the Rasch model, after estimating the 2PL model parameters, we may calculate the probabilities to endorse a certain item by a given individual on the basis of Equation 3.8. The results shown in Table 3.3 refer to low, intermediate, and high levels of job satisfaction and to the parameters for items 2 and 3, that is, low difficulty ($\hat{\beta}_2 = -1.392$) and high discriminating power ($\hat{\lambda}_2 = 0.913$) for item 2 and high difficulty ($\hat{\beta}_3 = 1.011$) and low discriminating power ($\hat{\lambda}_3 = 0.482$) for item 3.

With respect to Table 3.2, where all items have the same discriminating power, we now observe that the trend of $\hat{p}_j(\hat{\theta}_i)$ is less predictable. For instance, for a subject with a low level of job satisfaction ($\hat{\theta}_i = -3.160$), the probability to endorse item 3, having $\hat{\beta}_3 = 1.011$, rather than item 2, having $\hat{\beta}_2 = -1.392$, decreases from 0.166 to 0.118, which is much less than the reduction we would observe if items 2 and 3 had the same discriminating power. Indeed, if both items had, for instance, $\hat{\lambda}_2 = \hat{\lambda}_3 = 0.913$, then the probability to endorse the item would decrease from 0.166 to 0.022, whereas if both items had $\hat{\lambda}_2 = \hat{\lambda}_3 = 0.482$, then the probability to endorse the item would decrease from 0.299 to 0.118.

Once the Rasch model and the 2PL model have been estimated, it may be of interest to compare them. Since the Rasch model is nested in the 2PL model, we may perform a likelihood ratio (LR) test based on the χ^2 distribution with 3 degrees of freedom. Note that this is the number of additional parameters estimated under the 2PL model with respect to the Rasch model.

```
. lrtest RM TWOPLM

Likelihood-ratio test              LR chi2(3)   =      53.60
(Assumption: RM nested in TWOPLM)  Prob > chi2 =     0.0000
```

According to these results, we reject the Rasch parameterization in comparison with the 2PL parameterization and, therefore, we conclude for a significant difference in the discriminating power between the items.

Finally, to better understand the consequences of using the Rasch model instead of the 2PL model or vice versa, we may compare the job satisfaction individual estimates. For this, we first convert the dataset to a wide shape:

```
. drop d1-d4
. drop nd1-nd4
. reshape wide Y, i(ind) j(item)
(note: j = 1 2 3 4)

Data                                long    ->   wide
-----------------------------------------------------------------
Number of obs.                        64    ->      16
Number of variables                   10    ->      12
j variable (4 values)               item    ->   (dropped)
xij variables:
                                       Y    ->   Y1 Y2 ... Y4
-----------------------------------------------------------------
```

Then, we calculate the raw score for each response pattern:

```
. gen score = Y1 + Y2 + Y3 + Y4
. sort score
```

Finally, we create a table to visualize, for each response pattern, the raw score and the job satisfaction estimates under the Rasch model and the 2PL model, respectively. Note that the items are ordered according to their difficulty, from the easiest to the most difficult.

```
. list Y1 Y2 Y4  Y3   score   thetaRMm1 theta2PLMm1, nolab clean

        Y1   Y2   Y4   Y3   score   thetaRMm1   theta2P~m1
  1.     0    0    0    0       0   -3.4198155    5.3009924
  2.     0    0    0    1       1   -1.3743979    3.1666255
  3.     1    0    0    0       1   -1.3743979    1.8140526
  4.     0    1    0    0       1   -1.3743979    2.0165103
  5.     0    0    1    0       1   -1.3743979    3.142685
  6.     0    0    1    1       2   -.04587998    1.8827706
  7.     1    1    0    0       2   -.04587998   -.39880887
  8.     0    1    1    0       2   -.04587998    .90321498
  9.     1    0    0    1       2   -.04587998    .71823475
 10.     0    1    0    1       2   -.04587998    .92080303
 11.     1    0    1    0       2   -.04587998    .70023846
 12.     1    1    0    1       3    1.2912326   -2.1009547
 13.     0    1    1    1       3    1.2912326   -.31292648
 14.     1    0    1    1       3    1.2912326   -.56886429
 15.     1    1    1    0       3    1.2912326   -2.1343595
 16.     1    1    1    1       4    3.364276    -4.9574602
```

The table shows similar results as in Table 3.1. Due to the property of sufficiency of the raw scores, individuals with the same raw score and different response patterns obtain the same job satisfaction level under the Rasch model. On the other hand, under the 2PL model, the location of individuals on the latent trait also depends on the response pattern.

Note that the property of sufficiency of raw scores, which holds under the Rasch model, may appear counterintuitive. For instance, one may think that a subject (say, subject 6 in the above output) who endorses the two most difficult items and fails the two easiest items (so that his or her response pattern is 0011) has a greater level of the latent trait than a subject (say, subject 7) who endorses the two easiest items and fails the two most difficult ones (so that his or her response pattern is 1100), even if their raw scores are equal. However, the behavior of subject 6 is really anomalous in terms of probabilities. Indeed, let us consider the response pattern probabilities shown in Table 3.4, calculated under the Rasch model and under the 2PL model according to Equations 3.6 and 3.9, respectively, and ordered as in the previous `Stata` output. We observe that under the Rasch model, the probability of observing the response pattern 1100 is equal to 0.294 (0.221 under the 2PL model), whereas response pattern 0011 is very uncommon, having probability 0.004 (0.015 under the 2PL model). In other words, the response pattern 6 may be considered anomalous and, as such, it can be taken off the dataset (Bond and Fox, 2007). Alternatively, it is possible to assume that another latent variable (e.g., a sort of *carelessness* as suggested by McDonald, 1999), different from the job satisfaction, influences the item responses. This last interpretation would suggest a more complex model than the Rasch one, such as a multidimensional IRT model (see Section 6.4).

TABLE 3.4

Estimated Response Pattern Probabilities under the Rasch and 2PL Models

	Pattern	Score	$\hat{p}(y_i|\hat{\theta}_i)$	
			Rasch	2PL
1	0000	0	0.809	0.852
2	0001	1	0.018	0.068
3	1000	1	0.224	0.176
4	0100	1	0.184	0.127
5	0010	1	0.034	0.084
6	0011	2	0.004	0.015
7	1100	2	0.294	0.221
8	0110	2	0.045	0.043
9	1001	2	0.029	0.051
10	0101	2	0.024	0.034
11	1010	2	0.054	0.064
12	1101	3	0.147	0.186
13	0111	3	0.022	0.018
14	1011	3	0.027	0.030
15	1110	3	0.272	0.239
16	1111	4	0.811	0.778

3.8.1.3 *Alternatives to* `gllamm` *for the Rasch Model*

The Rasch model may be estimated through other functions than `gllamm`, which differ for the estimation method and/or for the output.

For the Rasch model, we may use the `Stata` function `xtlogit` by specifying option `re`, which is devoted to the estimation of logit models with random effects. Different from the `gllamm` function, `xtlogit` uses the MML with nonadaptive (rather than adaptive) quadrature. From other perspectives, the two functions are very similar. First, it is necessary to convert the original dataset in the long shape and to create as many dummies as items.

```
. use "RLMS_bin.dta", clear
. keep Y1-Y4
. gen ind=_n
. reshape long Y, i(ind) j(item)
. qui tab item, gen(d)
. forvalues j=1/4 {
    generate nd`j'=-d`j'
  }
```

Then, we run function `xtlogit` as follows:

```
. * Rasch model:
. xtlogit Y nd1-nd4, nocons i(ind) re
```

or, alternatively, we may constrain one difficulty parameter to 0 and introduce the constant term

```
. * Rasch model:
. xtlogit Y nd1-nd3, i(ind) re
```

After showing some details about the estimation process, the following output is obtained from the first command:

```
[...] Output omitted
```

```
Random-effects logistic regression       Number of obs     =      5672
Group variable: ind                       Number of groups  =      1418

Random effects u_i ~ Gaussian             Obs per group: min =         4
                                                         avg =       4.0
                                                         max =         4

                                          Wald chi2(4)      =    486.63
Log likelihood  = -3018.321               Prob > chi2       =    0.0000
```

Y	Coef.	Std. Err.	z	P>|z|	[95% Conf. Interval]	
nd1	-1.189871	.1184092	-10.05	0.000	-1.421948	-.9577929
nd2	-.9958962	.1168471	-8.52	0.000	-1.224912	-.7668801
nd3	1.307379	.1189261	10.99	0.000	1.074288	1.54047
nd4	.6939007	.1147761	6.05	0.000	.4689437	.9188577
/lnsig2u	2.147626	.0843003			1.982401	2.312852

```
-------------+----------------------------------------------------------
   sigma_u |   2.926517    .1233531                     2.694467    3.178552
       rho |   .7224764    .0169026                     .6881648    .7543603
-------------------------------------------------------------------------
Likelihood-ratio test of rho=0: chibar2(01) =  1476.46 Prob >= chibar2 = 0.000
```

Some details about the hierarchical structure of the data are displayed. Then, the estimates of the item difficulty parameters β_j (column Coef. in correspondence of rows nd1 to nd4 for items 1 to 4, respectively) and of the standard deviation of the latent variable (denoted by sigma_u) are provided. The results are very similar to those obtained by gllamm (see also Table 3.5).

Moreover, with xtlogit, the empirical Bayesian estimates of random effects are obtained through function predict, which creates a new variable:

```
. * Prediction of ability levels
. predict u
(option xb assumed; linear prediction)
. list u in 1/10

     +-----------+
     |         u |
     |-----------|
  1. |  1.189871 |
  2. |  .9958962 |
  3. | -1.307379 |
  4. | -.6939007 |
  5. |  1.189871 |
     |-----------|
  6. |  .9958962 |
  7. | -1.307379 |
  8. | -.6939007 |
  9. |  1.189871 |
 10. |  .9958962 |
```

Other than by the MML approach, the Rasch model may be estimated by the CML method (see Section 3.7 and Chapter 5 for details), which relies on a fixed-effects approach and, therefore, does not require any assumption for the distribution of Θ_i. In Stata, this approach is implemented in two functions, clogit and xtlogit, specifying option fe (staying for fixed effects) instead of option re:

```
. clogit Y nd1-nd4, group(ind)
. xtlogit Y nd1-nd3, nocons i(ind) fe
```

TABLE 3.5

Item Difficulty Parameters and Variance of Job Satisfaction Estimated under the Rasch Model by gllamm, xtlogit, and raschtest

	gllamm	xtlogit,re	raschtest
$\hat{\beta}_1$	−1.1916	−1.1899	−1.1397
$\hat{\beta}_2$	−0.9977	−0.9959	−0.9477
$\hat{\beta}_3$	1.3051	1.3074	1.3529
$\hat{\beta}_4$	0.6919	0.6939	0.7569
$\hat{\sigma}^2$	8.4114	8.5645	8.3029

These functions also allow us to use some traditional diagnostic tools, for the description of which we refer to Chapter 5.

Finally, we remind the routine `raschtest`, developed by Hardouin (2007) and freely downloadable by typing directly on the `Stata` console, the following commands:

```
. ssc install raschtest
. help  raschtest
. ssc install elapse
. ssc install genscore
. * Also the  function ghquadm is required, which is available in the sg158_1 program that
    you find typing:
. findit ghquadm
```

This routine is specialized in estimating Rasch-type models with different estimation methods, that is, MML, CML, and generalized estimating equations. Here, we refer to MML to obtain results that are directly comparable with those from `gllamm`. Different from the other routines, `raschtest` does not require the long shape of data, so that it can be directly run after loading the initial dataset:

```
. * Load data
. use "RLMS_bin.dta", clear
. * Define an identificative for each subject
. gen ind=_n
. * Rasch model
. raschtest Y1-Y4, id(ind) method(mml) icc genlt(thetaRM2) genscore(score)
```

Function `raschtest` requires the definition of an identificative variable for the individuals, through option `id`; if the identificative at issue is absent, as in our case, it can be previously defined in the usual way, through command `gen`. In addition, several options allow us to choose the estimation method (`mml`, `cml`, or `gee`) and some graphical outputs (e.g., option `icc`). It is also possible to generate new variables containing the latent variable predictions (option `genlt`) and the raw scores (option `genscore`) for each individual. The output is as follows:

```
Estimation method: Marginal maximum likelihood (MML)
Number of items: 4
Number of groups: 5 (5 of them are used to
compute the statistics of test)
Number of individuals: 1418 (0 individuals removed
for missing values)
Number of individuals with null or perfect score: 699
Marginal log-likelihood: -3020.3290
Log-likelihood: -1492.3565
```

	Difficulty					Standardized	
Items	parameters	Std.Err.	R1m	df	p-value	Outfit	Infit
Y1	-1.13968	0.11797	74.956	2	0.0000	-15.317	-4.680
Y2	-0.94769	0.11740	16.038	2	0.0003	-18.278	-6.142

```
Y3      1.35285    0.11921   20.894    2   0.0000  4.273   1.554
Y4      0.75691    0.11735   45.980    2   0.0000  4.105  -0.129
-------------------------------------------------------------------
R1m test              R1m=  74.954    8   0.0000
-------------------------------------------------------------------
Sigma  2.88147    0.11587
-------------------------------------------------------------------
```

| | Ability | | | Expected |
Group Score	parameters	Std. Err.	Freq.	Score
0 0	-3.23332	1.35798	361	0.23
1 1	-1.53502	0.93016	190	0.90
2 2	-0.00081	1.25040	244	2.00
3 3	1.53710	0.93259	285	3.09
4 4	3.25197	1.35872	338	3.77

After printing a summary about the estimation process and about the data structure, the item difficulty parameter estimates and the standard deviation of the latent variable are presented. As shown in Table 3.5, results are similar to those obtained by gllamm and xtlogit (option re).

Function raschtest also provides some additional output. First, the estimates of job satisfaction (denoted as ability parameters) are printed for each raw score (note that we may also define different ways of grouping the raw scores) with the corresponding standard errors and frequencies. Moreover, graphical outputs may be obtained (see also Chapter 5 for more details about the use of graphical tools for diagnostic purposes). Option icc allows us to plot the observed and expected ICC values for each item and ability estimate, corresponding to a score different from 0 and J. As shown in Figure 3.6, items clearly have different slopes; items 1 and 2 have higher slopes than items 3 and 4. This result rises questions on the adequacy of the Rasch model in favor of the 2PL model.

3.8.2 INVALSI Mathematics Data: Rasch, 2PL, and 3PL Models in R

In order to analyze the INVALSI reduced dataset, we consider four R packages. The first three, named ltm (Rizopoulos, 2006), irtoys (Partchev, 2014), and mirt (Chalmers, 2012), are particularly suitable for an MML estimation when we assume that the ability has a normal distribution. The fourth package, named MultiLCIRT (Bartolucci et al., 2014), is suitable for estimating the Rasch model and the 2PL model under the assumption that the ability has a discrete distribution, and then the population is shared in latent classes corresponding to different ability levels.

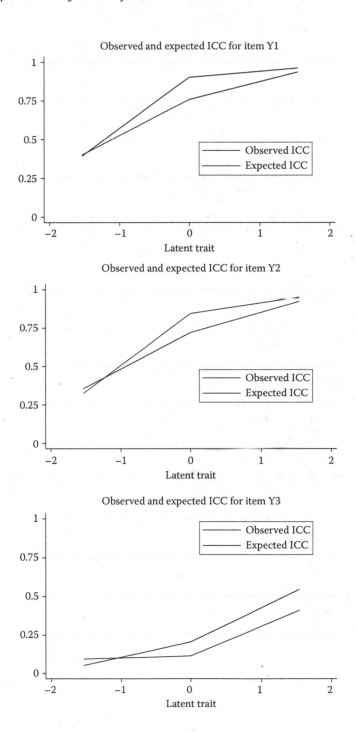

FIGURE 3.6
ICCs for items Y1–Y4.

(*Continued*)

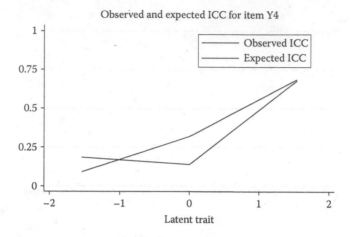

FIGURE 3.6 (Continued)
ICCs for items Y1–Y4.

3.8.2.1 Data Organization

First, the data that are in the file `Invalsi_reduced.RData` can be loaded through the following command:

```
> # Load data
> load("INVALSI_reduced.RData")
> # Show objects in memory
> ls()
"Y"
```

and a preliminary description of these data can be obtained by the following command:

```
> str(Y)
'data.frame': 1786 obs. of  27 variables:
 $ Y1 : int  1 1 1 1 1 0 1 1 1 ...
 $ Y2 : int  0 1 0 1 0 0 1 1 1 1 ...
 $ Y3 : int  1 1 1 1 0 1 1 1 1 ...
 $ Y4 : int  1 1 1 1 0 1 1 1 1 ...
 $ Y5 : int  1 1 1 1 1 1 1 1 1 ...
 $ Y6 : int  0 1 1 1 1 1 1 1 1 ...
 $ Y7 : int  1 0 0 1 0 0 0 0 0 ...
 $ Y8 : int  1 0 1 1 0 1 1 1 1 ...
 $ Y9 : int  1 1 1 1 0 1 0 1 1 1 ...
 $ Y10: int  1 0 1 1 0 1 1 1 1 ...
 $ Y11: int  0 1 0 1 1 1 1 1 1 ...
 $ Y12: int  1 1 1 1 1 1 1 1 1 ...
 $ Y13: int  1 1 0 1 0 1 1 1 1 ...
 $ Y14: int  1 1 1 1 0 0 0 1 1 ...
 $ Y15: int  1 1 1 1 1 1 1 1 1 ...
 $ Y16: int  1 1 1 1 0 1 1 1 1 ...
 $ Y17: int  1 1 1 1 1 1 1 1 1 ...
 $ Y18: int  1 1 0 1 0 1 1 1 1 ...
 $ Y19: int  1 1 0 1 0 1 1 1 1 ...
 $ Y20: int  1 1 0 1 0 1 0 1 1 1 ...
 $ Y21: int  0 1 1 1 1 1 1 1 1 ...
```

```
$ Y22: int  1 1 1 1 1 1 1 1 1 ...
$ Y23: int  1 1 0 1 0 0 1 1 1 1 ...
$ Y24: int  0 1 0 1 0 0 1 1 1 1 ...
$ Y25: int  0 0 0 1 0 0 1 1 0 1 ...
$ Y26: int  1 0 1 1 0 1 1 1 1 0 ...
$ Y27: int  1 1 0 1 0 0 1 1 1 0 ...
```

The dataset, which is a subset of the more general INVALSI one described in
Section 1.8.1, comprises the responses to $J = 27$ binary items in mathematics
provided by 1786 examinees.

3.8.2.2 Analysis under Normal Distribution for the Ability

We first load the package ltm through the standard R command after that
this package has been installed from the CRAN website:

```
> require(ltm)
Loading required package: ltm
Loading required package: MASS
Loading required package: msm
Loading required package: mvtnorm
Loading required package: polycor
Loading required package: sfsmisc
```

Different functions in the package may be used to fit the Rasch model based
on the assumption that the ability has a normal distribution. One possibility
is to simply use function rasch. In the following, we show how to use this
function, extract the estimates of the item difficulty levels, and display these
estimates together with the proportion of correct responses for each item.
Note that since this function estimates easiness item parameters $(-\beta_j)$, we
obtain difficulty parameter estimates by changing the sign.

```
> # Rasch model
> out = rasch(Y)
> # Extract difficulty parameters and compare them with success rates
> out$beta = -out$coefficients[,1]
> cbind(out$beta,prop=colMeans(Y))
          beta       prop
Y1   -2.13370026 0.8600224
Y2   -2.09741641 0.8561030
Y3   -0.71437646 0.6444569
Y4   -2.63961969 0.9064950
Y5   -4.49857442 0.9826428
Y6   -3.12882702 0.9384099
Y7    0.06050001 0.4876820
Y8   -1.36569829 0.7586786
Y9   -2.11814045 0.8583427
Y10  -1.90829607 0.8342665
Y11  -1.92676201 0.8365062
Y12  -1.50531362 0.7799552
Y13  -1.14319465 0.7222844
Y14  -1.45306637 0.7721165
Y15  -1.78792661 0.8191489
Y16  -2.44350327 0.8902576
Y17  -1.66216328 0.8023516
Y18  -1.60996394 0.7950728
Y19  -0.52798599 0.6080627
```

```
Y20 -1.41608993 0.7665174
Y21 -1.49022725 0.7777156
Y22 -2.13895742 0.8605823
Y23  0.65213031 0.3678611
Y24 -1.55892789 0.7877940
Y25  0.01725048 0.4966405
Y26 -1.09747921 0.7144457
Y27 -1.73670840 0.8124300
```

As expected, there is a correspondence between the difficulty level and the proportion of correct responses. This may be checked by comparing an easy item, such as the fifth (having the highest success rate and then the lowest difficulty estimate), with a more difficult item, such as the seventh (having a lower success rate and then a higher difficulty estimate).

It is important to clarify that these estimates are obtained under the assumption that the normal distribution of the ability has mean 0 but an arbitrary variance σ_θ^2. Indeed, function ltm also provides the maximum likelihood estimate of this variance that may be extracted as follows:

```
> (out$si2 = out$coefficients[1,2]^2)
[1] 0.9778139
```

In order to impose the constraint that this variance is equal to 1, equivalent to assuming that the ability has a standard normal distribution, the rasch function must be used with the additional argument constraint:

```
> out1 = rasch(Y, constraint = cbind(ncol(Y)+1, 1))
```

where ncol(Y)+1 denotes the parameter estimated after the 27 item difficulties, that is, the variance. In this way, a slightly worse value of the log-likelihood is obtained. In particular, the LR statistic between the two models is

```
> (out1$lr = -2*(out1$log.Lik-out$log.Lik))
[1] 0.2238517
```

leading to the conclusion that the hypothesis that $\sigma_\theta^2 = 1$ cannot be rejected or, in other words, this may be considered as a reliable assumption.

The plot of the ICCs under the estimated model, with free variance for the latent distribution, is obtained through the command plot(out), the result of which is given in Figure 3.7. This plot shows which is the easiest item, that is, the 5th for which a high probability of responding correctly is also obtained with a low ability level, and the most difficult, that is, the 23rd.

Finally, a prediction of the ability of every examinee, also called the score, may be obtained by the command factor.scores. This is how to use this function, show the scores for the first few individuals and the corresponding standard errors, and represent the histogram of all the scores, which is reported in Figure 3.8:

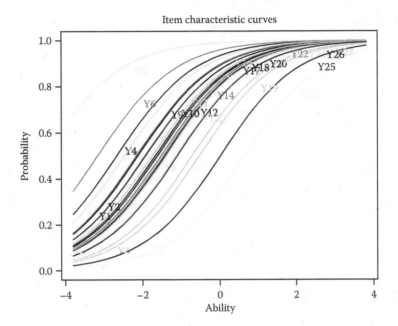

FIGURE 3.7
Estimated ICCs under the Rasch model with free variance for the ability.

```
> out2 = factor.scores(out)
> head(out2$score.dat[,30:31])
        z1       se.z1
1 -1.4484558 0.3899495
2 -1.5983151 0.3889885
3 -2.2057819 0.3984118
4 -2.7017164 0.4220854
5 -0.9870973 0.4007989
6 -2.0506308 0.3939532
> out2$score = out2$score.dat[,30]
> hist(out2$score)
```

Within the same `ltm` package used to fit the Rasch model, the estimation of the 2PL model can be performed as illustrated in the following. Note that we also include suitable commands to transform the provided estimates into those on which the parameterization in (3.8) is based under the assumption that the ability has a standard normal distribution. This assumption is required to identify all item parameters. An alternative is to constrain the difficulty of a reference item, say, the first, to be equal to 0, and its discriminating parameter to be equal to 1, while leaving free the mean and the variance of the ability distribution.

```
> # 2PL model
> out3 = ltm(Y ~z1)
> # Extract parameters and express them in the usual parametrization
> out3$beta = -out3$coefficients[,1]/out3$coefficients[,2]
> out3$lambda = out3$coefficients[,2]
```

```
> cbind(beta=out3$beta,lambda=out3$lambda)
          beta      lambda
Y1   -4.28830890 0.4391291
Y2   -2.01977753 1.0582537
Y3   -1.03463585 0.6250867
Y4   -2.42284853 1.1314395
Y5   -6.19468533 0.6880141
Y6   -4.42893372 0.6564641
Y7    0.20996942 0.2384152
Y8   -1.77611282 0.7130850
Y9   -1.87090231 1.2024179
Y10  -1.67705722 1.2137985
Y11  -1.40898921 1.6780182
Y12  -1.64841198 0.8870010
Y13  -1.01142887 1.2059979
Y14  -1.39444101 1.0648288
Y15  -1.76463399 1.0224066
Y16  -1.97701301 1.3796989
Y17  -1.86103958 0.8618793
Y18  -1.23116246 1.5454522
Y19  -0.37953076 1.7662966
Y20  -1.44787210 0.9733635
Y21  -1.68278292 0.8519483
Y22  -2.65601198 0.7584335
Y23   0.58081568 1.1971695
Y24  -1.29068887 1.3412312
Y25   0.01432826 1.5526372
Y26  -1.49692540 0.6722522
Y27  -1.76811841 0.9790551
```

In this case, the items may also be compared in terms of discriminating power. For instance, it is possible to state that items 19 and 7 have the

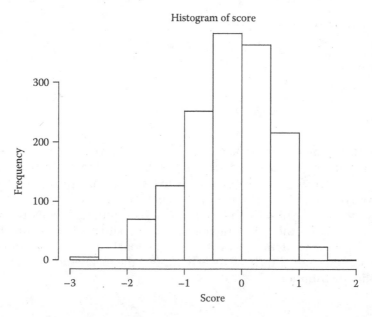

FIGURE 3.8
Histogram of the predicted ability levels.

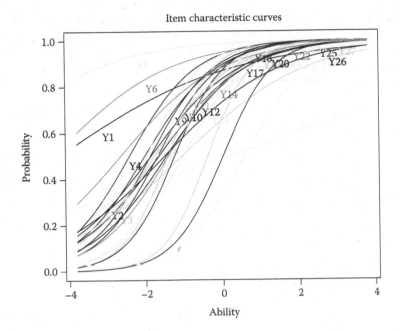

FIGURE 3.9
ICCs estimated under the 2PL model.

highest and lowest discriminating power, respectively. These results are also confirmed by the plots in Figure 3.9, which are obtained by the command `plot(out3)`. The prediction of the ability may be performed, as for the Rasch model, through the command `factor.scores`.

Finally, it may be of interest to compare the Rasch model with the 2PL model by an LR statistic, which is computed, together with the corresponding p-value, as follows:

```
> (out3$lr = -2*(out$log.Lik-out3$log.Lik))
[1] 477.9166
> (out3$pv = 1-pchisq(out3$lr,26))
[1] 0
```

Note that the asymptotic distribution used to compute the p-value is a χ^2 distribution with 26 degrees of freedom. This p-value leads us to reject the Rasch parameterization in comparison to the 2PL parameterization.

3.8.2.3 Alternatives to `ltm` for Rasch and 2PL Models

Alternatively to package `ltm`, one can use packages `irtoys` and `mirt` to estimate the Rasch and 2PL models.

Function `est` of package `irtoys` requires as main inputs the data, the specification of the item parameterization (`model`), and the basic routine for the estimation process (option `engine`). Besides, when `model='1PL'` and `rasch=T`, the discrimination parameters are forced to be equal to 1.

```
> require(irtoys)
> # Rasch model
> out_irtoys = est(Y, model="1PL", engine="ltm", rasch=T)
> # Display difficulty parameters (column 2)
> out_irtoys$est

      [,1]        [,2] [,3]
Y1       1 -2.13673680    0
Y2       1 -2.10048428    0
Y3       1 -0.71561961    0
Y4       1 -2.64306059    0
Y5       1 -4.50377232    0

[...] Output omitted

> # 2PL model
> out3_irtoys = est(Y, model="2PL", engine="ltm")
> # Display discriminating (column 1) and difficulty (column 2) parameters
> out3_irtoys$est

         [,1]        [,2] [,3]
Y1  0.4397658 -4.28303344    0
Y2  1.0551586 -2.02469833    0
Y3  0.6250031 -1.03514498    0
Y4  1.1308923 -2.42451557    0
Y5  0.6902617 -6.17733567    0

[...] Output omitted
```

Function `est` provides an output aligned to parameterizations used in (3.4) for the Rasch model and in (3.8) for the 2PL model. Note that column 2 of `out_irtoys$est` is immediately comparable to output `out$beta` obtained through `ltm` after a suitable reparameterization; similarly, estimates in columns 1 and 2 of `out3_irtoys$est` are the same as in `out3$lambda` and `out3$beta`.

Function `mirt` of the homonymous package requires the following inputs: the data, the number of latent traits (`model=1` in case of unidimensional IRT models), and the specification of the item parameterization (option `itemtype`):

```
> require(mirt)
> # Rasch model
> out_mirt = mirt(data=Y, model=1, itemtype="Rasch")
> # Display easiness item  parameters (column d)
> coef(out_mirt)

$Y1
    a1    d g u
par  1 2.134 0 1

$Y2
    a1    d g u
par  1 2.098 0 1

$Y3
```

```
       a1      d g u
par  1 0.714 0 1

$Y4
       a1      d g u
par  1 2.64 0 1

$Y5
       a1      d g u
par  1 4.499 0 1

[...] Output omitted

> # 2PL model
> out3_mirt <-  mirt(data=Y,model=1,itemtype="2PL")
> # Display discrimination (column a1) and easiness (column d) item parameters
> (coef=coef(out3_mirt))
$Y1
         a1      d g u
par 0.439 1.884 0 1

$Y2
         a1      d g u
par 1.057 2.138 0 1

$Y3
         a1      d g u
par 0.625 0.647 0 1

$Y4
         a1      d g u
par 1.132 2.743 0 1

$Y5
         a1      d g u
par 0.689 4.264 0 1

[...] Output omitted
```

One can note that function `mirt` provides a type of output similar to that of `ltm`, with the estimates of the easiness item parameters $-\beta_j$, instead of the difficulty item parameters, that enclose the discriminating indices. The output is displayed through function `coef()` that provides discrimination parameters in column `a1` and easiness parameters in column `d`. Parameterizations adopted in (3.4) and in (3.8) are obtained by dividing the coefficients in column `d` by the values in column `a1`. For instance, in case of 2PL model, parameter estimate $\hat{\beta}_1$ for item `Y1` is given by

```
> -coef$Y1[2]/coef$Y1[1]
[1] -4.291572
```

where the first element of `coef$Y1` corresponds to the discrimination parameter and the second element to the easiness item parameter.

3.8.2.4 3PL Model

Packages `irtoys` and `mirt` allow us to easily estimate a 3PL model, specifying options `model="3PL"` in function `est` of `irtoys` or `itemtype="3PL"` in function `mirt` of the homonymous package.

```
> require(irtoys)
> # 3PL model
> out4_irtoys = est(Y, model="3PL", engine="ltm")
> out4_irtoys$est
         [,1]          [,2]           [,3]
Y1  1.2085505   0.40886471  7.643604e-01
Y2  1.0452346  -2.03785136  4.183285e-07
Y3  0.6223125  -1.03594425  4.973354e-04
Y4  1.1152379  -2.45091774  4.283870e-08
Y5  0.6564699  -6.16268740  1.773024e-01

[...] Output omitted

> require(mirt)
> # 3PL model
> out4_mirt =  mirt(data=Y,model=1,itemtype="3PL")
Iteration: 500, Log-Lik: -21675.281, Max-Change: 0.00062EM cycles terminated
    after 500 iterations.
> coef(out4_mirt)
$Y1
        a1      d     g u
par 1.219 -0.502 0.765 1

$Y2
        a1     d     g u
par 1.048 2.116 0.011 1

$Y3
        a1     d     g u
par 0.661 0.514 0.076 1

$Y4
       a1     d     g u
par 1.12 2.725 0.009 1

$Y5
        a1     d     g u
par 0.661 3.916 0.275 1

[...] Output omitted
```

Outputs are very similar to those previously illustrated, with the difference that now the estimates of pseudo-guessing parameters δ_j are also provided: see column 3 of output `out4_irtoys $est` and column denoted by `g` of `coef()`.

Package `mirt` is particularly flexible, as it allows us to easily constrain one or more item parameters. For instance, observing estimates of δ_j in the previous output, it is reasonable to assume that just a few pseudo-guessing parameters differ from zero; therefore, we constrain all the other ones specifying `itemtype='2PL'` instead of `itemtype='3PL'` as follows:

```
> # 3PL model with some pseudo-guessing parameters constrained to 0
> out4bis_mirt =  mirt(data=Y,model=1,itemtype=c("3PL","2PL","2PL",
+ "2PL","3PL","2PL","3PL","2PL","2PL","2PL","2PL","2PL","2PL","3PL", "3PL",
+ "2PL","3PL","3PL","2PL","3PL","3PL","2PL","2PL","3PL","2PL","3PL","2PL"))
Iteration: 500, Log-Lik: -21675.302, Max-Change: 0.00066EM cycles terminated
    after 500 iterations.
> coef(out4bis_mirt)
$Y1
        a1     d     g u
par 1.209 -0.492 0.764 1
```

```
$Y2
        a1    d g u
par 1.047 2.13 0 1

$Y3
        a1     d g u
par 0.623 0.645 0 1

$Y4
        a1     d g u
par 1.118 2.735 0 1

$Y5
        a1     d     g u
par 0.657 4.022 0.195 1

[...] Output omitted
```

3.8.2.5 Analysis under Discrete Distribution for the Ability

In order to fit IRT models with the MML-LC method, that is, under the assumption that the ability has a discrete distribution, we use package MultiLCIRT that is loaded in the usual way:

```
> require(MultiLCIRT)
Loading required package: MultiLCIRT
Loading required package: MASS
Loading required package: limSolve
```

Then, a crucial point in applying this estimation method is the choice of the number of support points of the latent distribution, denoted by k. For instance, we may assume $k = 5$ so as to identify five latent classes of examinees in the population and estimate the Rasch model as shown in the following. Note that we use option fort=TRUE so as to use the Fortran implementation of the estimation algorithm in order to reduce the computing time.

```
> # Create data matrix
> Y = as.matrix(Y)
> # LC-Rasch model
> out4 = est_multi_poly(Y,k=5,link=1,fort=T)
*---------------------------------------------------------------------*
Model with multidimensional structure
            [,1] [,2] [,3] [,4] [,5] [,6] [,7] [,8] [,9] [,10] [,11] [,12]
Dimension 1    1    2    3    4    5    6    7    8    9   10    11    12
            [,13] [,14] [,15] [,16] [,17] [,18] [,19] [,20] [,21] [,22] [,23]
Dimension 1    13    14    15    16    17    18    19    20    21    22    23
            [,24] [,25] [,26] [,27]
Dimension 1    24    25    26    27
Link of type =            1
Discrimination index =    0
Constraints on the difficulty = 0
Type of initialization =  0
*---------------------------------------------------------------------*
```

The estimates of the parameters of the latent distribution may be obtained as follows:

```
> cbind(theta=t(out4$Th),pi=out4$piv)
        Dimension 1          pi
Class 1  -0.5700836 0.008982661
Class 2   0.4255201 0.097877221
Class 3   1.4310556 0.305729299
Class 4   2.4961228 0.430405753
Class 5   3.6950892 0.157005065
```

We observe that the ability levels (column `Dimension 1` in the previous output) are increasing in order and that the fourth class has the highest probabilities (column `pi`). However, in order to compare these results with those obtained previously under the normal distribution of the ability, we need to center this distribution on 0. In particular, mean and variance of the estimated latent distribution are obtained through a single command:

```
> # Standardize ability distribution
> out5 = standard.matrix(t(out4$Th),out4$piv)
> out5$mu
[1] 2.128537
> out5$si2
Dimension 1
   0.941478
> # Centered support points
> (out4$ths = as.vector(out4$Th-out5$mu))
[1] -2.6986145 -1.7030116 -0.6974745  0.3675974  1.5665707
```

Vector `out4$ths` contains the centered support points. Also note that the estimated variance is close to the variance estimated under the normal distribution, which is contained in `out$si2`. The two distributions, continuous and discrete, may be represented through the following commands, obtaining the plot in Figure 3.10:

```
> x = seq(-5,5,0.01)
> f = dnorm(x,0,sqrt(out$si2))
> plot(x,f,xlab="ability",ylab="density/probability",type="l",ylim=c(0,0.5))
> for(c in 1:5) lines(c(out4$th[c],out4$th[c]),c(0,out4$piv[c]),type="b")
```

Taking into account the mean of the estimated latent distribution, the estimates of the item parameters may be shifted so that they are directly comparable with those obtained under the assumption of a normal distribution centered on 0. This shift and the comparison between parameter estimates can be performed as follows:

```
> out4$betas = as.vector(out4$Be-out5$mu)
> cbind(LC=out4$betas,normal=out$beta,diff=out4$betas-out$beta)
            LC      normal        diff
Y1 -2.12853682 -2.13370026 0.005163441
Y2 -2.09231278 -2.09741641 0.005103638
Y3 -0.70794808 -0.71437646 0.006428381
Y4 -2.63445080 -2.63961969 0.005168885
Y5 -4.49181814 -4.49857442 0.006756286
Y6 -3.12331285 -3.12882702 0.005514179
Y7  0.06639843  0.06050001 0.005898427
Y8 -1.35986307 -1.36569829 0.005835223
Y9 -2.11291857 -2.11814045 0.005221883
```

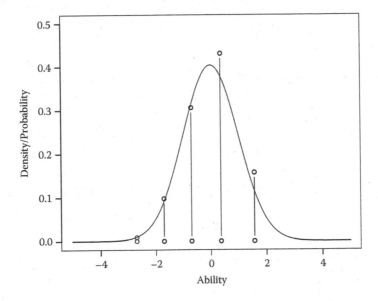

FIGURE 3.10
Estimated distribution of the ability under the Rasch model with the assumption of normality (curve) and discreteness (bars).

```
Y10 -1.90306198 -1.90829607 0.005234092
Y11 -1.92158972 -1.92676201 0.005172292
Y12 -1.49966427 -1.50531362 0.005649349
Y13 -1.13711051 -1.14319465 0.006084143
Y14 -1.44720356 -1.45306637 0.005862811
Y15 -1.78251266 -1.78792661 0.005413953
Y16 -2.43841491 -2.44350327 0.005088357
Y17 -1.65664685 -1.66216328 0.005516433
Y18 -1.60438272 -1.60996394 0.005581213
Y19 -0.52148759 -0.52798599 0.006498405
Y20 -1.41043069 -1.41608993 0.005659238
Y21 -1.48455471 -1.49022725 0.005672538
Y22 -2.13377481 -2.13895742 0.005182612
Y23  0.65602795  0.65213031 0.003897644
Y24 -1.55334911 -1.55892789 0.005578784
Y25  0.02324759  0.01725048 0.005997112
Y26 -1.09134912 -1.09747921 0.006130094
Y27 -1.73122940 -1.73670840 0.005479003
```

We observe that the difference between the estimates obtained through the two models is very small for all items.

As mentioned at the end of Section 3.7, it is typically necessary to choose the number of support points of the latent distribution on the basis of the data, and we base this choice on BIC. A suitable function, search.model(), may be used for this; this function also tries different starting values for each value of k in order to deal with the problem of multimodality of the likelihood function.

```
> # Search the optimal number of support points
> out6 = search.model(Y,kv=1:6,link=1,fort=TRUE)

[...] Output omitted

[1] "********************************************************************************"
[1]  6 10
*---------------------------------------------------------------------------------*
Model with multdimensional structure
              [,1] [,2] [,3] [,4] [,5] [,6] [,7] [,8] [,9] [,10] [,11] [,12]
Dimension 1    1    2    3    4    5    6    7    8    9   10    11    12
              [,13] [,14] [,15] [,16] [,17] [,18] [,19] [,20] [,21] [,22] [,23]
Dimension 1    13    14    15    16    17    18    19    20    21    22    23
              [,24] [,25] [,26] [,27]
Dimension 1    24    25    26    27
Link of type =                   1
Discrimination index =           0
Constraints on the difficulty = 0
Type of initialization =         1
*---------------------------------------------------------------------------------*
lktrace =  -21925.37 -21925.37 -21925.37 -21925.37 -21925.37 -21925.37 -21925.37
-21925.37 -21925.37 -21925.37 -21925.37 -21925
lk =   -23606.97 -22144.78 -21958.11 -21930.48 -21925.37 -21925
aic =   47267.94 44347.56 43978.22 43926.97 43920.75 43924
bic =   47416.11 44506.7 44148.34 44108.06 44112.82 44127.05
ent =   3.119727e-14 253.4769 559.0794 908.7215 979.9466 1067.154
nec =   1 0.1733544 0.3390701 0.5420395 0.582748 0.6344677
```

The last rows of the output show the values of some information criteria, among which that of BIC, for each tried value of k. We observe that the lowest value of the BIC index is for $k = 4$ that, according to this approach, is the proper number of latent classes to be used if there is no substantial reason to use a different value of k. Moreover, the complete output for the selected model may be extracted by the command out6$out.single[[4]].

In order to estimate the 2PL model under the assumption of the discrete distribution for the ability, it is possible to use the same functions previously described including the option disc=1, so that the discriminant indices are also estimated for each item.

```
> # LC-2PL model
> out7 = est_multi_poly(Y,k=5,link=1,disc=1,fort=T)
*---------------------------------------------------------------------------------*
Model with multidimensional structure
              [,1] [,2] [,3] [,4] [,5] [,6] [,7] [,8] [,9] [,10] [,11] [,12]
Dimension 1    1    2    3    4    5    6    7    8    9   10    11    12
              [,13] [,14] [,15] [,16] [,17] [,18] [,19] [,20] [,21] [,22] [,23]
Dimension 1    13    14    15    16    17    18    19    20    21    22    23
              [,24] [,25] [,26] [,27]
Dimension 1    24    25    26    27
Link of type =                   1
Discrimination index =           1
Constraints on the difficulty = 0
Type of initialization =         0
*---------------------------------------------------------------------------------*
```

Then, in order to compare the results with those obtained under the assumption of normality, the estimated support points must be standardized, and the item parameter estimates must be transformed accordingly.

In the following, it is shown how to perform these transformations and to make a comparison with the corresponding parameter estimates under the assumption that the ability has a standard normal distribution.

```
> out8 = standard.matrix(t(out7$Th),out7$piv)
> out8$mu
[1] 1.883717
> out8$si2
Dimension 1
  0.1964785
> # Standardized support points
> out7$ths = as.vector(out7$Th-out8$mu)/sqrt(out8$si2)
> # Standardized difficulties
> out7$betas = as.vector(out7$Be-out8$mu)/sqrt(out8$si2)
> # Standardized discriminating indices
> out7$lambdas = as.vector(out7$gac)*sqrt(out8$si2)
> cbind("LC-beta"=out7$betas,"normal-beta"=out3$beta,
+ "LC-lambda"=out7$lambdas,"normal-lambda"=out3$lambda)
           LC-beta normal-beta LC-lambda normal-lambda
Y1  -4.249698767 -4.28830890 0.4432590     0.4391291
Y2  -1.965700119 -2.01977753 1.0942650     1.0582537
Y3  -1.041875369 -1.03463585 0.6220671     0.6250867
Y4  -2.343915241 -2.42284853 1.1789426     1.1314395
Y5  -6.105237964 -6.19468533 0.6979933     0.6880141
Y6  -4.329361577 -4.42893372 0.6724970     0.6564641
Y7   0.201037494  0.20996942 0.2475506     0.2384152
Y8  -1.745151440 -1.77611282 0.7283540     0.7130850
Y9  -1.808272998 -1.87090231 1.2583517     1.2024179
Y10 -1.628675247 -1.67705722 1.2621487     1.2137985
Y11 -1.374757305 -1.40898921 1.7398367     1.6780182
Y12 -1.618920536 -1.64841198 0.9068984     0.8870010
Y13 -0.986972702 -1.01142887 1.2513374     1.2059979
Y14 -1.360208338 -1.39444101 1.1008443     1.0648288
Y15 -1.713327180 -1.76463399 1.0617187     1.0224066
Y16 -1.944083428 -1.97701301 1.4040901     1.3796989
Y17 -1.823025133 -1.86103958 0.8835170     0.8618793
Y18 -1.217503878 -1.23116246 1.5620223     1.5454522
Y19 -0.378234633 -0.37953076 1.7901269     1.7662966
Y20 -1.429200406 -1.44787210 0.9890928     0.9733635
Y21 -1.657181897 -1.68278292 0.8678778     0.8519483
Y22 -2.640874694 -2.65601198 0.7620991     0.7584335
Y23  0.543797153  0.58081568 1.2596200     1.1971695
Y24 -1.268105206 -1.29068887 1.3721860     1.3412312
Y25  0.005153885  0.01432826 1.6119179     1.5526372
Y26 -1.448633464 -1.49692540 0.7000342     0.6722522
Y27 -1.738770953 -1.76811841 0.9985428     0.9790551
```

We observe some differences in the discriminant indices that are typically smaller under the MML method based on the normal distribution for the ability. However, the comparison among the items in terms of difficulty and discriminating power leads to the same results under both approaches.

Finally, even by adopting an approach based on a discrete latent distribution, we can compare the Rasch model with the 2PL model:

```
> (out7$lr = -2*(out4$lk-out7$lk))
[1] 480.0054
> (out7$pv = 1-pchisq(out7$lr,26))
[1] 0
```

We come to the same conclusion on the basis of the MML method, that is, the Rasch model is too restrictive for these data and the inclusion of the discriminant indices in the model is unavoidable. In fact, we have a very similar value of the LR statistic that has a p-value equal to 0.

Exercises

1. Using the dataset `lsat.dta` (downloadable from `http://www.gllamm.org/lsat.dta`), perform the following analyses through `Stata`:

   (a) Estimate a Rasch model constraining the difficulty parameter of item 1 to 0.

   (b) Estimate a Rasch model constraining the average ability to 0 and compare the results with those obtained at point (a).

   (c) On the basis of the model considered at point (b), detect the most difficult item and the easiest item.

   (d) On the basis of the model estimated at point (b), predict the individual abilities and detect the extreme values, that is, the minimum and the maximum.

   (e) Compute the probabilities of endorsing each of the five items for an intermediate level of ability and for the two extreme values.

2. Using the dataset `mislevy.dat` (downloadable from `http://www.gllamm.org/books/mislevy.dat`), perform the following analyses through `Stata`:

   (a) Reshape the data in long format (variables cwm, cwf, cbm, and cbf denote the absolute frequencies of each item responses pattern for white males, white females, black males, and black females, respectively).

   (b) Estimate the Rasch model.

   (c) Estimate the 2PL model and detect the item with the highest discriminating power and that with the lowest discriminating power.

   (d) Compare the two models estimated at points (b) and (c) through an LR test and detect the one with the best fit.

   (e) Predict the individuals' abilities both under the Rasch model and under the 2PL model and compare these predictions.

3. Using the dataset `Scored` available in the R package `irtoys`, perform the following analyses in R using the preferred package:

(a) Estimate a Rasch model by the CML method, the MML method based on the normal distribution for the ability, and that based on three latent classes.

(b) Compare the estimates of the item parameters obtained by the three methods.

(c) Predict the ability level of each respondent in a suitable way when the MML method is applied.

(d) Perform the analysis based on the MML method for the 2PL model.

(e) Perform a suitable comparison between the difficulty parameter estimates obtained under the Rasch model and the 2PL model with the MML method.

(f) Comment about the estimates of the discrimination indices obtained under the 2PL model.

4. Consider the binary version of dataset Science obtained to solve Exercise 4 of Chapter 1. For this dataset, repeat the same analyses required in the previous exercise using two latent classes (instead of three) for the MML estimation method.

4

Item Response Theory Models for Polytomous Items

4.1 Introduction

In the previous chapter, item response theory (IRT) models for dichotomously scored items were illustrated. However, it is very common that psychological and educational tests are characterized by polytomous items to supply a more precise piece of information about the response process. The most relevant case, to which we will typically refer, concerns items with ordered categories. In this context, it is necessary to provide a suitable generalization of the models for binary data to account for the categorical nature of the item responses.

Similar to the dichotomous case, IRT models for polytomously scored items, or more briefly, *polytomous IRT models*, assume unidimensionality of the latent trait and local independence, whereas the assumption of monotonicity has to be reformulated as will be clarified in the following. In fact, monotonicity is now expressed with reference to the probability that the response provided by an examinee is at least equal to a certain category; this probability must be a nondecreasing function of the ability level for each item. This assumption is formulated by using different types of logit that are also commonly used in the literature on categorical and ordinal data analysis. Once the type of logit has been chosen, the conditional distribution of each response variable is parameterized by extending the formulations for dichotomous items illustrated in the previous chapter. These formulations involve a different number of parameters to describe the characteristics of each item.

Concerning the nature of ability parameters, the fixed-effects and the random-effects approaches are still valid. We recall that the latter approach may be based on a parametric assumption on the distribution of the ability, which is considered as a continuous latent variable, or may be based on the assumption that this latent variable has a discrete distribution with a certain number of support points. These support points correspond to latent classes of individuals being internally homogeneous in terms of ability level.

The assumptions about the ability parameters have implications on the possible estimation methods for the models considered here, which closely resemble those outlined for the models for dichotomous items. Therefore, we refer the reader to Section 3.7 (in Chapter 3) for an illustration of the principles behind these methods, whereas a detailed illustration is postponed to Chapter 5.

This chapter is organized as follows. In Section 4.2, we illustrate in detail the assumptions of IRT models for polytomous items, and then, in Section 4.3, we illustrate the main criteria used to classify these models, with special attention to their statistical properties. Then, the description of the structure of the most common models for ordered items is provided in Section 4.4, whereas the case of unordered items is dealt with in Section 4.5. The chapter concludes with some illustrative examples in Stata and R software.

4.2 Model Assumptions

In the polytomous case, each response variable Y_{ij}, $i = 1, \ldots, n, j = 1, \ldots, J$, has l_j response categories denoted by $y = 0, \ldots, l_j - 1$. In this context, the conditional probability that examinee i with latent trait level θ_i responds with category y to item j is denoted by

$$p_{jy}(\theta_i) = p(Y_{ij} = y|\theta_i).$$

We also let $p_j(\theta_i)$ denote the probability vector $(p_{j0}(\theta_i), \ldots, p_{j,l_j-1}(\theta_i))'$, the elements of which sum up to 1.

Function $p_{jy}(\theta_i)$, which has a similar meaning as $p_j(\theta_i)$ defined in the case of binary responses, is hereafter named as *item response category characteristic curve* (IRCCC, Baker and Kim, 2004). Alternatively, the term item response function, already introduced in Chapter 3, may be used, which encompasses the characteristic curves for the items independently of the scoring process (i.e., dichotomous or polytomous items). Note that for each test item, there are as many IRCCCs as the response categories. In addition, the assumption of monotonicity of the item characteristic curves (ICCs), which characterizes the item response models for binary items, is now formulated differently. Indeed, as shown in Figure 4.1, IRCCCs of a given item do not share a common shape: on one hand, the extreme response categories show a monotonic shape, which is decreasing for category $y = 0$ and increasing for category $y = l_j - 1$. On the other hand, IRCCCs for the intermediate categories $(y = 1, \ldots, l_j - 2)$ are bell-shaped unimodal functions.

The monotonicity assumption is still validly formulated with reference to

$$p_{jy}^*(\theta_i) = p(Y_{ij} \geq y|\theta_i), \quad y = 1, \ldots, l_j - 1,$$

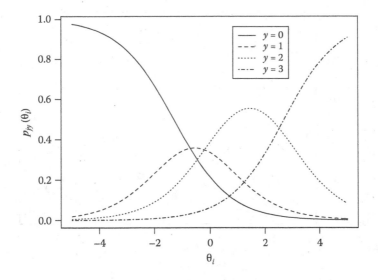

FIGURE 4.1
IRCCCs for a given item j with four categories under an IRT model for ordered items.

which corresponds to a cumulative probability; for an illustration see Figure 4.2. In particular, this is the probability that the response category is equal to y or to a higher category. This function is named in different ways: *category boundary curve*, *boundary characteristic curve*, *threshold curve*, *cumulative probability curve*, or *operating characteristic curve*. It should be noted that Samejima (1969) used a slightly different terminology, referring to $p_{jy}(\theta_i)$ as the operating characteristic curve and to $p_{jy}^*(\theta_i)$ as the *cumulative operating characteristic curve*. Naturally, there are as many category boundary curves as response categories minus one, because $p_{j0}^*(\theta_i) = 1$. Moreover, $p_{j,l_j-1}^*(\theta_i) = p_{j,l_j-1}(\theta_i)$.

Regarding the nature of θ_i, we recall that, according to the fixed-effects approach, θ_i is a fixed parameter, whereas, according to the random-effects approach, θ_i is considered as a realization of the random variable Θ_i on which several distributive hypotheses are possible. The choice about the type of approach and about the distributive nature of the latent trait has important consequences mainly on the estimation process, as shown in Section 3.7.

Concerning the random-effects approach, it is important to remember that, as outlined in Section 3.6.1 for the dichotomous case, certain models can be conceptualized as nonlinear mixed models. More precisely, it has been shown that the Rasch model and the 2PL model can be derived in terms of two-level random intercept logistic models (with loadings for the 2PL model), where each latent variable Θ_i is normally distributed. The same framework may also be adopted in the case of polytomously scored items (Rijmen et al., 2003; Tuerlinckx and Wang, 2004). In such a setting, IRT models may be

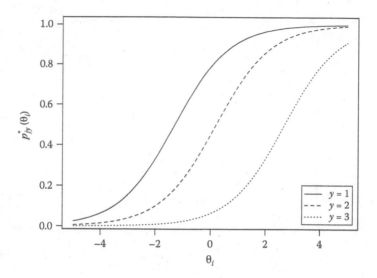

FIGURE 4.2
Boundary characteristic curves for a given item j with four categories under an IRT model for ordered items.

conceptualized as two-level random intercept ordinal or multinomial logistic models, since $l_j - 1$ nonredundant logits are defined, and each of them is similar to the logit defined in the case of binary responses; see assumption (3.14). The main difference with the dichotomous case is that, with polytomous items, parameters are defined for each item-by-category combination rather than for each item. Moreover, there are different ways to specify the logits according to the two disjoint subsets, say, A_y and B_y, of item categories that are compared:

$$\log \frac{p(Y_{ij} \in A_y|\theta_i)}{p(Y_{ij} \in B_y|\theta_i)}, \quad y = 1, \ldots, l_y - 1. \tag{4.1}$$

As will be explained in Section 4.3.1, in the case of ordered items, the logit specification can be based on different types of comparison: (1) between cumulative probabilities, so that the probability that the response is in category y or higher is compared with the probability that the response is in a smaller one ($A_y = \{y, \ldots, l_j - 1\}$ and $B_y = \{0, \ldots, y - 1\}$); (2) between the conditional probabilities referred to adjacent item categories ($A_y = \{y\}$ and $B_y = \{y - 1\}$), given that one of them has been chosen; and (3) between the conditional probabilities of a category with respect to a higher one ($A_y = \{y, \ldots, l_j - 1\}$ and $B_y = \{y - 1\}$). Finally, $l_j - 1$ logits derived from the comparison of each category with respect to another one (say, the first one) may be defined in the case of unordered (or *nominally scored*) items ($A_y = \{y\}$ and $B_y = \{0\}$). All these types of logit have the same expression in the case of dichotomously scored items, so that the IRT models described in Chapter 3 represent special cases of the more general models for polytomous items that are of interest in this chapter.

As illustrated in the next sections, the type of logit link function is not the only criterion to classify IRT models for polytomous data. Also other criteria may be adopted, such as that based on the constraints imposed on the item parameters. The choice among the wide range of models obtained as a result of the latter classification is not easy in practical situations, because they often present similar goodness-of-fit. Rather, the different statistical properties that characterize the models represent a complementary tool for selecting a suitable model for a certain application.

4.3 Taxonomy of Models for Polytomous Responses

IRT models for polytomous responses may be expressed through the general formulation

$$g_y[p_j(\theta_i)] = \lambda_j(\theta_i - \beta_{jy}), \quad j = 1, \ldots, J, \ y = 1, \ldots, l_j - 1, \tag{4.2}$$

where $g_y(\cdot)$ is a link function specific for category y based on a logit of type (4.1) and, as for the dichotomous case, λ_j and β_{jy} are the item parameters that, in most cases, identify the discriminating power and the difficulty level of item j, respectively, and on which suitably constraints may be assumed. The item difficulty parameter β_{jy} is also named as *step (difficulty) parameter*, *threshold (difficulty) parameter*, or *category boundary (difficulty) parameter*, so as to account for the specificity of this parameter for item j and response category y. Note that, in principle, it might also be conceived a discrimination parameter that is item and category specific; however, this type of generalization is not conceivable in some cases, as for the graded response models described in Section 4.4.1, whereas it may result in a too complex model to interpret in other cases.

In the following, we provide the different taxonomies of IRT models for polytomous items based on the mathematical function assumed for $p_{jy}(\theta_i)$; see Van der Linden and Hambleton (1997) for an alternative classification of over 100 IRT models. Then, the focus will be on the statistical properties characterizing the models at issue.

4.3.1 Classification Based on the Formulation of the Item Response Characteristic Curves

Many IRT models for items with polytomous responses have been proposed in the psychometric and statistical literature, and several taxonomies can be adopted, which account for different mathematical specifications of $p_{jy}(\theta_i)$.

Among the first ones, Samejima (1972) proposes to distinguish between *homogeneous models* and *heterogeneous models*. Homogeneous models are characterized by the identical shape of $p_{jy}^*(\theta_i)$ for all categories ($y = 0, \ldots, l_j - 1$) of the same item j; models that do not satisfy this requirement are instead heterogeneous. Two subcategories are conceivable: one for models that can be naturally expanded to deal with continuous responses (as, for instance, the *acceleration model* by Samejima, 1995) and the other for models that are discrete in nature.

Another well-known taxonomy is due to Thissen and Steinberg (1986), who distinguish among

- *difference models* that are appropriate for ordered responses and obtain $p_{jy}(\theta_i)$ as the difference between the probability of a response in category y or higher and the probability of a response in category $y + 1$ or higher
- *divide-by-total models* that may be used for either ordinal or nominal responses and in which $p_{jy}(\theta_i)$ is based on exponential functions suitably normalized
- *left-side added models* that add an asymptote on the left side of the item curve defined by the model to accommodate random guessing in the presence of polytomous items

A very useful criterion to classify IRT models for polytomous data relies on the works of Molenaar (1983a), Mellenbergh (1995), Van der Ark (2001), and Tuerlinckx and Wang (2004). This criterion is based on the logit function describing the transition from a response category to another. More precisely, it is possible to define three main specifications of $g_y[\boldsymbol{p}_j(\theta_i)]$ distinguishing among (Agresti, 2002; Bartolucci et al., 2007): (1) cumulative or global logits, (2) adjacent category or local logits, and (3) continuation ratio logits.

The link function defined in terms of *global logits* (or cumulative odds logits in ascending order) compares the probability that an item response is in category y or higher with the probability that it is in a lower category:

$$g_y[\boldsymbol{p}_j(\theta_i)] = \log \frac{p(Y_{ij} \geq y|\theta_i)}{p(Y_{ij} < y|\theta_i)} = \log \frac{p_{jy}(\theta_i) + \cdots + p_{j,l_j-1}(\theta_i)}{p_{j0}(\theta_i) + \cdots + p_{j,y-1}(\theta_i)},$$

$$y = 1, \ldots, l_j - 1.$$

On the other hand, *local logits* are adopted when the interest is in comparing the probability of each category y with the probability of the previous category, $y - 1$, so that

$$g_y[\boldsymbol{p}_j(\theta_i)] = \log \frac{p(Y_{ij} = y|\theta_i)}{p(Y_{ij} = y - 1|\theta_i)} = \log \frac{p_{jy}(\theta_i)}{p_{j,y-1}(\theta_i)}, \quad y = 1, \ldots, l_j - 1.$$

Finally, *continuation ratio logits* are considered to compare the probability of a response in category y or higher with the probability of the previous category $y - 1$:

$$g_y[\boldsymbol{p}_j(\theta_i)] = \log \frac{p(Y_{ij} \geq y | \theta_i)}{p(Y_{ij} = y - 1 | \theta_i)} = \log \frac{p_{jy}(\theta_i) + \cdots + p_{j,l_j-1}(\theta_i)}{p_{j,y-1}(\theta_i)},$$

$$y = 1, \ldots, l_j - 1.$$

Global logits are typically used when the trait of interest is assumed to be continuous but latent, so that it can be observed only when each subject reaches a given threshold on the latent continuum; see also definition (3.5). On the contrary, local logits are used to identify one or more intermediate levels of performance on an item and to award a partial credit for reaching such intermediate levels. Finally, continuation ratio logits are useful when sequential cognitive processes are involved (e.g., problem solving or repeated trials), as typically happens in the educational context. Note that the interpretation of continuation ratio logits is very different from that of local logits and models based on these two types of logit rely on different notions of *item step*. Local logits describe the transition from one category to the adjacent one, given that one of these two categories is chosen. Thus, each of these logits excludes any other categories and the steps are locally evaluated. On the other hand, continuation ratio logits describe the transition between adjacent categories, given that at least the smallest between the two has been reached. Therefore, the item steps are actually performed in a consecutive order. Typical examples of stepwise response mechanisms, in the sense of continuation ratio logits, are encountered in responding to items in mathematics or in the presence of psychomotor skills (Tutz, 1990).

IRT models based on global logits are also known as *graded response models* and those based on local logits are known as *partial credit models*. Moreover, IRT models based on continuation ratio logits are called *sequential models* or *stepwise models*.

Finally, Hambleton et al. (2010) distinguish between *indirect models* and *direct models*. In the former, each IRCCC is obtained by a two-step process: first, the cumulative probability $p_{jy}^*(\theta_i)$ is modeled and then $p_{jy}(\theta_i)$ is defined in terms of difference between two consecutive cumulative probabilities. In contrast, in direct models, the IRCCCs are directly formulated as functions of θ_i.

Note that the classification criteria described previously converge to the same direction (Ostini and Nering, 2010). In fact, homogeneous models, difference models, models based on global logits, and indirect models represent the same class of models, and the same holds for heterogeneous models that are discrete in nature, divide-by-total models, models based on local logits, and direct models. Moreover, heterogeneous models that can be naturally expanded to continuous response models belong to the class of models based on continuation ratio logits (Hemker et al., 2001).

4.3.2 Statistical Properties of Models for Polytomous Responses

The different ways of classifying the polytomous IRT models just illustrated are based on the mathematical form of the curves that describe the conditional probability of a given response, but they do not take explicitly into account the substantial principles on which these models are based. This implies that the level of fit to the observed data is usually the basis for model selection. However, it is well known that different IRT models may provide almost identical sets of curves that approximate equally well the true ones (Maydeu-Olivares et al., 1994; Samejima, 1996). In these cases, the goodness-of-fit of the models based on different curves may be considered as a necessary but not a sufficient condition to justify the use of a specific formulation. Other criteria have to be considered, which are based on some interesting and useful properties satisfied by the different IRT models, as we discuss in the following.

Samejima (1996) proposed a classification that distinguishes between models that satisfy one or more of the following measurement properties: *additivity* (A), *generalization to a continuous response (GCR) model, unique maximum condition* (UMC), and *orderliness of modal points* (O).

Property A means that the mathematical form of the conditional response probabilities to a given item is the same for both more finely categorized responses (e.g., right or wrong is changed to 0, 1, 2, 3) and combined category responses (e.g., 0, 1, 2, 3 are collapsed in right and wrong). If A does not hold, then the model will be affected by some incidental factors, such as the number of response categories. Note that A is always encountered in the item response models belonging to the homogeneous class defined by Samejima (1996). GCR implies that, as the number of ordinal responses increases, the phenomenon may be adequately described by a continuous model. UMC means that the likelihood function used to estimate the latent variable levels has a unique modal point, implying that only a unique ability estimate is obtained for every subject. This property is linked to property O, which implies that, for each item, the curves describing the conditional probabilities of the different response categories provide ordered modal points in accordance with the item scores.

Further relevant properties may be considered other than the previous ones, among which are the *monotonicity of likelihood ratio (MLR) of the total score* (Hemker et al., 1996), the *stochastic ordering of the manifest (SOM) variable* (Hemker et al., 1997), the *stochastic ordering of the latent (SOL) variable* (Hemker et al., 1997; Sijtsma and Hemker, 2000), and *invariant item ordering* (IIO) for all individuals (Sijtsma and Hemker, 1998, 2000). For a review and a classification of 20 different polytomous IRT models according to these properties, see Van der Ark (2001).

In particular, recalling that $Y_{i.} = \sum_{j=1}^{J} Y_{ij}$ is the raw score and $y_{i.}$ denotes one of its possible realizations, MLR holds when for any pair of subjects i_1 and i_2, we have that

$$\frac{p\left(Y_{i_1.} = y_{i_1.}|\theta\right)}{p\left(Y_{i_2.} = y_{i_2.}|\theta\right)} \geq \frac{p\left(Y_{i_1.} = y_{i_1.}|\theta'\right)}{p\left(Y_{i_2.} = y_{i_2.}|\theta'\right)},$$

provided that $y_{i_1.} > y_{i_2.}$ and $\theta < \theta'$. On the other hand, SOM is referred to the order of individuals on the latent trait given a stochastically correct ordering of individuals on the manifest variable $Y_{i.}$, that is,

$$p(Y_{i_1.} \geq y_{i.}|\theta_{i_1}) \leq p(Y_{i_2.} \geq y_{i.}|\theta_{i_2}),$$

whenever $\theta_{i_1} < \theta_{i_2}$. On the contrary, SOL means that the total score gives a stochastically correct ordering of the individuals on the latent trait, that is,

$$p(\Theta_{i_1} \geq \theta|y_{i_1.}) \leq p(\Theta_{i_2} \geq \theta|y_{i_2.})$$

whenever $y_{i_1.} < y_{i_2.}$. When this property does not hold, the common use of the raw scores $y_{i.}$ for ordering the individuals on the latent trait is not justified.

Finally, IIO means that all items have the same ordering difficulty for all subjects. If a model does not satisfy this property, then several different item orderings exist and these orderings may be much different from one to another. Therefore, results pertaining to item ordering are more difficult to interpret and, for many applications, a test functioning may not be fully understood.

Another relevant property already outlined in Section 3.3 (in Chapter 3) and still valid in the context of polytomously scored items is represented by the *parameter separability* (PS), which means that the estimation equations of item parameters do not involve the person parameters and, vice versa, the estimation equations of person parameters do not involve the item parameters. This property results in measurements that are specifically objective. On the basis of the specific objectivity, it is possible to discern between (1) Rasch-type models, which give specifically objective measurements and represent extensions of the Rasch model to the polytomous case and (2) non-Rasch-type models, which give no specifically objective measurements. While in the dichotomous case the fulfillment of PS and, therefore, of the specific objectivity is obtained in a single case, which corresponds to the Rasch model illustrated in Section 3.3, in the polytomous case there exist different ways to achieve PS, according to the type of logit. However, the mainstream of the literature agrees about including in the class of the properly named Rasch-type models only those for which PS is obtained through sufficient statistics based on the raw scores.

4.4 Models for Ordinal Responses

In this section, more details about the most well-known IRT models for ordinal responses are provided; good reviews and comparisons of the main models of interest in this section are also provided in Wright and Masters (1982), Andrich (1988), and Verhelst et al. (2005). The distinction based on the type of logit link function (i.e., global, local, or continuation ratio) is taken into account in the following. Moreover, a further distinction has to account for the constraints that may be imposed on the item parameters λ_j and β_{jy}.

First, one can consider a general situation in which each item may discriminate differently from the others and a special case in which all the items discriminate in the same way, that is, $\lambda_j = 1, j = 1, \ldots, J$, similar to what was discussed in Section 3.3 about the Rasch model. In any case, it is assumed that, within each item, all response categories share the same λ_j instead of a category-specific discrimination parameter. This is to keep the boundary characteristic curves of the same item away from crossing and so avoiding degenerate conditional response probabilities.

Second, the item difficulty parameters β_{jy} may be formulated in a completely general way, so that for each item j, there are as many difficulty parameters as the number of response categories (l_j) minus 1. Note that β_{jy} may also be expressed in an additive way as $\beta_{jy} = \beta_j + \tau_{jy}$, where β_j summarizes the difficulty of item j and τ_{jy} is a cutoff point between categories, whose interpretation depends on the specific model. An alternative and more parsimonious formulation, known as *rating scale parameterization*, consists in assuming that the distance between cutoff points is the same for each item, so that $\tau_{jy} = \tau_y$ for all items. Therefore, we have that

$$\beta_{jy} = \beta_j + \tau_y, \quad j = 1, \ldots, J, \ y = 1, \ldots, l_j - 1. \tag{4.3}$$

Obviously, this constrained version of the difficulty parameters makes only sense when all items have the same number of response categories, that is, $l_j = l, j = 1, \ldots, J$.

By combining the mentioned constraints, four different specifications of the item parameterization are obtained, based on free or constrained discrimination parameters and on a rating scale or a free parameterization for the difficulty parameters. Therefore, also according to the type of link function, 12 different types of unidimensional IRT models for ordinal responses result. These models are listed in Table 4.1; see also Bacci et al. (2014).

Abbreviations used for the models specified in Table 4.1 refer to the way the corresponding models are named in the following. Thus, GRM denotes the *graded response model* (Samejima, 1969) and RS-GRM indicates its rating scale version (Muraki, 1990), whereas 1P-GRM (Van der Ark, 2001) and

TABLE 4.1

List of Unidimensional IRT Models for Ordinal Polytomous Responses That Result from the Different Choices of the Logit Link Function, Constraints on the Discrimination Indices, and Constraints on the Difficulty Levels

λ_j	β_{jy}	Resulting Parameterization	Logit Link		
			Global	Local	Continuation
Free	Free	$\lambda_j(\theta_i - \beta_{jy})$	GRM	GPCM	2P-SM
Free	Constrained	$\lambda_j[\theta_i - (\beta_j + \tau_y)]$	RS-GRM	RS-GPCM	2P-RS-SM
Constrained	Free	$\theta_i - \beta_{jy}$	1P-GRM	PCM	SRM
Constrained	Constrained	$\theta_i - (\beta_j + \tau_y)$	1P-RS-GRM	RSM	SRSM

1P-RS-GRM (Van der Ark, 2001) are the equally discriminating versions of GRM and RS-GRM, respectively. In addition, GPCM (Muraki, 1992, 1993) denotes the *generalized partial credit model* and RS-GPCM (Muraki, 1992) is the corresponding rating scale formulation, PCM is the *partial credit model* (Masters, 1982), and RSM is the *rating scale model* (Andersen, 1977; Andrich, 1978b). Finally, 2P-SM indicates the *two-parameter sequential model* (Hemker et al., 2001; Van der Ark, 2001), which represents a constrained version of the acceleration model of Samejima (1995), 2P-RS-SM is the corresponding rating scale version, and SRM and SRSM are the *sequential Rasch model* (also known as the *sequential logit model*) and the *sequential rating scale model* by Tutz (1990), respectively.

Table 4.1 identifies a hierarchy of models in correspondence with each type of link function (Van der Ark, 2001), which is useful for the model selection process, because each restrictive model may be compared with a more general one by means of a likelihood ratio (LR) test. Naturally, models attaining different types of logit link functions are not nested, and therefore, they cannot be simply compared on the basis of this test. In addition, the item parameters that pertain to a class of models cannot be written as a function of those of the other classes, and consequently, they give rise to different interpretations.

It is also possible to classify the polytomous IRT models with respect to all the properties mentioned in Section 4.3.2, as shown in Table 4.2. It is worth observing that almost all desirable properties hold for RSM. However, since RSM is the most restrictive model, it usually does not fit data properly. GRM and PCM satisfy different properties, and therefore, the choice depends on the researcher's priorities. On the other hand, GPCM satisfies the least number of properties, but it can be useful to make a direct comparison with GRM, as both models have the same number of parameters. Finally, PS is ensured by 1P-GRM, PCM, RSM, SRM, and SRSM. However, the mainstreams of the literature agree on encompassing only PCM and RSM in the class of Rasch-type

TABLE 4.2

Properties of Some IRT Models for Ordinal Items

	Polytomous IRT Model						
	GRM	1P-GRM	PCM	GPCM	RSM	SRM	SRSM
A	Yes	Yes	No	No	No	—	—
GCR	Yes	Yes	No	No	No	—	—
UMC	Yes	Yes	Yes	Yes	Yes	—	—
O	Yes	Yes	Yes	Yes	Yes	—	—
MLR	No	No	Yes	No	Yes	No	No
SOM	Yes	Yes	Yes	Yes	Yes	Yes	Yes
SOL	No	No	Yes	No	Yes	No	No
IIO	No	No	No	No	Yes	No	Yes
PS	No	Yes	Yes	No	Yes	Yes	Yes

Note: — for not established.

models, because only these models admit sufficient statistics of the latent trait based on the raw scores. Also note that for the sequential IRT models, certain properties have not been established and this is highlighted in the table.

In the following, the mathematical structure of the models listed in Table 4.2 is described. To conclude, it is important to outline that the models illustrated in the subsequent sections can be adapted in a straightforward way to the latent class (LC) framework (Section 3.6.2), assuming that Θ_i has a discrete distribution with support points $\xi_1, \ldots, \xi_k$ and corresponding weights $\pi_1, \ldots, \pi_k$. Accordingly, Equations 3.16 and 3.17 must be modified to define the conditional and the manifest distributions of y_i.

4.4.1 Graded Response Models

Samejima's GRM (1969, 1972) is formulated through the global logit link function based on the probability that an item response is observed in category y or higher. It is an adaptation of the 2PL model to ordered responses. The general expression for the probability of scoring y on item j is the following:

$$p_{jy}(\theta_i) = p_{jy}^*(\theta_i) - p_{j,y+1}^*(\theta_i), \tag{4.4}$$

where $p_{jy}^*(\theta_i)$ is the probability that a generic person responds by category y or higher. In the present formulation, we have

$$p_{jy}^*(\theta_i) = \frac{e^{\lambda_j(\theta_i - \beta_{jy})}}{1 + e^{\lambda_j(\theta_i - \beta_{jy})}} = \frac{1}{1 + e^{-\lambda_j(\theta_i - \beta_{jy})}}, \tag{4.5}$$

so that, according to general Equation 4.2, GRM is formulated as follows:

$$\log \frac{p(Y_{ij} \geq y|\theta_i)}{p(Y_{ij} < y|\theta_i)} = \lambda_j(\theta_i - \beta_{jy}), \quad j = 1,\ldots,J, \; y = 1,\ldots,l_j - 1. \quad (4.6)$$

To better understand this expression, the elementary specifications of $p_{jy}(\theta_i)$ for an item j with $l_j = 4$ possible response categories (0, 1, 2, 3) are here shown:

$$p_{j0}(\theta_i) = 1 - \frac{e^{\lambda_j(\theta_i - \beta_{j1})}}{1 + e^{\lambda_j(\theta_i - \beta_{j1})}} = \frac{1}{1 + e^{\lambda_j(\theta_i - \beta_{j1})}}, \quad (4.7)$$

$$p_{j1}(\theta_i) = \frac{e^{\lambda_j(\theta_i - \beta_{j1})}}{1 + e^{\lambda_j(\theta_i - \beta_{j1})}} - \frac{e^{\lambda_j(\theta_i - \beta_{j2})}}{1 + e^{\lambda_j(\theta_i - \beta_{j2})}},$$

$$p_{j2}(\theta_i) = \frac{e^{\lambda_j(\theta_i - \beta_{j2})}}{1 + e^{\lambda_j(\theta_i - \beta_{j2})}} - \frac{e^{\lambda_j(\theta_i - \beta_{j3})}}{1 + e^{\lambda_j(\theta_i - \beta_{j3})}},$$

$$p_{j3}(\theta_i) = \frac{e^{\lambda_j(\theta_i - \beta_{j3})}}{1 + e^{\lambda_j(\theta_i - \beta_{j3})}}. \quad (4.8)$$

The results of the estimation process for the GRM are similar to those for the 2PL model. An ability parameter estimate is obtained for each person and a difficulty parameter is obtained for each threshold and item. The total number of thresholds for each item is given by the number of response modalities minus 1, as each of them is referred to the step from a given modality (or smaller) to the following one (or higher). Finally, a discriminant parameter λ_j is estimated for each item; however, within the same item, all responses share the same discriminating power so that the boundary characteristic curves do not cross. In addition, since $p_{jy}(\theta_i)$ is defined as the difference between cumulative probabilities, then inequality constraints must be imposed on the threshold difficulty parameters, that is,

$$\beta_{j1} < \cdots < \beta_{j,l_j-1}, \quad j = 1,\ldots,J.$$

It should be noted that no constraint of equal distance between the β_{jy} parameters is required.

IRCCCs defined in this section are nonmonotonic curves, except for the first and last response categories; see Figures 4.3 and 4.4 for a comparison with the corresponding boundary characteristic curves. Indeed, for the first category ($y = 0$), $p_{jy}(\theta_i)$ has the expression defined in (4.7), which follows

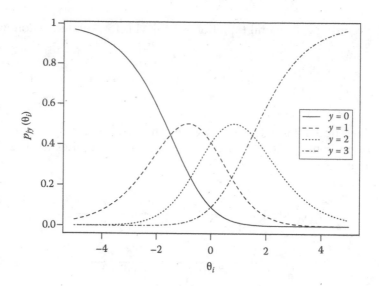

FIGURE 4.3
IRCCCs for a given item j with four categories under GRM ($\lambda_j = 1$, $\beta_{j1} = -2.0$, $\beta_{j2} = 0.0$, $\beta_{j3} = 1.5$).

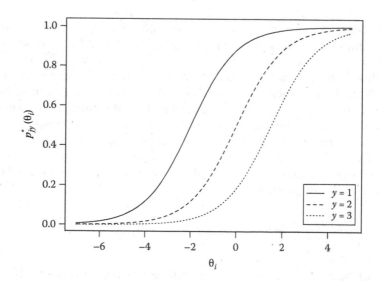

FIGURE 4.4
Boundary characteristic curves for a given item j with four categories under GRM ($\lambda_j = 1$, $\beta_{j1} = -2.0$, $\beta_{j2} = 0.0$, $\beta_{j3} = 1.5$).

a monotonically decreasing logistic function in θ_i. For the last category $(y = l_j - 1)$, $p_{jy}(\theta_i)$ follows a monotonically increasing logistic function:

$$p_{j,l_j-1}(\theta_i) = \frac{e^{\lambda_j(\theta_i - \beta_{j,l_j-1})}}{1 + e^{\lambda_j(\theta_i - \beta_{j,l_j-1})}};$$

this becomes equal to the expression in (4.8) with four response categories. The intermediate categories present a mode located at the midpoint of the two adjacent threshold difficulty parameters, that is, at $(\beta_{jy} + \beta_{j,y+1})/2$, $y = 1, \ldots, l_j - 2$.

Finally, it is important to outline that the item parameters β_{j0} and β_{j,l_j-1} express the value of θ_i at which the probability of choosing category $y = 0$ and $y = l_j - 1$, respectively, equals 0.50, as happens for IRT models for binary items (see Section 3.3). In the other cases, that is, for the nonextreme response categories, β_{jy} simply contributes to locating the modal point of the corresponding IRCCC. It also corresponds to the difficulty to pass the yth step (for item j), that is, the difficulty of choosing a response category greater than or equal to the yth with respect to choosing a previous category. In more detail, parameters β_{jy} represent the value of θ_i at which the probability of answering by category y or higher equals the probability of answering by category $y - 1$ or smaller, that is,

$$p_{jy}^*(\theta_i) = 1 - p_{jy}^*(\theta_i)$$

when $\theta_i = \beta_{jy}$. Besides, if $\theta_i > \beta_{jy}$, then $p_{jy}^*(\theta_i) > 1 - p_{jy}^*(\theta_i)$, whereas if $\theta_i < \beta_{jy}$, then $p_{jy}^*(\theta_i) < 1 - p_{jy}^*(\theta_i)$.

An interesting type of model is represented by 1P-GRM (Table 4.1), which is obtained as a special case of GRM when all the discrimination indices λ_j are constrained to 1. In this case, the cumulative probabilities $p_{jy}^*(\theta_i)$ have the same form as for the Rasch model:

$$p_{jy}^*(\theta_i) = \frac{e^{\theta_i - \beta_{jy}}}{1 + e^{\theta_i - \beta_{jy}}}; \tag{4.9}$$

see Equation 3.4. This has important consequences on the estimation process of 1P-GRM, since approaches based on the conditional maximum likelihood (see Section 3.7 and Chapter 5 for details) may be adopted, so realizing a person-free item calibration and an item-free person calibration, which are typical of Rasch-type models.

4.4.2 Partial Credit Models

In the setting of partial credit models, a local logit link function is adopted, as reported in Table 4.1. The most well known are PCM and RSM, which are illustrated in the following.

4.4.2.1 Basic Formulation

For items with two or more ordered response categories, Masters (1982) proposed PCM within the Rasch model framework. This model is suitable when the same questionnaire includes items with ordered categories, which may be different in number and/or for the meaning (e.g., strongly disagree/disagree/agree/strongly agree, always/often/hardly/never).

To understand PCM, this model may be conceived as a simple adaptation of the Rasch model for dichotomous items. In the latter one, answering to an item involves only one step: choosing category 1 rather than 0 and then Equation 3.4 expresses the corresponding probabilities, being only two response categories possible. More in general, the response process involves a choice between as many steps as response categories of each item minus 1: for instance, if item j has 3 response categories (0, 1, 2), then person i responds with one of these categories and his or her choice process can be thought as a choice between modality 1 rather than 0 and modality 2 rather than 1. Hence, the probability of person i scoring 1 rather than 0 on item j (conditional on responding by one of these two categories) is identical to the dichotomous case:

$$\frac{p_{j1}(\theta_i)}{p_{j0}(\theta_i) + p_{j1}(\theta_i)} = \frac{e^{\theta_i - \beta_{j1}}}{1 + e^{\theta_i - \beta_{j1}}},$$

with the only difference that now $p_{j0}(\theta_i) + p_{j1}(\theta_i) < 1$, because more than two ordered performance levels are considered. While β_{j1} still governs the probability of completing the first step (choosing category 1 rather than 0), this is not the only step in responding to item j; in other words, β_{j1} does not design the difficulty of item j, but only the difficulty of passing the first threshold (i.e., step from score 0 to score 1) of item j. In a similar way, the second step from score 1 to score 2 designs the probability of a generic person choosing modality 2 rather than 1 for item j, and it is given by

$$\frac{p_{j2}(\theta_i)}{p_{j1}(\theta_i) + p_{j2}(\theta_i)} = \frac{e^{\theta_i - \beta_{j2}}}{1 + e^{\theta_i - \beta_{j2}}}.$$

In general, the probability of responding by modality y rather than modality $y - 1$ to item j is given by

$$\frac{p_{jy}(\theta_i)}{p_{j,y-1}(\theta_i) + p_{jy}(\theta_i)} = \frac{e^{\theta_i - \beta_{jy}}}{1 + e^{\theta_i - \beta_{jy}}}, \quad y = 1, \ldots, l_j - 1.$$

This implies that, the probability of a generic person scoring y on item j may be computed as follows:

$$p_{jy}(\theta_i) = \frac{e^{\sum_{h=1}^{y}(\theta_i - \beta_{jh})}}{\sum_{m=0}^{l_j-1} e^{\sum_{h=1}^{m}(\theta_i - \beta_{jh})}} = \frac{e^{y\theta_i - \sum_{h=1}^{y}\beta_{jh}}}{1 + \sum_{m=1}^{l_j-1} e^{m\theta_i - \sum_{h=1}^{m}\beta_{jh}}}, \quad (4.10)$$

$$y = 0, 1, \ldots, l_j - 1,$$

where the sum $\sum_{h=1}^{y}$ at the numerator vanishes for $y = 0$ and the sum $\sum_{h=1}^{m}$ at the denominator behaves in a similar way for $m = 0$; this convention will be adopted even in the following. Consequently, the probability of scoring 0 on item j is given by

$$p_{j0}(\theta_i) = \frac{1}{1 + \sum_{m=1}^{l_j-1} e^{m\theta_i - \sum_{h=1}^{m}\beta_{jh}}}.$$

Note that the denominator of Equation 4.10 is the sum of the numerators for all the possible l_j response categories. The dichotomous Rasch model can be considered as a special case of PCM where all items have only two response categories; hence, only one threshold difficulty is estimated.

To better understand, the elementary specifications of $p_{jy}(\theta_i)$ for an item j with $l_j = 4$ possible response categories are here shown. On the basis of Equation 4.10, we have

$$p_{j0}(\theta_i) = \frac{1}{\Psi(\theta_i)},$$

$$p_{j1}(\theta_i) = \frac{e^{\theta_i - \beta_{j1}}}{\Psi(\theta_i)},$$

$$p_{j2}(\theta_i) = \frac{e^{2\theta_i - \beta_{j1} - \beta_{j2}}}{\Psi(\theta_i)},$$

$$p_{j3}(\theta_i) = \frac{e^{3\theta_i - \beta_{j1} - \beta_{j2} - \beta_{j3}}}{\Psi(\theta_i)},$$

where $\Psi(\theta_i)$ is the sum of the numerators, that is,

$$\Psi(\theta_i) = 1 + e^{\theta_i - \beta_{j1}} + e^{2\theta_i - \beta_{j1} - \beta_{j2}} + e^{3\theta_i - \beta_{j1} - \beta_{j2} - \beta_{j3}}.$$

The results of the estimation process for PCM are similar to those for dichotomous IRT models. An ability parameter estimate is obtained for each person and a difficulty parameter is obtained for each threshold of each item. As the results from Table 4.2 show, the PCM violates the property of IIO, that is, the ranking of items does not necessarily coincide with the ranking of the total scores, although they are sufficient statistics (Bertoli-Barsotti, 2005).

IRCCCs under PCM and the corresponding boundary characteristic curves are very similar to those of models based on a graded response formulation, as represented in Figures 4.5 and 4.6. In particular, IRCCCs are nonmonotonic curves, except for the first and last response categories: for the first category the curve is monotonically decreasing, and for the last category it is monotonically increasing. The corresponding boundary characteristic curves are S-shaped curves that do not intersect, independently of the values of the threshold parameters β_{jy}.

Moreover, the item parameters represent the point on the latent continuum where two consecutive IRCCCs cross. In particular, β_{jy} is the value of θ_i at which the probability of answering by category y equals the probability of answering by category $y - 1$. In fact, if $\theta_i = \beta_{jy}$, then

$$p_{jy}(\theta_i) = p_{j,y-1}(\theta_i).$$

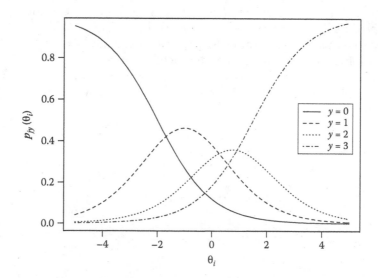

FIGURE 4.5
IRCCCs for a given item j with four categories under PCM ($\beta_{j1} = -2.5$, $\beta_{j2} = 0.0$, $\beta_{j3} = 1.3$).

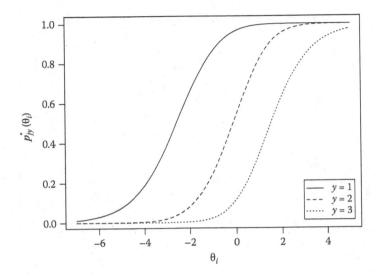

FIGURE 4.6
Boundary characteristic curves for a given item j with four categories under PCM ($\beta_{j1} = -2.5$, $\beta_{j2} = 0.0$, $\beta_{j3} = 1.3$).

In addition, if $\theta_i > \beta_{jy}$, then $p_{jy}(\theta_i) > p_{j,y-1}(\theta_i)$, whereas if $\theta_i < \beta_{jy}$, then $p_{jy}(\theta_i) < p_{j,y-1}(\theta_i)$. Therefore, β_{jy} cannot be interpreted as the difficulty of category y of item j; it must be interpreted as the relative difficulty of category y compared to category $y - 1$ of the same item. See Verhelst et al. (1997) for an extension of PCM that allows for a nicer interpretation of the item parameters in terms of difficulty of the response categories. On the basis of the aforementioned arguments, the parameters β_{jy} under the partial credit formulation cannot be directly compared with parameters β_{jy} under the graded response formulation, because they have a different meaning.

Differently from the GRMs, the item difficulty parameters β_{jy} of PCM (and of the other partial credit models) are defined with respect to just two adjacent categories, and then, they can take any order. Therefore, reversed thresholds may be conceived, as illustrated in Figure 4.7. In such a case $\beta_{j3} < \beta_{j2}$, and as a consequence, IRCCC corresponding to category $y = 2$ is not the most likely at any value of θ_i. According to some authors (see, for instance, Andrich, 2010), this is a clear symptom of a problem with the data and with the empirical ordering of the categories. On the contrary, other authors (e.g., Tuerlinckx and Wang, 2004) believe that this is not a problem in itself, but only that some theories requiring that each response category has to be the most likely for certain ability levels are too restrictive.

Another possibility, which represents a further example of disordered thresholds, consists in response steps with the same difficulty level, as shown in Figure 4.8, where all β_{jy} parameters are equal to 0. In this case, all IRCCCs

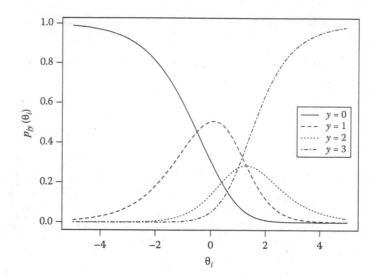

FIGURE 4.7
IRCCCs for a given item j with four categories under PCM with the assumption of reversed thresholds ($\beta_{j1} = -0.5$, $\beta_{j2} = 1.3$, $\beta_{j3} = 1.0$).

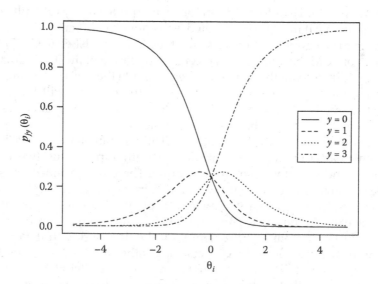

FIGURE 4.8
IRCCCs for a given item j with four categories under PCM with the assumption of equal difficulties ($\beta_{j1} = \beta_{j2} = \beta_{j3} = 0.0$).

cross at the same value of θ_i, so that any intermediate category is never the most likely.

A straightforward generalization of the PCM is represented by GPCM (Muraki, 1992, 1993), in which the discrimination parameters are not equal for all items:

$$p_{jy}(\theta_i) = \frac{e^{\sum_{h=1}^{y} \lambda_j(\theta_i - \beta_{jy})}}{\sum_{m=0}^{l_j-1} e^{\sum_{h=1}^{m} \lambda_j(\theta_i - \beta_{jh})}}, \quad j = 1, \dots, J, \; y = 0, \dots, l_j - 1.$$

As usual, parameter λ_j rules the slope of $p_{jy}(\theta_i)$; as this parameter decreases, IRCCCs become flatter. It is clear from examining Table 4.2 that GPCM does not satisfy the properties characterizing the Rasch-type models, due to the presence of this discrimination parameter.

In conclusion, an LC version of PCM is given by von Davier and Rost (1995) and Molenaar (1997), which represents a generalization to ordered items of the mixed Rasch model defined in Equation 3.15:

$$p_{jy}(\xi_v) = \frac{e^{\sum_{h=1}^{y} (\xi_v - \beta_{jhv})}}{\sum_{m=0}^{l_j-1} e^{\sum_{h=1}^{m} (\xi_v - \beta_{jhv})}}, \quad v = 1, \dots, k, \; y = 0, \dots, l_j - 1.$$

A constrained version with item parameters constant over all classes (i.e., $\beta_{jhv} = \beta_{jh}$) is illustrated in Bacci et al. (2014), who also consider a general LC framework to encompass all the other IRT models here illustrated.

4.4.2.2 Rating Scale Model

RSM was independently proposed by Andersen (1977) and Andrich (1978a,b) before PCM, but it can be derived from this one as a special case. Indeed, RSM differs from PCM in that the distance between step difficulties from category to category is assumed to be the same across all items. Therefore, RSM is suitable when all items of the questionnaire have the same kind of response categories (i.e., the same number $l_j = l, j = 1, \dots, J$, of categories, which also have the same meaning). RSM includes a difficulty parameter for each item and a difficulty parameter for each threshold, and therefore, it is more parsimonious than PCM. The following general expression for the probability of person i responding in category l for item j is derived:

$$p_{jy}(\theta_i) = \frac{e^{\sum_{h=1}^{y} \theta_i - (\beta_j + \tau_h)}}{\sum_{m=0}^{l-1} e^{\sum_{h=1}^{m} \theta_i - (\beta_j + \tau_h)}} = \frac{e^{y\theta_i - \sum_{h=1}^{y} (\beta_j + \tau_h)}}{1 + \sum_{m=1}^{l-1} e^{m\theta_i - \sum_{h=1}^{m} (\beta_j + \tau_h)}},$$
$$y = 0, 1, \dots, l-1.$$

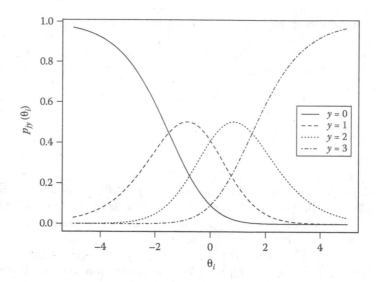

FIGURE 4.9
IRCCCs for a given item j with four categories under RSM with the assumption of equally spaced thresholds ($\beta_{j1} = -1.5$, $\beta_{j2} = 0.0$, $\beta_{j3} = 1.5$).

Note that $l - 1$ thresholds, $\tau_1, \dots, \tau_{l-1}$, are estimated for the l response categories. This general expression implies that the probability of responding by category 0 to a generic item j is equal to

$$p_{j0}(\theta) = \frac{1}{1 + \sum_{m=1}^{l-1} e^{\sum_{h=1}^{m} \theta_i - (\beta_j + \tau_h)}}.$$

The graphical representation of IRCCCs for RSM is very similar to that for PCM. Note that, in all models here illustrated, the possible presence of equally spaced thresholds of a given item makes all the corresponding IRCCCs for intermediate categories to have the same maximum height (Figure 4.9).

4.4.3 Sequential Models

In the sequential model setting, SRM is the most well-known model (Tutz, 1990). Relying on the framework of discrete choice models (Wooldridge, 2002), the model may be justified assuming the existence of an underlying continuous variable Y_{ijy}^* defined as follows:

$$Y_{ijy}^* = \Theta_i + \varepsilon_{ijy},$$

where ε_{ijy} is a random error with standard logistic distribution, which drives the influence of accidental factors on the latent variable of interest. Then, the response mechanism for item j is described as follows:

$$Y_{ij} > y \quad \text{if} \quad Y^*_{ijy} > \beta_{jy} \quad \text{given} \quad Y_{ij} \geq y, \quad y = 0, \ldots, l_j - 2.$$

The assumption of logistically distributed errors ε_{ijy} implies that the probability of overcoming the yth item step (i.e., transiting from category y to category $y + 1$ given that category y has been reached) is obtained as

$$p(Y_{ij} > y | Y_{ij} \geq y) = \frac{e^{\theta_i - \beta_{jy}}}{1 + e^{\theta_i - \beta_{jy}}},$$

which is the same as Equation 3.4 defining the Rasch model. This fact is particularly relevant because it leads to PS and, then, to apply the CML approach (Section 3.7 and Chapter 5) for parameter estimation.

The probability of scoring $Y_{ij} = y$ depends on the probability of performing successfully the first $y - 1$ steps and of being unsuccessful at the yth step. In particular, for $y = 1, \ldots, l_j - 2$, we have

$$p_{jy}(\theta_i) = p(Y^*_{ij0} > \beta_{j0}) p(Y^*_{ij1} > \beta_{j1} | Y^*_{ij0} > \beta_{j0}) \cdots$$

$$p(Y^*_{ij,y-1} > \beta_{j,y-1} | Y^*_{ij,y-2} > \beta_{j,y-2}) p(Y^*_{ijy} \leq \beta_{jy} | Y^*_{ij,y-1} > \beta_{j,y-1})$$

$$= \left[\prod_{h=0}^{y-1} p(Y_{ij} > h | Y_{ij} \geq h) \right] p(Y_{ij} = y | Y_{ij} \geq y)$$

$$= \left(\prod_{h=0}^{y-1} \frac{e^{\theta_i - \beta_{jh}}}{1 + e^{\theta_i - \beta_{jh}}} \right) \frac{1}{1 + e^{(\theta_i - \beta_{jy})}},$$

and for $y = l_j - 1$, we have

$$p_{j,l_j-1}(\theta_i) = \prod_{h=0}^{l_j-2} \frac{e^{\theta_i - \beta_{jh}}}{1 + e^{\theta_i - \beta_{jh}}}.$$

The interpretation of the item parameters β_{jy} is similar to that of the Rasch model: β_{jy} is the difficulty of solving the step from y to $y + 1$ if the previous steps were successfully completed. Therefore, β_{jy} represents the value of θ_i at which the probability of answering by category $y + 1$ or higher equals the probability of answering by category y. In particular, if $\theta_i = \beta_{jy}$, then

$$p^*_{j,y+1}(\theta_i) = p_{jy}(\theta_i).$$

Besides, if $\theta_i > \beta_{jy}$, then $p^*_{j,y+1}(\theta_i) > p_{jy}(\theta_i)$, whereas if $\theta_i < \beta_{jy}$, then $p^*_{j,y+1}(\theta_i) < p_{jy}(\theta_i)$.

Similar to PCM and related models, sequential models do not require the threshold parameters to have the same ordering of the item categories.

4.5 Models for Nominal Responses

The previous models allow for items with ordered polytomous response categories. However, in the presence of nominally scored items, another type of models is more appropriate, which is based on *baseline-category logits*, also known as *multinomial logits* (Agresti, 2002). These logits are formulated comparing the probability of answering by category $y = 1, \ldots, l_j - 1$ versus a baseline category, say, category $y = 0$. However, the reference category is arbitrary and then any other category may be chosen.

The most well-known IRT model for nominal responses is the *nominal response model* proposed by Bock (1972); for a more flexible solution for multiple choice items, see Thissen and Steinberg (1984). The model assumes that

$$p_{jy}(\theta_i) = \frac{e^{\lambda_{jy}\theta_i + \beta_{jy}}}{1 + \sum_{h=1}^{l_j-1} e^{\lambda_{jh}\theta_i + \beta_{jh}}}, \quad y = 1, \ldots, l_j - 1, \tag{4.11}$$

with

$$p_{j0}(\theta_i) = \frac{1}{1 + \sum_{h=1}^{l_j-1} e^{\lambda_{jh}\theta_i + \beta_{jh}}}.$$

Up to a reparameterization, the model at issue is then an extension of the 2PL model where each response category y has a discriminant index λ_{jy}, which is also item specific.

The relevance of the nominal response model is not only due to the possibility to account for nominal categories, which however are not so particularly widespread in the educational and psychological settings, but, mainly, to the possibility of deriving partial credit models as special cases of Equation 4.11. Indeed, considering the following basic relation:

$$\log \frac{p_{jy}(\theta_i)}{p_{j0}(\theta_i)} = \log \frac{p_{jy}(\theta_i)}{p_{j,y-1}(\theta_i)} + \log \frac{p_{j,y-1}(\theta_i)}{p_{j,y-2}(\theta_i)} + \cdots + \log \frac{p_{j1}(\theta_i)}{p_{j0}(\theta_i)},$$

$$y = 1, \ldots, l_j - 1,$$

it is possible to pass from local logits to baseline-category logits and vice versa. In particular, with $\lambda_{jy} = 1$ for all j and y in expression (4.11) and reparameterizing in a suitable way this expression, PCM based on Equation 4.10 is obtained.

4.6 Examples

In the following, we illustrate the estimation of IRT models for ordered polytomous items in `Stata` and R. First, we illustrate the `Stata` commands through the analysis of the items that drive the measurement of one of the two latent dimensions of the HADS dataset (see Section 1.8.3 for the dataset description). Then, we provide an analysis in R of the RLMS dataset, about job satisfaction, illustrated in Section 1.8.2.

4.6.1 Depression Dimension of HADS: Analysis in `Stata`

In the following, the commands for the estimation of models based on a graded response formulation and a partial credit formulation in `Stata` are provided through the analysis of HADS data.

4.6.1.1 Data Organization

The subset of items that is taken into account concerns measurement of depression. For sake of simplicity, these items are renamed as follows:

```
. use "HADS.dta", clear
. keep Y1 Y3-Y5 Y9 Y13-Y14
. rename Y3 Y2
. rename Y4 Y3
. rename Y5 Y4
. rename Y9 Y5
. rename Y13 Y6
. rename Y14 Y7
```

As described in Section 3.8.1, we first collapse the data and then we convert them from the wide shape to the long shape:

```
. gen cons=1
. collapse (sum) wt2=cons, by (Y1-Y7)
. gen ind=_n
. reshape long Y, i(ind) j(item)
. qui tab item, gen(d)
```

4.6.1.2 Analysis in `gllamm` under the Assumption of Normality for the Latent Trait

The estimation of models based on the graded response formulation may be easily implemented in `Stata` by means of the `gllamm` function, specifying link `ologit`, which accounts for the ordinal nature of the responses through cumulative logits, rather than link `logit`, which is used with binary responses.

By default, function `gllamm` with `ologit` link allows us to estimate the twice constrained GRM, which is named as 1P-RS-GRM:

```
. * 1P-RS-GRM
. gllamm Y d2-d7, i(ind) weight(wt) l(ologit) f(binom) adapt dots
```

The output is similar to that obtained in the case of models for binary items:

```
[…] Output omitted
```

Y	Coef.	Std. Err.	z	P>|z|	[95% Conf. Interval]
Y					
d2	.3499265	.2053634	1.70	0.088	-.0525783 .7524313
d3	1.14251	.1960366	5.83	0.000	.7582858 1.526735
d4	2.091689	.2005032	10.43	0.000	1.69871 2.484668
d5	-.7944309	.2100929	-3.78	0.000	-1.206205 -.3826564
d6	-.1269114	.1998025	-0.64	0.525	-.5185172 .2646943
d7	.7292569	.1977444	3.69	0.000	.3416849 1.116829
_cut11					
_cons	-.5418186	.1780707	-3.04	0.002	-.8908307 -.1928065
_cut12					
_cons	1.994202	.1887562	10.56	0.000	1.624246 2.364157
_cut13					
_cons	4.525094	.2382391	18.99	0.000	4.058154 4.992034

```
Variances and covariances of random effects
------------------------------------------------------------------
***level 2 (ind)
   var(1): 2.2484597 (.32714146)
------------------------------------------------------------------
```

In the previous output, the estimated item parameters $\hat{\beta}_j$ are printed out for each item $j = 2, \ldots, 7$ (column `Coef.` in correspondence of row d*j*) with the corresponding standard errors, *z*-values, *p*-values, and inferior and superior limits of the confidence intervals at the 95% level, being $\beta_1 \equiv 0$ to ensure model identifiability. Besides, three constant terms (`_cons`) are displayed, which correspond to the estimates of the parameters τ_y in Equation 4.3. Each of these thresholds corresponds to the difficulty of passing from category $y - 1$ or smaller to category y or higher, and they are the same for all items. Therefore, $\hat{\tau}_1 = -0.542$ denotes the difficulty of answering 1, 2, or 3 rather

TABLE 4.3

Estimates of the Item Difficulty Parameters for 1P-RS-GRM, 1P-GRM, and GRM

Item	1P-RS-GRM			1P-GRM			GRM		
j	$\hat{\beta}_{j1}$	$\hat{\beta}_{j2}$	$\hat{\beta}_{j3}$	$\hat{\beta}_{j1}$	$\hat{\beta}_{j2}$	$\hat{\beta}_{j3}$	$\hat{\beta}_{j1}$	$\hat{\beta}_{j2}$	$\hat{\beta}_{j3}$
1	−0.542	1.994	4.525	−0.891	2.824	4.315	−0.978	3.099	4.682
2	−0.192	2.344	4.875	−0.268	1.039	3.105	−0.358	1.327	3.975
3	0.601	3.137	5.668	−2.047	1.091	3.690	−2.341	1.250	4.216
4	1.550	4.086	6.617	−3.539	−0.160	3.404	−3.859	−0.181	3.739
5	−1.336	1.200	3.731	0.431	2.612	4.061	0.553	3.454	5.402
6	−0.669	1.867	4.398	−0.479	2.398	4.295	−0.564	2.831	5.059
7	0.187	2.723	5.254	−1.186	1.042	4.601	−1.540	1.362	5.990

than 0 to any item; $\hat{\tau}_2 = 1.994$ is the difficulty of answering 2 or 3 rather than 0 or 1; $\hat{\tau}_3 = 4.525$ is the difficulty of answering 3 rather than 0, 1, or 2. The high value of $\hat{\tau}_3$ denotes that only subjects with a very high level of depression likely answer by category 3. It is also possible to formulate a single item-by-step parameter estimate, $\hat{\beta}_{jy}$, by summing $\hat{\beta}_j$ and $\hat{\tau}_y$ according to Equation 4.3; see the output shown in Table 4.3 (columns 2–4).

As usual, the random intercept is assumed to be normally distributed with mean equal to 0. Its estimated variance is equal to 2.248 with a standard error of 0.327.

To remove the constraint on the discrimination parameters, we may proceed as for the 2PL model (Section 3.8.1):

```
. * RS-GRM
. eq load: d1-d7
. gllamm Y d2-d7, i(ind) weight(wt) l(ologit) f(binom) eqs(load)  adapt dots
```

The resulting model is RS-GRM and the output is similar to that described earlier for 1P-RS-GRM (with values for $\hat{\beta}_j$ displayed as usual in column Coef. in correspondence of row dj), with the addition of the estimates of the discrimination parameters $\hat{\lambda}_j$, named in the following output as loadings for random effect 1.

```
[…] Output omitted

-----------------------------------------------------------------------
         Y |    Coef.   Std. Err.     z    P>|z|    [95% Conf. Interval]
-----------+-----------------------------------------------------------
Y          |
        d2 |   .1935127   .2307244    0.84   0.402   -.2586987    .6457242
        d3 |   1.152311   .1977451    5.83   0.000    .7647372    1.539884
        d4 |   2.102084   .2033418   10.34   0.000    1.703541    2.500626
        d5 |  -.8576675   .2252811   -3.81   0.000   -1.29921    -.4161246
        d6 |  -.1271112   .2037297   -0.62   0.533   -.5264141    .2721916
        d7 |   .7550588   .2004671    3.77   0.000    .3621506    1.147967
```

```
-------------+-------------------------------------------------------------
_cut11       |
       _cons |   -.5565025    .1787694    -3.11   0.002    -.9068841   -.2061208
-------------+-------------------------------------------------------------
_cut12       |
       _cons |    2.013437    .1899539    10.60   0.000     1.641134    2.38574
-------------+-------------------------------------------------------------
_cut13       |
       _cons |    4.537729    .2385088    19.03   0.000     4.070261    5.005198
-------------------------------------------------------------------------

Variances and covariances of random effects
-------------------------------------------------------------------------
***level 2 (ind)

    var(1): 2.1769015 (.58194825)

    loadings for random effect 1
    d1: 1 (fixed)
    d2: 1.6815462 (.2513525)
    d3: .94788743 (.15451226)
    d4: .75985425 (.13793175)
    d5: 1.1602182 (.21581503)
    d6: .98639441 (.17728993)
    d7: .90481723 (.15982056)
-------------------------------------------------------------------------
```

Concerning the item difficulty estimates, we remind the reader that gllamm uses the multiplicative factor only on the latent trait (i.e., the left side of Equation 4.2 is specified as $\lambda_j\theta_i - \beta_{jy}^*$ rather than as $\lambda_j(\theta_i - \beta_{jy})$). In any case, there are no sensible differences in the estimates of β_{jy} and τ_y between 1P-RS-GRM and RS-GRM, as can be seen from the corresponding Stata outputs. Concerning the item discrimination estimates, items 2 and 5 have the highest discriminating power, whereas item 4 discriminates to a lower extent than the other items.

Another generalization of 1P-RS-GRM consists in a model with item-specific thresholds. To specify unconstrained item difficulties, we proceed by defining an equation containing the names of the items with different thresholds and, then, retrieving them in the gllamm command using option thresh. For instance, the estimation of 1P-GRM is performed as follows:

```
. * Model without  item specification
. quietly gllamm Y, i(ind) weight(wt) l(ologit) f(binom) adapt
. * 1P-GRM
. eq thr: d2-d7
. matrix a=e(b)
. gllamm Y, i(ind) weight(wt) l(ologit) f(binom) from(a) thresh(thr) adapt dots
```

Note that, in this script, we first run a naive model without item specification, and then, we add item-specific thresholds in a new model by using the estimates of the naive model as initial values of the thresholds. For this, option from is used in function gllamm, which retrieves the values specified through command matrix. A much faster estimation process results.

The output for 1P-GRM is as follows:

```
[...] Output omitted

---------------------------------------------------------------------------
         Y |    Coef.   Std. Err.      z    P>|z|    [95% Conf. Interval]
-----------+---------------------------------------------------------------
_cut11     |
        d2 |   .6236129   .2426081    2.57   0.010    .1481099    1.099116
        d3 |  -1.15592    .2708092   -4.27   0.000   -1.686696   -.6251437
        d4 |  -2.647893   .3571926   -7.41   0.000   -3.347978   -1.947808
        d5 |   1.322385   .2474904    5.34   0.000    .8373125    1.807457
        d6 |   .4124845   .2442411    1.69   0.091   -.0662194    .8911883
        d7 |  -.2945652   .250236    -1.18   0.239   -.7850187    .1958883
     _cons |  -.8911466   .2095076   -4.25   0.000   -1.301774   -.4805193
-----------+---------------------------------------------------------------
_cut12     |
        d2 |  -1.785294   .3080694   -5.80   0.000   -2.389099   -1.181489
        d3 |  -1.732556   .3092309   -5.60   0.000   -2.338638   -1.126475
        d4 |  -2.983554   .3081367   -9.68   0.000   -3.587491   -2.379617
        d5 |  -.2118015   .339294    -0.62   0.532   -.8768054    .4532025
        d6 |  -.4255455   .3336814   -1.28   0.202   -1.079549    .2284581
        d7 |  -1.781838   .3093432   -5.76   0.000   -2.38814    -1.175536
     _cons |   2.823921   .2791752   10.12   0.000    2.276749    3.371095
-----------+---------------------------------------------------------------
_cut13     |
        d2 |  -1.209769   .4746721   -2.55   0.011   -2.14011    -.2794293
        d3 |  -.6244952   .5002234   -1.25   0.212   -1.604915    .3559246
        d4 |  -.9110467   .4848148   -1.88   0.060   -1.861266    .0391727
        d5 |  -.2532284   .5342031   -0.47   0.635   -1.300247    .7937904
        d6 |  -.0196197   .5482167   -0.04   0.971   -1.094105    1.054865
        d7 |   .2862327   .5867831    0.49   0.626   -.863841     1.436306
     _cons |   4.314556   .4151129   10.39   0.000    3.50095     5.128163
---------------------------------------------------------------------------

Variances and covariances of random effects
---------------------------------------------------------------------------
***level 2 (ind)

   var(1): 2.5514232 (.36929087)
---------------------------------------------------------------------------
```

In the present extended version of the model, an item difficulty parameter is provided for each item-by-threshold combination (column Coef.), with $\beta_{1y} \equiv 0$, $y = 1, 2, 3$. To compare the output of gllamm with the parameterization in Equation 4.9 (see also Table 4.1), $\hat{\beta}_{jy}$ is obtained by summing dj and the corresponding constant term _cons for each _cut1y. For instance, $\hat{\beta}_{32} = -1.733 + 2.824 = 1.091$. The item difficulty parameters reformulated in this way are shown in Table 4.3 (columns 5–7), together with the corresponding parameters of 1P-RS-GRM. Note that the requirement of ordered thresholds is met in all cases.

Finally, the most general GRM may be estimated by removing the constraint of equal discrimination parameters that characterize 1P-GRM. In practice, the estimation process is performed by specifying both options eqs and thresh in command gllamm. In addition, to accelerate the algorithm, we suggest retrieving the starting values of the parameters from the previously

estimated 1P-GRM and to initialize the newly added discrimination parameters starting from 1. For this, we need to run command `matrix` twice: the
first time to define object a containing output from the previously estimated
1P-GRM and the second time to define object b containing all the elements of
a and a vector of ones to initialize the additional parameters of GRM (i.e., the
discrimination parameters). Then, b is retrieved in function `gllamm` through
option `from`. In more detail, we observe that object a is composed of 22 elements (see the previously displayed output of 1P-GRM): 21 estimates of difficulty parameters (one value for each item by threshold) and one estimate for
the second-level variance. All these elements are contained in object b, where
6 elements equal to 1 are added, one for each item minus one (recall that the
discrimination parameter of the first item equals 1 by default).

```
. * Define starting values to accelerate the algorithm
. matrix a=e(b)
. matrix b=a[1,1..21],1,1,1,1,1,1,a[1,22]
. * GRM
. gllamm Y, i(ind) weight(wt) l(ologit) f(binom)  eqs(load) thresh(thr) adapt from(b) copy
dots
```

The output for the item difficulties of GRM is similar to that for 1P-
GRM; see also Table 4.3 (columns 8–10) for a direct comparison with the two
constrained versions of GRM.

```
[…] Output omitted
```

Y	Coef.	Std. Err.	z	P>|z|	[95% Conf. Interval]
_cut11					
d2	.710587	.2617497	2.71	0.007	.1975671	1.223607
d3	-1.149101	.323967	-3.55	0.000	-1.784065	-.5141375
d4	-2.833721	.4949601	-5.73	0.000	-3.803825	-1.863617
d5	1.368546	.266588	5.13	0.000	.8460428	1.891049
d6	.4950536	.2718618	1.82	0.069	-.0377857	1.027893
d7	-.1433805	.2819435	-0.51	0.611	-.6959796	.4092187
_cons	-.9776687	.244778	-3.99	0.000	-1.457425	-.4979126
_cut12					
d2	-2.108669	.4135504	-5.10	0.000	-2.919213	-1.298125
d3	-1.963451	.4289268	-4.58	0.000	-2.804132	-1.12277
d4	-3.278121	.4193439	-7.82	0.000	-4.10002	-2.456223
d5	-.6596388	.4792632	-1.38	0.169	-1.598977	.2796997
d6	-.6784715	.5025338	-1.35	0.177	-1.66342	.3064766
d7	-2.107299	.434651	-4.85	0.000	-2.959199	-1.255399
_cons	3.098897	.4051963	7.65	0.000	2.304727	3.893067
_cut13					
d2	-1.715874	.6152319	-2.79	0.005	-2.921707	-.5100419
d3	-.8508396	.6862234	-1.24	0.215	-2.195813	.4941335
d4	-.9891516	.7495075	-1.32	0.187	-2.458159	.4798562
d5	-.8667488	.7069264	-1.23	0.220	-2.252299	.5188014
d6	-.3561436	.777832	-0.46	0.647	-1.880666	1.168379
d7	-.3214376	.7692255	-0.42	0.676	-1.829092	1.186217
_cons	4.681836	.5758304	8.13	0.000	3.553229	5.810443

```
Variances and covariances of random effects
---------------------------------------------------------------------
***level 2 (ind)

   var(1): 3.6332959 (1.2275823)

   loadings for random effect 1
   d1: 1 (fixed)
   d2: .74620893 (.15045823)
   d3: .90867852 (.18944338)
   d4: .98764904 (.2457114)
   d5: .7062869 (.1712189)
   d6: .85499897 (.21187592)
   d7: .7279227 (.17783315)
---------------------------------------------------------------------
```

Based on the GRM estimates, function `gllapred` may be used to plot IRCCCs of the items. First, the range to display latent trait values is defined as follows:

```
. quietly egen N=max(ind)
. generate trait1 = (-5) + (ind-1)*(5-(-5))/(N-1)
```

and the cumulative probability functions $\hat{p}_{jy}^*(\theta_i)$, $y = 1, 2, 3$, are calculated using command `gllapred` as follows:

```
. gllapred probcum0, mu above(0) us(trait)
(mu will be stored in probcum0)
. gllapred probcum1, mu above(1) us(trait)
(mu will be stored in probcum1)
. gllapred probcum2, mu above(2) us(trait)
(mu will be stored in probcum2)
```

Then, probability functions $\hat{p}_{jy}(\theta_i)$, $y = 1, 2, 3$, are computed according to Equation 4.4:

```
. gen probcat0 = 1-probcum0
. gen probcat1 = probcum0-probcum1
. gen probcat2 = probcum1-probcum2
. gen probcat3 = probcum2
```

and function `twoway` is applied for each item as follows:

```
. * Plot IRCCC for item 3 (repeat similarly for any other item, substituting in a suitable
way item==3)
. twoway (line probcat0 trait1 if Y==0, sort) (line probcat1 trait1 if Y ==1, sort lpatt("."))
(line probcat2 trait1 if Y ==2, sort lpatt("-")) (line probcat3 trait1 if Y ==3,
sort lpatt("_")) if item==3,  legend(order(1 "category 0" 2 "category 1"
3 "category 2" 4 "category 3"))
```

Note that through command `twoway`, we define a curve for each probability as previously computed (`probcat0`, `probcat1`, ...) and the corresponding

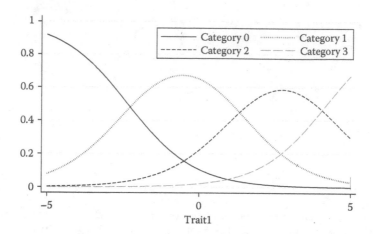

FIGURE 4.10
IRCCCs for item Y3 under GRM.

item response category (Y==0, Y==1, ...); then, the item that the plot refers to is declared (e.g., item==3), and a legend is specified. The resulting IRCCCs for item 3 are shown in Figure 4.10.

As the four GRMs applied to HADS data define a nested structure, an LR test may be used for model selection (use function lrtest, as shown in the application in Section 3.8.1). As shown in Table 4.4, 1P-GRM is preferable to 1P-RS-GRM and GRM.

In addition to graded response models, the nonlinear mixed framework allows us to also estimate PCM and related models. The main difficulty with these models is the correct specification of the link function. Indeed, certain statistical software, such as Stata, usually consider only the global logit link and the baseline-category logit link. The problem may be bypassed by relying on the basic relation between baseline-category logits and local logits (see Section 4.5). For this, we need to format the long-shaped data in a suitable way so as to have one row for each category-by-item-by-person combination.

TABLE 4.4

LR Tests for the Selection of Graded Response Models

	df	Test Statistic	p-Value
Compared models			
1P-RS-GRM vs. RS-GRM	6	26.25	0.0002
1P-RS-GRM vs. 1P-GRM	12	119.33	<0.0001
1P-RS-GRM vs. GRM	18	124.86	<0.0001
RS-GRM vs. GRM	12	98.61	<0.0001
1P-GRM vs. GRM	6	5.53	0.4784

A very clear explanation of the suitable Stata commands is given in Zheng and Rabe-Hesketh (2007), to which we refer the reader for details.

First, we manipulate the data as follows:

```
. gen obs=_n
. * Expand data to have one row for each response category
. expand 4
(3234 observations created)
. sort ind item obs
. * Generate variable cat to contain all  possible scores (0,1,2,3) for each  item-by-person
combination
. by obs, sort: gen cat = _n-1
. * Specify the actually chosen category by each person
. gen chosen = Y == cat
. * Define dummies for each item-by-category combination
. * Category y=0 as reference
. forvalues j=1/7{
    forvalues y=1/3{
    gen d`j'_`y' = -1*d`j'*(cat>=`y')
    }
  }
```

Then, PCM is estimated through function gllamm, where link mlogit (standing for multinomial logit) is adopted and constant factor loadings are specified for each item (cat denotes the response category chosen by the subject):

```
. * PCM
. eq slope: cat
. gllamm cat d1_1-d7_3, i(ind) eqs(slope) link(mlogit) expand(obs chosen o) weight(wt) adapt
trace nocons
```

Note that option expand is used to specify the expanded structure of the original data. More precisely, obs denotes the original row of data, chosen is an indicator for the corresponding response category, and o provides only one set of item parameters (instead of a separate set for each category). The resulting output displays the item difficulty estimates, $\hat{\beta}_{jy} = dj\_y$ ($dj\_0 \equiv 0$, $j = 1, \ldots, 7$) in column Coef., and, as usual, the variance of the latent trait with the corresponding standard error. Note that the thresholds of some item parameters are reversed, as those of item 1 ($\hat{\beta}_{12} > \hat{\beta}_{13}$).

```
[...] Output omitted
```

cat	Coef.	Std. Err.	z	P>|z|	[95% Conf. Interval]	
d1_1	-.6604956	.1896326	-3.48	0.000	-1.032169	-.2888225
d1_2	2.517428	.297879	8.45	0.000	1.933596	3.10126
d1_3	2.132152	.4787884	4.45	0.000	1.193744	3.07056
d2_1	.4278411	.2121418	2.02	0.044	.0120507	.8436314
d2_2	.4457956	.2416711	1.84	0.065	-.027871	.9194622
d2_3	1.952805	.3157078	6.19	0.000	1.334029	2.57158
d3_1	-1.508519	.2268007	-6.65	0.000	-1.95304	-1.063998
d3_2	.958704	.2067007	4.64	0.000	.5535781	1.36383
d3_3	2.447397	.352464	6.94	0.000	1.756581	3.138214
d4_1	-2.692205	.3300794	-8.16	0.000	-3.339149	-2.045262

```
 d4_2 |   -.1096149    .1874571    -0.58   0.559     -.4770242     .2577943
 d4_3 |    2.581059    .3093178     8.34   0.000      1.974807    3.187311
 d5_1 |     .581521    .1925655     3.02   0.003      .2040996     .9589425
 d5_2 |    1.956474    .3009003     6.50   0.000       1.36672    2.546227
 d5_3 |    2.169831    .4538006     4.78   0.000      1.280399    3.059264
 d6_1 |   -.2127048    .1885815    -1.13   0.259     -.5823177     .1569082
 d6_2 |    1.937638    .2677713     7.24   0.000      1.412816     2.46246
 d6_3 |    2.466591    .4552288     5.42   0.000      1.574359    3.358823
 d7_1 |   -.6284935    .2073518    -3.03   0.002     -1.034896    -.2220914
 d7_2 |    .6392459    .2094321     3.05   0.002      .2287665    1.049725
 d7_3 |    3.474814    .4647641     7.48   0.000      2.563893    4.385735
----------------------------------------------------------------------------

Variances and covariances of random effects
----------------------------------------------------------------------------
***level 2 (ind)
   var(1): 1.3336163 (.21389782)
```

As for GRM, we may plot IRCCCs for the items, under PCM (Figure 4.11). First, function `gllapred` is invoked for the estimation of the conditional probabilities $p_{jy}(\theta_i)$, which, differently from the GRM, are now directly provided:

```
. * quietly egen N=max(ind)
. * generate trait1 = (-5) + (ind-1)*(5-(-5))/(N-1)
. gllapred prob1, mu us(trait)
(mu will be stored in prob1)
```

Then, function `twoway` is applied to these probabilities:

```
. * Plot IRCCC for item 5
. twoway (line prob1 trait1 if cat==0, sort) (line prob1 trait1 if cat ==1, sort lpatt("."))
  (line prob1 trait1 if cat ==2,  sort lpatt("-")) (line prob1 trait1 if cat ==3,
  sort lpatt("_"))   if item==5,  legend(order(1 "category 0" 2 "category 1"  3
  "category 2" 4 "category 3"))
```

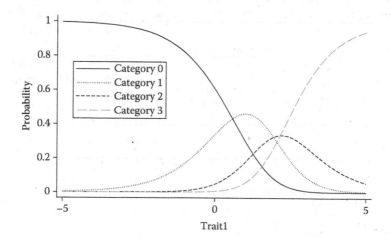

FIGURE 4.11
IRCCCs for item Y5 under PCM.

RSM and GPCM are estimated according to the same procedure for PCM, after ad hoc variable generation. More precisely, RSM requires fewer new variables than PCM, as it has fewer item parameters. Therefore, new variables are generated to identify the common item thresholds, as follows:

```
. * Generate dummies to identify the thresholds (one threshold for each response category
minus 1)
. gen step1 = -1*(cat>=1)
. gen step2 = -1*(cat>=2)
. gen step3 = -1*(cat>=3)
. foreach var of varlist d*{
    gen n'var' = -1*'var'*cat
    }
. sort ind item cat
```

which are used in function g11amm instead of the previously created dj_y:

```
. * RSM
. eq slope: cat
. gllamm cat nd1-nd7 step2 step3, i(ind) eqs(slope) link(mlogit) expand(obs chosen o)
weight(wt) adapt trace nocons
```

The displayed output is quite clear, being $\hat{\beta}_j =$ ndj and $\hat{\tau}_y =$ stepy ($y = 2, 3$ and $\tau_1 = 0$).

```
[...] Output omitted
```

cat	Coef.	Std. Err.	z	P>|z|	[95% Conf. Interval]	
nd1	-.2183407	.1452534	-1.50	0.133	-.5030322	.0663507
nd2	-.5406346	.1461998	-3.70	0.000	-.827181	-.2540883
nd3	-1.050304	.1516018	-6.93	0.000	-1.347438	-.7531696
nd4	-1.748899	.1653371	-10.58	0.000	-2.072953	-1.424844
nd5	.2840981	.1481139	1.92	0.055	-.0061997	.5743959
nd6	-.1335526	.1453541	-0.92	0.358	-.4184414	.1513361
nd7	-.7474362	.1478506	-5.06	0.000	-1.037218	-.4576543
step2	1.686612	.1368632	12.32	0.000	1.418365	1.954859
step3	3.346269	.2104802	15.90	0.000	2.933736	3.758803

```
Variances and covariances of random effects
------------------------------------------------------------------------
***level 2 (ind)
    var(1): 1.2422302 (.19748846)
```

Finally, GPCM is a generalization of PCM, where different discrimination parameters are added. Therefore, additional data manipulation with respect to PCM is needed, so that a new variable is defined for each item and it is then retrieved in the g11amm function through option eqs:

```
. * Note that variables cat and d1_1 to d7_3 are defined as for PCM
. * Generate dummies to accommodate free item discrimination parameters
. forvalues j=1/7{
    gen cat_d`j'= cat*d`j'
  }
. sort ind item cat
. * GPCM
. eq load: cat_d1-cat_d7
. gllamm cat d1_1-d7_3, i(ind) eqs(load) link(mlogit) expand(obs chosen o)
  weight(wt) adapt trace nocons

[...] Output omitted
```

To conclude, we outline that an alternative to `gllamm` for estimating PCM and RSM is provided by the `Stata` module `pcmodel`, which is very easy to use; see Section 5.10.1 for an example.

4.6.2 RLMS Data: Analysis in R

In order to illustrate the estimation of the IRT models for ordinal data dealt with in this chapter, we consider the RLMS dataset about Russian job satisfaction in its original version based on $J = 4$ items with $l_j = 5$ response categories for every item j; see Section 1.8.2 for a detailed description of the dataset. However, in order to have a clearer interpretation of the results, we arrange the response categories in increasing order and then from absolutely unsatisfied (0) to absolutely satisfied (4).

In the following, we first show how to analyze the dataset at issue by using the R packages `ltm` (Rizopoulos, 2006) and `MultiLCIRT` (Bartolucci et al., 2014) that were already considered in Chapter 3 with reference to dichotomous IRT models. We recall that the first package is suitable to estimate IRT models under the assumption of normality of the latent trait, whereas the second package is used when this trait is represented by a latent variable having a discrete distribution. Among other packages for polytomous IRT models, we recall `mirt` (used in Section 3.8.2 for estimating binary item response models) and `eRm` (which will be used in Section 5.10.2 for the implementation of some diagnostics tools); see also `http://cran.r-project.org/web/views/Psychometrics.html` for a complete review of the main R packages devoted to polytomous IRT models.

4.6.2.1 Data Organization

The RLMS dataset is organized as shown in Section 2.12.2. First, we load the data and we retain only the item responses. Then, we drop records with at least one missing response and we relabel the response categories in reverse order for an easier interpretation:

```
> load("RLMS.RData")
> # Keep item responses
> data = data[,6:9]
```

```
> # Drop records with missing observations
> ind = which(apply(is.na(data),1,any))
> data = data[-ind,]
> # Reverse response categories
> data = 4-data
> (n = nrow(data))
[1] 1418
```

4.6.2.2 Analysis under Normal Distribution for the Ability

We first load the package ltm by the command require, as already shown in Section 3.8.2. Then, we fit models based on the graded response formulation by function grm, which also allows for the constraint of equal discriminant parameters across items. It is also important to note that this function uses a type of cumulative logits defined in a different way with respect to what was shown in Section 4.4.1; see Equation 4.6. Moreover, by default, this function uses a different parameterization. In order to obtain estimates based on the parameterization here adopted, it is necessary to use option IRT.param. In any case, the latent trait is assumed to be normally distributed with mean 0 and variance 1, so that identifiability constraints on the item parameters are not necessary.

In order to fit GRM in the standard formulation with free discrimination parameters, we use the following command:

```
> # GRM
> out = grm(data,IRT.param=TRUE)
> summary(out)

Call:
grm(data = data, IRT.param = TRUE)

Model Summary:
   log.Lik     AIC       BIC
  -6967.09 13974.18 14079.32

Coefficients:
$Y1
         value
Extrmt1 -1.911
Extrmt2 -1.063
Extrmt3 -0.388
Extrmt4  0.993
Dscrmn   4.155

$Y2
         value
Extrmt1 -1.872
Extrmt2 -1.035
Extrmt3 -0.340
Extrmt4  1.040
Dscrmn   3.717

$Y3
         value
Extrmt1 -1.515
Extrmt2 -0.348
Extrmt3  0.445
Extrmt4  2.139
Dscrmn   1.559
```

```
$Y4
          value
Extrmt1  -1.665
Extrmt2  -0.601
Extrmt3   0.201
Extrmt4   1.689
Dscrmn    1.904


Integration:
method: Gauss-Hermite
quadrature points: 21

Optimization:
Convergence: 0
max(|grad|): 3.7
quasi-Newton: BFGS
```

This output reports the estimates of the discriminant index (λ_j), denoted by Dscrmn, for each item, together with the estimates of the four threshold parameters (β_{jy}), denoted by Extrmty ($y = 1, 2, 3, 4$). On the basis of this output, we conclude that the first item has the highest discriminating power. Nevertheless, this item concerns the general satisfaction about work; see Section 1.8.2. On the other hand, the last two items are those with the lowest discriminating power.

Regarding the threshold parameters, we note that for almost all response categories, the estimate for the third item is the highest with respect to the other items. This means that this is the most difficult item, as confirmed by the frequency distributions previously reported. We recall that this item is about earnings, meaning that this is the aspect of smallest satisfaction of the interviewees. The easiest items are instead the first two.

The previous conclusions are confirmed by the probability plots that may be obtained by command

```
plot(out)
```

In particular, we can compare the plot which refers to the first item with the one which refers to the third item; these plots are reported in Figure 4.12. For instance, taking an ability level of 2, we observe that the probability of the last response category is much higher for the first item (upper panel) than for the third item (lower panel).

In order to fit 1P-GRM, we have to constrain the discriminant indices to be equal across items through option constrained as follows:

```
> # 1P-GRM
> out2 = grm(data,IRT.param=TRUE,constrained=TRUE)
> summary(out2)

Call:
grm(data = data, constrained = TRUE, IRT.param = TRUE)

Model Summary:
   log.Lik      AIC       BIC
 -7090.891 14215.78 14305.15
```

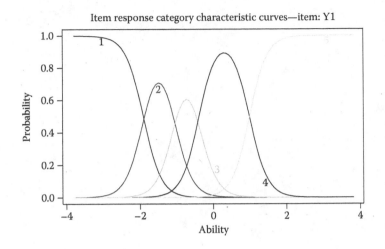

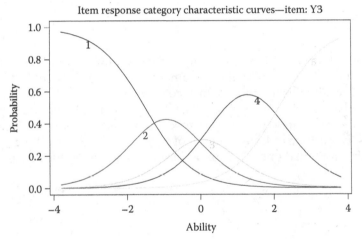

FIGURE 4.12

IRCCCs for the first and third items estimated under GRM.

```
Coefficients:
$Y1
          value
Extrmt1 -2.256
Extrmt2 -1.182
Extrmt3 -0.403
Extrmt4  1.164
Dscrmn   2.443

$Y2
          value
Extrmt1 -2.134
Extrmt2 -1.129
Extrmt3 -0.343
Extrmt4  1.187
Dscrmn   2.443
```

```
$Y3
             value
Extrmt1    -1.191
Extrmt2    -0.268
Extrmt3     0.373
Extrmt4     1.738
Dscrmn      2.443

$Y4
             value
Extrmt1    -1.473
Extrmt2    -0.518
Extrmt3     0.206
Extrmt4     1.533
Dscrmn      2.443


Integration:
method: Gauss-Hermite
quadrature points: 21

Optimization:
Convergence: 0
max(|grad|): 0.12
quasi-Newton: BFGS
```

The interpretation of this output is similar to the previous one considering the constraint on the discrimination parameters. These results confirm that the most difficult item is the third, which concerns earnings.

On the basis of the previous results, it is also possible to test the hypothesis that the discrimination parameters are equal to each other by an LR statistic that has asymptotic null distribution of type $\chi^2(3)$, where 3 is the number of constraints assumed under this hypothesis. This test is performed as follows:

```
> (dev = -2*(out2$log.Lik-out$log.Lik))
[1] 247.6034
> (pvalue = 1-pchisq(dev,3))
[1] 0
```

These results provide strong evidence against the hypothesis of interest, given the very high value of the LR statistic and the corresponding *p*-value that is equal to 0.

The same analysis illustrated previously may be performed under partial credit models. In this case we use function gpcm, paying attention to option constraint. We start from GPCM, which has free discrimination parameters and may be fitted as follows:

```
> # GPCM
> out3 = gpcm(data,IRT.param=TRUE)
> summary(out3)

Call:
gpcm(data = data, IRT.param = TRUE)

Model Summary:
  log.Lik      AIC       BIC
 -7063.46 14166.92 14272.06
```

```
Coefficients:
$Y1
          value std.err z.value
Catgr.1 -1.893   0.071 -26.518
Catgr.2 -1.010   0.042 -24.140
Catgr.3 -0.444   0.036 -12.240
Catgr.4  0.988   0.045  21.819
Dscrmn   3.971   0.355  11.184

$Y2
          value std.err z.value
Catgr.1 -1.843   0.073 -25.322
Catgr.2 -0.995   0.045 -21.863
Catgr.3 -0.420   0.039 -10.689
Catgr.4  1.062   0.050  21.450
Dscrmn   3.055   0.235  12.998

$Y3
          value std.err z.value
Catgr.1 -1.336   0.121 -11.053
Catgr.2  0.092   0.109   0.843
Catgr.3 -0.056   0.110  -0.510
Catgr.4  2.609   0.184  14.165
Dscrmn   0.752   0.049  15.282

$Y4
          value std.err z.value
Catgr.1 -1.531   0.105 -14.611
Catgr.2 -0.389   0.082  -4.753
Catgr.3 -0.105   0.079  -1.323
Catgr.4  1.929   0.118  16.284
Dscrmn   1.036   0.065  16.047


Integration:
method: Gauss-Hermite
quadrature points: 21

Optimization:
Convergence: 0
max(|grad|): 0.043
optimizer: nlminb
```

This output is similar to that for graded response models, with parameters $\hat{\beta}_{jy}$ denoted by `Catgr.y` ($y = 1, 2, 3, 4$). These results confirm that the first item has the highest discriminating power, whereas the third has the lowest discriminating power.

In order to fit PCM, which includes the constraint of constant discrimination parameters across items, we use in a different way the command already used:

```
> # PCM
> out4 = gpcm(data, IRT.param=TRUE, constraint="1PL")
```

In order to compare PCM with GPCM and then to test the hypothesis of equal discriminating power of the items, we can again use an LR statistic:

```
> (dev2 = -2*(out4$log.Lik-out3$log.Lik))
[1] 286.5034
> (pvalue2 = 1-pchisq(dev2,3))
[1] 0
```

This test confirms that the hypothesis of equal discriminating power across items must be rejected.

A final point concerns the comparison between the graded response formulation and the partial credit formulation for the data at issue. Though a formal test between models based on these different formulations cannot be performed in a simple way, GRM has undoubtedly a better fit than GPCM with the same number of free parameters. In fact, for the first model we have a maximum log-likelihood of -6967.09, which is equal to -7063.46 for the second. This is also confirmed by the values of indices on which the Akaike and the Bayesian information criteria are based (see Section 5.7.1 for details) that can be displayed by command `summary` applied with argument equal to the object collecting all the model fitting results (e.g., `summary(out)`).

In conclusion, we can state that GRM is the most suitable among the models here considered for the data at issue, at least when the latent trait is assumed to be normally distributed. The same conclusion can be reached on the basis of substantial considerations about the phenomenon that is considered in the study and the formulation of the test items.

4.6.2.3 Analysis under Discrete Distribution for the Ability

Under the assumption that the latent variable has a discrete distribution, we can fit models based on GRM and PCM formulations by package `MultiLCIRT` in a similar way as illustrated in Section 3.8.2. This package is loaded by the usual command `require`. We recall that in this framework, the choice of the number of support points of the latent distribution, which identify latent classes in the population of interest, is crucial.

We start from fitting GRM with free discrimination parameters and $k = 3$ latent classes by the following commands:

```
> # Aggregate data
> out5 = aggr_data(data)
> S = out5$data_dis
> yv = out5$freq
> # LC-GRM
> out6 = est_multi_poly(S,yv,k=3,link=1,disc=1)
*------------------------------------------------------------------------------*
Model with multidimensional structure
            [,1] [,2] [,3] [,4]
Dimension 1   1    2    3    4
Link of type =                1
Discrimination index =        1
Constraints on the difficulty = 0
Type of initialization =      0
*------------------------------------------------------------------------------*
> summary(out6)

Call:
est_multi_poly(S = S, yv = yv, k = 3, link = 1, disc = 1)

Log-likelihood:
[1] -7117.17
```

```
AIC:
[1] 14280.34

BIC:
[1] 14401.25

Class weights:
[1] 0.3480 0.5376 0.1144

Conditional response probabilities:
, , class = 1

         item
category     1      2      3      4
       0 0.1078 0.1263 0.3373 0.2727
       1 0.3567 0.3508 0.3899 0.4165
       2 0.4461 0.4169 0.1671 0.2133
       3 0.0883 0.1040 0.0966 0.0908
       4 0.0011 0.0020 0.0092 0.0067

, , class = 2

         item
category     1      2      3      4
       0 0.0011 0.0026 0.0713 0.0336
       1 0.0069 0.0136 0.2154 0.1370
       2 0.0786 0.1155 0.2737 0.2914
       3 0.8087 0.7668 0.3817 0.4703
       4 0.1047 0.1016 0.0580 0.0678

, , class = 3

         item
category     1      2      3      4
       0 0.0000 0.0000 0.0070 0.0017
       1 0.0000 0.0001 0.0287 0.0084
       2 0.0002 0.0009 0.0695 0.0310
       3 0.0228 0.0522 0.4947 0.3666
       4 0.9769 0.9468 0.3999 0.5922
```

Note that in command `est_multi_poly`, option `link=1` is necessary to select logits of type global, whereas option `disc=1` indicates that there are no constraints on the discrimination parameters. Moreover, the output directly shows the estimated weight of each latent class and the conditional response probabilities given this class.

The estimates of the model parameters may be displayed as follows:

```
> # Abilities and class weights
> rbind(out6$Th,pi=out6$piv)
             Class 1   Class 2    Class 3
Dimension 1 2.1133995 6.7911661 12.6815072
pi          0.3479746 0.5376487  0.1143767
> # Item difficulty parameters
> out6$Bec
          [,1]     [,2]     [,3]       [,4]
[1,]  0.0000000 1.971138 4.434774  8.936721
[2,] -0.1384556 2.006430 4.595056  9.329225
[3,]  0.4429256 4.536849 7.390848 13.684742
[4,]  0.1840715 3.680074 6.491005 11.947701
> # Item discrimination parameters
> out6$ga
[1] 1.0000000 0.8589115 0.4044001 0.5084211
```

In order to properly interpret these results, consider that the identifiability constraint adopted in this case is that, for the first item, the discrimination parameter is equal to 1 and the first threshold is equal to 0. In any case, these results confirm that the first item has the highest discriminating power, whereas the third has the lowest discriminating power. These two items are also different in terms of difficulty level. It is also of interest to consider the distribution of the latent variable based on three ordered latent classes, with the last one corresponding to the most satisfied subjects; this class has the smallest weight.

It is possible to express the estimates of the model parameters under the alternative identifiability constraint that the latent distribution is standardized, that is, it has mean 0 and variance 1. This is obtained through the following simple commands:

```
> out7 = standard.matrix(t(out6$Th),out6$piv)
> out7$mu
[1] 5.83714
> out7$si2
Dimension 1
   10.67247
> # Standardized abilities
> (ths = (out6$Th-out7$mu)/out7$si)
               Class 1    Class 2   Class 3
Dimension 1 -1.139848 0.2920302 2.095081
> # Standardized difficulties
> (bes = (out6$Bec-out7$mu)/out7$si)
           [,1]       [,2]        [,3]      [,4]
[1,] -1.786766 -1.1833947  -0.4292683 0.948791
[2,] -1.829148 -1.1725917  -0.3802055 1.068937
[3,] -1.651185 -0.3980230   0.4755946 2.402174
[4,] -1.730421 -0.6602843   0.2001499 1.870461
> # Standardized discrimination parameters
> (gas = out6$gac*out7$si)
[1] 3.266875 2.805956 1.321124 1.660948
```

The estimates of the item parameters are now directly comparable with those obtained under the assumption that the latent trait has a normal distribution (Section 4.6.2.2). This comparison shows a certain agreement between the two sets of parameter estimates obtained under the different assumptions about the latent distribution.

In the same package MultiLCIRT, it is possible to fit some constrained GRM versions. In particular, we can fit the model with constrained discrimination parameters by omitting the option disc=1 as follows:

```
> # LC-1P-GRM
> out8 = est_multi_poly(S,yv,k=3,link=1)
*------------------------------------------------------------------------------*
Model with multidimensional structure
            [,1] [,2] [,3] [,4]
Dimension 1   1    2    3    4
Link of type =                1
Discrimination index =        0
Constraints on the difficulty = 0
Type of initialization =      0
*------------------------------------------------------------------------------*
> summary(out8)
```

```
Call:
est_multi_poly(S = S, yv = yv, k = 3, link = 1)

Log-likelihood:
[1] -7202.57

AIC:
[1] 14445.13

BIC:
[1] 14550.27

Class weights:
[1] 0.3638 0.5476 0.0887

Conditional response probabilities:
, , class = 1

        item
category      1      2      3      4
       0 0.0921 0.1111 0.4022 0.2871
       1 0.3034 0.3088 0.4308 0.4489
       2 0.3846 0.3838 0.1201 0.1985
       3 0.2112 0.1881 0.0450 0.0622
       4 0.0087 0.0083 0.0019 0.0034

, , class = 2

        item
category      1      2      3      4
       0 0.0051 0.0062 0.0327 0.0199
       1 0.0268 0.0289 0.1679 0.1031
       2 0.1195 0.1356 0.3051 0.2947
       3 0.6994 0.6862 0.4577 0.5195
       4 0.1492 0.1431 0.0367 0.0628

, , class = 3

        item
category      1      2      3      4
       0 0.0001 0.0001 0.0006 0.0004
       1 0.0005 0.0006 0.0040 0.0022
       2 0.0027 0.0031 0.0139 0.0105
       3 0.0921 0.0959 0.3084 0.2031
       4 0.9046 0.9003 0.6731 0.7838
```

The two graded response formulations may be compared by an LR statistic:

```
> (dev3 = -2*(out8$lk-out6$lk))
[1] 170.7949
> (pvalue3 = 1-pchisq(dev3,3))
[1] 0
```

This statistic confirms the conclusion that the hypothesis of constant discriminating power of the items must be strongly rejected.

We can also fit RS-GRM, that is, GRM under the constraint in (4.3). In this case, we have to use option difl as follows:

```
> # LC-1P-RS-GRM
> out9 = est_multi_poly(S,yv,k=3,link=1,disc=1,difl=1)
```

```
*-------------------------------------------------------------------------------*
Model with multidimensional structure
            [,1] [,2] [,3] [,4]
Dimension 1    1    2    3    4
Link of type =                 1
Discrimination index =         1
Constraints on the difficulty = 1
Type of initialization =       0
*-------------------------------------------------------------------------------*
> summary(out9)

Call:
est_multi_poly(S = S, yv = yv, k = 3, link = 1, disc = 1, difl = 1)

Log-likelihood:
[1] -7160.93

AIC:
[1] 14349.86

BIC:
[1] 14423.46

Class weights:
[1] 0.3433 0.5500 0.1067

Conditional response probabilities:
, , class = 1

          item
category       1      2      3      4
       0 0.0650 0.0854 0.4279 0.3170
       1 0.3796 0.3930 0.3478 0.4121
       2 0.4207 0.3869 0.1516 0.1942
       3 0.1319 0.1310 0.0665 0.0721
       4 0.0027 0.0035 0.0062 0.0045

, , class = 2

          item
category       1      2      3      4
       0 0.0016 0.0027 0.0646 0.0292
       1 0.0160 0.0229 0.1774 0.1194
       2 0.1079 0.1297 0.2988 0.2899
       3 0.7652 0.7340 0.3960 0.4960
       4 0.1094 0.1108 0.0633 0.0654

, , class = 3

          item
category       1      2      3      4
       0 0.0000 0.0000 0.0037 0.0010
       1 0.0002 0.0003 0.0130 0.0049
       2 0.0012 0.0020 0.0422 0.0203
       3 0.0689 0.0894 0.3817 0.3038
       4 0.9298 0.9083 0.5594 0.6700


> out9$Bec
$difficulties
[1] 0.0000000 0.1300641 2.2023827 1.5987719

$cutoffs
[1] 0.000000 2.443275 4.525888 8.565204

> out9$gac
[1] 1.0000000 0.9350058 0.6268061 0.7194276
```

In the new output, the difficulty parameter of each item is collected in the vector `out9$Bec$difficulties`, whereas the common thresholds are collected in `out9$Bec$cutoffs`. These results confirm that the most difficult item is the third. However, even the rating scale hypothesis must be rejected according to the following LR test, which considers that this hypothesis is formulated by nine independent constraints on the item parameters of GRM:

```
> (dev4 = -2*(out9$lk-out6$lk))
[1] 87.52847
> (pvalue4 = 1-pchisq(dev4,9))
[1] 5.107026e-15
```

All the previous analyses may be performed in a similar way on the basis of a partial credit formulation. It is enough to use option `link=2`, so that conditional response probabilities are modeled by local logits. For instance, for PCM (characterized by constrained discrimination parameters and free difficulty parameters), we have

```
> # LC-PCM
> out10 = est_multi_poly(S = S, yv = yv, k = 3, link = 2)
> summary(out10)

Call:
est_multi_poly(S = S, yv = yv, k = 3, link = 2)

Log-likelihood:
[1] -7249.21

AIC:
[1] 14538.41

BIC:
[1] 14643.55

Class weights:
[1] 0.3322 0.6049 0.0628

Conditional response probabilities:
, , class = 1

        item
category     1      2      3      4
       0 0.1120 0.1324 0.4251 0.3177
       1 0.3344 0.3357 0.4398 0.4522
       2 0.3161 0.3206 0.1088 0.1789
       3 0.2273 0.2024 0.0259 0.0500
       4 0.0102 0.0089 0.0003 0.0012

, , class = 2

        item
category     1      2      3      4
       0 0.0015 0.0019 0.0323 0.0144
       1 0.0262 0.0290 0.2014 0.1235
       2 0.1493 0.1666 0.3002 0.2945
       3 0.6472 0.6340 0.4314 0.4959
       4 0.1758 0.1685 0.0346 0.0718

, , class = 3
```

```
         item
category     1       2       3       4
       0 0.0000 0.0000 0.0000 0.0000
       1 0.0000 0.0000 0.0000 0.0000
       2 0.0003 0.0003 0.0024 0.0012
       3 0.0639 0.0652 0.1874 0.1134
       4 0.9358 0.9345 0.8101 0.8853

> # Abilities and class weights
> rbind(out10$Th,pi=out10$piv)
               Class 1    Class 2     Class 3
Dimension 1 1.0939909 2.8902987 6.87753595
pi          0.3322454 0.6049417 0.06281291
> # Item difficulties
> out10$Bec
          [,1]      [,2]      [,3]      [,4]
[1,] 0.0000000 1.150213 1.423499 4.193535
[2,] 0.1635137 1.140120 1.554026 4.215250
[3,] 1.0600193 2.491167 2.527734 5.413607
[4,] 0.7407284 2.021320 2.369192 4.822845
> # Item discrimination parameters
> out10$gac
[1] 1 1 1 1
```

However, as for the case of a normally distributed latent trait, models of partial credit type show a worse fit with respect to those of graded response type for the data at issue.

Finally, it is important to stress that all the previous analyses are performed under the assumption that there are $k = 3$ latent classes. If we want to base this choice on the data and then avoid an a priori choice, we can use the command search.model, as already shown in Section 3.8.2, where options link, disc, and difl are specified according to the chosen model.

```
> # Search the optimal number of support points by BIC
> out11 = search.model(S,yv,kv=1:7,link=1,disc=1)
*****************************************************************************
7 12
*--------------------------------------------------------------------------*
Model with multidimensional structure
            [,1] [,2] [,3] [,4]
Dimension 1    1    2    3    4
Link of type =                1
Discrimination index =        1
Constraints on the difficulty = 0
Type of initialization =      1
*--------------------------------------------------------------------------*
lktrace   -6974.314 -6974.314 -6941.269 -6920.098 -6920.098 -6920.098 -6920.098 -6920.098
          -6920.098 -6920.098 -6920.098 -6920.098 -6886.701
lk =  -8209.439 -7414.972 -7117.168 -6974.314 -6920.098 -6886.701 -6886.701
aic =  16450.88 14871.94 14280.34 13998.63 13894.2 13831.4 13835.4
bic =  16534.99 14982.34 14401.25 14130.05 14036.14 13983.85 13998.37

[…] Output omitted

> out11$bicv
[1] 16534.99 14982.34 14401.25 14130.05 14036.14 13983.85 13998.37
```

On the basis of the list of the BIC values so obtained, $k = 6$ is the most suitable number of classes for the RLMS dataset. The estimation results for this value of k may be extracted from the output provided by `search.model` through the command

```
> out12 = out11$out.single[[6]]
```

These results may be displayed by the command `summary`. Clearly, the model has a much better fit with 6 latent classes than with 3 classes, as proved by the much higher value of the log-likelihood. The fit is also clearly superior than that under the assumption that the latent trait is normally distributed. However, it is clear that with a higher number of latent classes, the results may be more difficult to interpret because more groups of subjects have to be described.

Exercises

1. Using the dataset `delinq.txt` (downloadable from `http://www.gllamm.org/delinq.txt`), perform the following analyses through `Stata`:

   (a) Reshape the data in long format and estimate the 1P-RS-GRM.

   (b) Using the parameter estimates obtained at point (a) as starting values, estimate the 1P-GRM, RS-GRM, and GRM.

   (c) Compare the models estimated at points (a) and (b) using, in a suitable way, the LR test and indicate the model with the best fit.

   (d) On the basis of the model selected at point (c), plot IRCCCs for each item.

   (e) Discuss the shape of IRCCCs for each item.

2. Using the first five items considered in dataset `aggression.dat` (downloadable from `http://www.gllamm.org/aggression.dat`), perform the following analyses through `Stata`:

   (a) After reshaping the data in a suitable way, estimate PCM and RSM and select the best model.

   (b) On the basis of the model selected at point (a), plot IRCCCs for each item.

   (c) On the basis of the aforementioned IRCCCs, detect the possible presence of items with reversed thresholds.

   (d) Collapse category 2 with category 1 and estimate the Rasch model.

   (e) Compare the estimates of the item difficulties under the Rasch model with those obtained under the model selected at point (a).

3. Consider dataset Science available in the R package ltm:

   (a) Convert the original dataset with nominal response categories into a matrix in which the categories are coded from 0 to 3.

   (b) Using the same package, estimate GRM under the assumption that the latent trait has a normal distribution and comment on the parameter estimates.

   (c) Using package MultiLCIRT, estimate the same model described at point (b) assuming that the latent distribution is discrete with 3 support points.

   (d) Compare the results obtained at point (c) with those obtained at point (b) in terms of parameter estimates and goodness-of-fit.

4. Using the R package MultiLCIRT, estimate the LC version of GRM and PCM for the HADS data. In this analysis, select the number of latent classes by BIC and compare the obtained results with those illustrated in Section 4.6.1.

5

Estimation Methods and Diagnostics

5.1 Introduction

Two fundamental steps in item response theory (IRT) analysis are those of parameter estimation and diagnostics. Regarding the main estimation methods, we have already provided a summary description in Section 3.7 with reference to IRT models for dichotomous items. We recall that the most well-known estimation strategies are based on the joint maximum likelihood (JML) method, on the conditional maximum likelihood (CML) method, and on the marginal maximum likelihood (MML) method. The first two methods rely on the so-called fixed-effects formulation of the model, in which a fixed parameter is used to represent the latent trait level of every sample unit. On the other hand, the MML method relies on the random-effects formulation that assumes that each of these individual levels is a realization of a latent variable, the distribution of which may be continuous or discrete. These methods have different advantages and disadvantages and a different degree of applicability.

Regarding diagnostics, both classical test theory (CTT) and IRT provide tools to understand how each item response depends on the latent trait of interest. Among the classical tools (Chapter 2), we recall Cronbach's α, the item-test correlations, the proportion of responses for the different modalities of each item, and the factor analysis to assess the strength of the association between an item and the latent variable. Corresponding tools are available within IRT.

It is important to recall that most statistical tools for IRT diagnostics are aimed at evaluating the validity of the Rasch paradigm against more general IRT models. Indeed, since Rasch-type models (i.e., the dichotomous Rasch model, the partial credit model [PCM], and the rating scale model [RSM]) provide a mathematical/probabilistic description of the measurement, the analysis of the discrepancy between the Rasch model and the collected data is necessary. Rather than dealing with fit in the usual way, that is, the model should fit the data, in the Rasch framework the aim is obtaining data that fit the model: only if data fit the measurement model, one can be sure that the resulting measurements are objective and unidimensional. For instance, the bad fit of an item to the Rasch-type model means that this item does not contribute to measure the same latent trait as the other items; therefore, it does not make sense to keep it in the questionnaire and the measurement process should be carried out by only relying on the remaining items. Similarly, if a person gives unexpected responses in comparison with his or her latent trait estimate, then the behavior of this person cannot be considered in agreement with the measurement process.

On the other hand, from a statistical point of view, we can talk of misfit of the model to the observed data and, therefore, abandon Rasch-type models in favor of more general IRT models having a better fit. Dropping items and/or persons' responses, or switching to another model, depends on the aim of the analysis. If the aim is the construction of a measurement instrument for a certain latent variable, then the Rasch paradigm should be followed and a set of items with a good fit to the Rasch model should be selected. However, if the aim is analyzing observed data, then a typical statistical point of view should be adopted by searching a suitable IRT model in terms of fit.

This chapter is organized as follows. The first part of the chapter is devoted to the illustration of the most important estimation methods. More precisely, JML, CML, and MML are described in Sections 5.2 through 5.4, respectively. These sections explicitly deal with the estimation of models for dichotomous items, whereas the extension to polytomous items is briefly described in Section 5.5. The second part of the chapter is devoted to the illustration of several diagnostic instruments. We mainly distinguish between graphical tools (Section 5.6), methods based on the goodness-of-fit measurement and on parametric and nonparametric hypothesis tests (Sections 5.7 and 5.8), and methods focused on differential item functioning (DIF) (Section 5.9). As usual, the chapter ends with illustrative examples in Stata and R (Section 5.10).

5.2 Joint Maximum Likelihood Method

As already mentioned in Section 3.7 (in Chapter 3), the JML method consists of maximizing, with respect to both ability and item parameters, the likelihood of the assumed model. The latter corresponds to the probability $p(Y|\theta)$, where Y is the observed matrix of responses. We recall that, being based on a fixed-effects approach, the JML method considers the person parameters as fixed parameters corresponding to the latent trait level of each individual.

In the following, this method is illustrated in detail for the Rasch model and for the two-parameter logistic (2PL) and three-parameter logistic (3PL) models. In this regard, we recall that the 2PL model may be obtained as a particular case of the 3PL model constraining the guessing parameter of each item to 0.

5.2.1 Rasch Model

For the Rasch model, the likelihood function to maximize is

$$L_J(\psi) = \frac{e^{\sum_{i=1}^{n} y_{i\cdot}\theta_i - \sum_{j=1}^{J} y_{\cdot j}\beta_j}}{\prod_{i=1}^{n} \prod_{j=1}^{J} \left(1 + e^{\theta_i - \beta_j}\right)},$$

where $y_{i\cdot} = \sum_{j=1}^{J} y_{ij}$ and $y_{\cdot j} = \sum_{i=1}^{n} y_{ij}$, which directly derives from (3.7). Moreover, ψ is used to denote the vector of all model parameters with respect to which the likelihood function has to be maximized. This notation is used throughout the chapter, specifying the structure of this vector for each case of interest. In the present case, this vector contains both ability parameters θ_i and difficulty parameters β_j. In practice, we proceed by maximizing the corresponding log-likelihood

$$\ell_J(\psi) = \log L_J(\psi) = \sum_{i=1}^{n} y_{i\cdot}\theta_i - \sum_{j=1}^{J} y_{\cdot j}\beta_j - \sum_{i=1}^{n}\sum_{j=1}^{J} \log\left(1 + e^{\theta_i - \beta_j}\right).$$

Note that both $L_J(\psi)$ and $\ell_J(\psi)$ are invariant with respect to translations of the parameters θ_i and β_j; this is related to the invariance of the probabilities in (3.4) already noted in Section 3.3. Then, a suitable constraint has to be put on the parameters in order to make the model identifiable. The following constraints may be used:

- $\beta_1 = 0$: The first item is taken as a reference item and then β_j, $j = 2, \ldots, J$, must be interpreted as the difficulty of item j with respect to the first one. This constraint is used when there is an item whose characteristics are well known, mainly because it has already been administered in previous applications.
- $\sum_{j=1}^{J} \beta_j = 0$: So that the average difficulty of the items is fixed and β_j, $j = 1, \ldots, J$, is interpreted as the difficulty of item j with respect to the average difficulty. This constraint is used when a reference item cannot be singled out.
- $\sum_{i=1}^{n} \theta_i = 0$: So that the average ability of the subjects is fixed and θ_i, $i = 1, \ldots, n$, is interpreted as the ability of subject i with respect to the average of the group.

It has to be clarified that the previous constraints are equivalent in the sense that the maximum value of the likelihood that is reached under each of them is the same; this value is also equal to the unconstrained maximum likelihood. Moreover, the parameter estimates obtained under each of these constraints may be simply transformed into the estimates that may be obtained under any other constraint. For this reason, we prefer to adopt the first one, that is,

$\beta_1 = 0$, and then this parameter is dropped from the overall parameter vector ψ. So, we can define this vector as $\psi = (\theta', \beta^{*\prime})'$, where $\beta^* = (\beta_2, \ldots, \beta_J)'$ is the vector of free item parameters.

To maximize $\ell_J(\psi)$, we can use an iterative algorithm that performs a series of steps until convergence. Let $\psi^{(h)}$ denote the estimate of ψ obtained at the end of the h-th step. Its elements $\theta_i^{(h)}$, $i = 1, \ldots, n$, are collected in the subvector $\theta^{(h)}$, and $\beta_j^{(h)}$, $j = 2, \ldots, J$, are collected in the subvector $\beta^{*(h)}$. At the $(h + 1)$-th step, the algorithm performs the following operations:

- Update the estimate of each parameter θ_i, $i = 1, \ldots, n$, by a first-step solution of the Newton–Raphson maximization of $\ell_J(\psi)$ with respect to this parameter, with all the other parameters held fixed at the value obtained from the previous step. In practice, we have

$$\theta_i^{(h+1)} = \theta_i^{(h)} - \frac{\partial \ell_J(\psi^{(h)})}{\partial \theta_i} \Big/ \frac{\partial^2 \ell_J(\psi^{(h)})}{\partial \theta_i^2}, \quad i = 1, \ldots, n, \qquad (5.1)$$

with

$$\frac{\partial \ell_J(\psi^{(h)})}{\partial \theta_i}$$

denoting the first derivative of $\ell_J(\psi)$ with respect to θ_i evaluated at $\psi = \psi^{(h)}$. A similar notation is used for the other derivatives.
- Update the estimate of β_j, $j = 2, \ldots, J$, by a first-step solution of the Newton–Raphson maximization of $\ell_J(\psi)$ with respect to this parameter, with all the other parameters fixed at the value obtained from the previous step. We have

$$\beta_j^{(h+1)} = \beta_j^{(h)} - \frac{\partial \ell_J(\psi^{(h)})}{\partial \beta_j} \Big/ \frac{\partial^2 \ell_J(\psi^{(h)})}{\partial \beta_j^2}, \quad j = 2, \ldots, J. \qquad (5.2)$$

Explicit expressions for the derivatives in (5.1) and (5.2) are given in Appendix 5.A.1 of this chapter; see also Baker and Kim (2004).

Obviously, the algorithm has to be initialized by choosing an initial guess $\psi^{(0)}$ of the parameter vector. This choice does not affect the point at convergence of the algorithm as, under mild conditions, $\ell_J(\psi)$ is a strictly concave function in the full parameter space. In any case, we suggest adopting the following rule for this initialization:

$$\theta_i^{(0)} = \log \frac{y_{i\cdot}}{J - y_{i\cdot}}, \quad i = 1, \ldots, n,$$

$$\beta_j^{(0)} = \log \frac{n - y_{\cdot j}}{y_{\cdot j}} - \log \frac{n - y_{\cdot 1}}{y_{\cdot 1}}, \quad j = 2, \ldots, J.$$

Another important point is how to check the convergence of the algorithm. A suitable rule consists in stopping the algorithm at step h when the following conditions are satisfied:

$$\ell_J\big(\psi^{(h)}\big) - \ell_J\big(\psi^{(h-1)}\big) < \text{tol}_1,$$

$$\max\big|\psi^{(h)} - \psi^{(h-1)}\big| < \text{tol}_2.$$

In these expressions, tol_1 and tol_2 are small constants and $\max|v|$ denotes the largest element (in absolute value) of the vector v. As a rule of thumb, we can choose these constants to be equal to 10^{-6}.

When the algorithm is properly initialized, it usually converges in a reasonable number of steps to the maximum of $\ell_J(\psi)$, which corresponds to the JML estimate of the parameters ψ, denoted by $\hat{\psi}_J = \big(\hat{\theta}_J', \hat{\beta}_J^{*\prime}\big)'$. It has to be clarified that a full Newton–Raphson algorithm, which at each step jointly updates the estimate of ψ until convergence, is in principle faster with respect to the algorithm previously described. However, in practical situations, this algorithm may not be viable as it uses the second derivative matrix having dimension $(n+J-1) \times (n+J-1)$, which may become huge. For this reason, we suggest the implementation based on (5.1) and (5.2).

The implementation previously described also allows us to easily estimate the parameters θ_i when the parameters β_j are assumed to be known. This typically happens when the same *bank* of items is used at several occasions, and then β^* is set equal to the estimate of this parameter vector based on a dataset obtained by merging the data deriving from previous administrations. In this case, only the steps based on (5.1) are used in order to update the ability parameters, whereas the steps to update the difficulty parameters, see (5.2), are skipped. More details about estimation ability are given in Thissen and Wainer (2001) and Baker and Kim (2004).

We also have to consider that the JML estimate is not ensured to exist, since it is not ensured that there exist vectors ψ with finite elements such that $\ell_J(\psi)$ corresponds to the supremum of this function. A set of conditions on the matrix Y that ensure the JML estimate to exist is provided in Fischer (1981). A necessary condition, which typically ensures that these conditions are satisfied, is

$$0 < y_{i\cdot} < J, \quad i = 1,\ldots,n,$$

$$0 < y_{\cdot j} < n, \quad j = 1,\ldots,J,$$

so that there are no subjects that respond correctly or incorrectly to all items and there are no items to which all subjects respond correctly or incorrectly. When such *anomalous* subjects or items exist, they must be eliminated from the dataset. Note that, in practical situations, it may happen that some anomalous subjects exist, but it is very unlikely that one or more anomalous items exist. This is because, typically, n is much larger than J. In practice, it is then sufficient to remove some rows from Y in order to ensure estimability of the Rasch model by the JML method.

Standard errors for both subject and item parameter estimates can be obtained through the square root of the diagonal elements of the inverse of the observed information matrix. However, since this matrix is very large, computing its inverse may be difficult. For this reason, in Appendix 5.A.1, we exploit the blocking structure of the information matrix to provide fast and stable estimates of the standard errors.

Finally, we recall that, for J fixed, the JML estimator is not consistent as n grows to infinity. This does not mean that the estimates obtained by using this method are unreliable. Indeed, if the number of subjects and that of items are large enough, the bias of the JML estimator may be low. Nevertheless, methods to reduce the bias may be used, as the one based on the general approach of McCullagh and Tibshirani (1990). The same may be said for the reliability of the standard errors.

5.2.2 2PL and 3PL Models

The 3PL model is very general as the 2PL model is a particular case of the first when $\delta_j = 0$ for $j = 1, \ldots, J$, where δ_j is the guessing parameter for item j. Moreover, the Rasch model is a particular case of the 2PL model with $\lambda_j = 1$ and $\delta_j = 0$ for $j = 1, \ldots, J$. The joint log-likelihood of the 3PL model has expression

$$\ell_J(\psi) = \sum_{i=1}^{n} \sum_{j=1}^{J} y_{ij} \log p_j(\theta_i) + \left(1 - y_{ij}\right) \log \left(1 - p_j(\theta_i)\right),$$

with $p_j(\theta_i)$ defined in (3.11), where now ψ contains, in addition to the ability parameters θ_i and the difficulty parameters β_j, the parameters λ_j and δ_j. Different from the Rasch model, the previous expression cannot be further simplified.

The JML method is based on the same steps described previously, in which the estimation of each single parameter is updated in order to avoid inversion of huge matrices as in the standard Newton–Raphson algorithm that updates all the parameters at the same time. In this case, we use the identifiability constraints $\lambda_1 = 1$ and $\beta_1 = 0$. The derivatives to be used are given in Appendix 5.A.1.

When applied to the 2PL and 3PL models, the aforementioned algorithm may suffer from an instability problem due to the discriminating indices assuming too large values in magnitude. This problem, known as the Heywood case in the context of factorial analysis, may be overcome by requiring these parameters to lie in a certain interval, say $[1/10; 10]$. Similarly, the algorithm must also include a constraint to ensure that each guessing parameter δ_j is between 0 and 1.

5.3 Conditional Maximum Likelihood Method

This method may be only applied with the Rasch model (Rasch, 1960, 1961) and it is typically used to estimate the item parameters in β. In this case, the method is based on the maximization of the conditional likelihood of the model given the sufficient statistics $y_{1.}, \ldots, y_{n.}$ for the ability parameters. When the interest is on the ability parameters θ only, this method may be applied to estimate these parameters by maximizing the conditional likelihood given the sufficient statistics for β, which correspond to $y_{.1}, \ldots, y_{.J}$.

First, when only β is estimated, consider that the conditional probability that subject i attains score $y_{i.}$, given the ability parameter θ_i, is equal to

$$p(y_{i.}|\theta_i) = \sum_{y \in \mathcal{Y}(y_{i.})} p(y_i = y|\theta_i) = \frac{e^{\theta_i y_{i.}}}{\prod_{j=1}^{J} \left(1 + e^{\theta_i - \beta_j}\right)} \omega_i,$$

where

$$\omega_i = \sum_{y \in \mathcal{Y}(y_{i.})} e^{-\sum_{j=1}^{J} y_j \beta_j},$$

with $\mathcal{Y}(y_{i.})$ denoting the set of all the binary vectors $y = (y_1, \ldots, y_J)'$ with elements having a sum $y_{i.}$. Obviously, these vectors have $y_{i.}$ elements equal to 1 and $J - y_{i.}$ elements equal to 0. Consequently, for each subject i, the conditional probability of the response configuration y_i given $y_{i.}$ is equal to

$$p(y_i|y_{i.}, \theta_i) = \frac{p(y_i|\theta_i)}{p(y_{i.}|\theta_i)} = \frac{e^{-\sum_{j=1}^{J} y_{ij} \beta_j}}{\omega_i}, \tag{5.3}$$

which does not depend on θ_i and is then denoted by $p_C(y_i)$.

The likelihood used for CML estimation is equal to

$$L_C(\psi) = \prod_{i=1}^{n} p_C(y_i).$$

Note that the subjects who respond correctly or incorrectly to all items do not contribute to this likelihood and can then be dropped. Therefore, with

$$y_{.j}^* = \sum_{i=1}^{n} y_{ij} 1\{0 < y_{i.} < J\},$$

the conditional log-likelihood may be expressed as

$$\ell_C(\psi) = \log L_C(\beta) = -\sum_{j=1}^{J} y_{\cdot j}^* \beta_j - \sum_{i=1}^{n} 1\{0 < y_{i\cdot} < J\} \log w_i. \qquad (5.4)$$

An identifiability problem also arises in this case, since we can add an arbitrary constant to each parameter β_j without changing the conditional probability in (5.3). For this reason, we use the constraint $\beta_1 = 0$ and then $\psi = \beta^*$; an equivalent constraint is $\sum_{j=1}^{J} \beta_j = 0$.

In order to maximize $\ell_C(\psi)$, we can rely on a Newton–Raphson algorithm. At step $h + 1$, it consists in updating the estimate of ψ as

$$\psi^{(h+1)} = \psi^{(h)} - \left(\frac{\partial^2 \ell_C(\psi^{(h)})}{\partial \psi \partial \psi'}\right)^{-1} \frac{\partial \ell_C(\psi^{(h)})}{\partial \psi},$$

where $\psi^{(h)}$ is the estimate obtained at the end of the previous step. A method to compute the conditional log-likelihood and its derivatives, in a way that is viable even when J is large, is illustrated in Appendix 5.A.2.

For initializing the algorithm and checking its convergence, we can use the procedure suggested in the previous section. Note that the algorithm usually converges in a very few steps, being a Newton–Raphson algorithm that updates the full parameter vector ψ at each step. The value at convergence is the CML estimate of ψ, denoted by $\hat{\psi}_C = \hat{\beta}_C^*$. It is also worth noting that standard errors for these estimates may be obtained as the square root of the elements in the main diagonal of the inverse of $I_C(\hat{\psi}_C)$, where

$$I_C(\psi) = -\frac{\partial^2 \ell_C(\psi)}{\partial \psi \partial \psi'}$$

is the observed information matrix.

For the existence of the CML estimate, certain conditions need to be fulfilled that are less restrictive in comparison to those mentioned in the previous section; see Fischer (1981). In most situations, a condition that ensures the existence of the estimate is that there do not exist items to which no subjects or all subjects respond correctly.

Finally, we have to stress that the main advantage of the method is that the resulting estimator is consistent for J fixed as n growing to infinity. For a detailed description of the asymptotic properties of the CML estimator, see Andersen (1970, 1972).

5.4 Marginal Maximum Likelihood Method

A relevant problem of the JML method is that it does not lead to consistent estimates of the item parameters as the sample size grows to infinity with the number of items remaining constant. We showed in the previous section that a solution to this problem is the CML method. However, this method may be applied only with the Rasch model, which is based on the one-parameter logistic parameterization.

A more general method, which is applicable to the 2PL and 3PL models, is the MML method. As already mentioned in Section 3.7, this method consists of maximizing the likelihood of the model after the ability parameters have been integrated out on the basis of a common distribution assumed on these parameters. For a sample of n subjects having response configurations y_i, $i = 1, \ldots, n$, this *marginal likelihood* may be expressed as

$$L_{\mathrm{M}}(\psi) = \prod_{i=1}^{n} p(y_i), \qquad (5.5)$$

where $p(y_i)$ is defined in (3.2) and the vector ψ now includes the item parameters and the parameters of the distribution of the ability. The corresponding log-likelihood is

$$\ell_{\mathrm{M}}(\psi) = \sum_{i=1}^{n} \log p(y_i), \qquad (5.6)$$

which needs to be maximized by suitable algorithms.

The first authors who considered this approach are Bock and Lieberman (1970). In particular, they dealt with an IRT model based on the normal ogive item characteristic curve (ICC) in which the ability parameters have a normal distribution. Their approach is based on computing the integral in $p(y_i)$ by a Gauss–Hermite quadrature method as follows:

$$p(y_i) = \sum_{v=1}^{k} p(y_i | \xi_v) \pi_v = \sum_{v=1}^{k} \left[\prod_{j=1}^{J} p_j(\xi_v)^{y_{ij}} \left(1 - p_j(\xi_v)\right)^{1-y_{ij}} \right] \pi_v, \qquad (5.7)$$

where $\xi_1, \ldots, \xi_k$ are the *quadrature points* and $\pi_1, \ldots, \pi_k$ the corresponding *weights*. Under a uniform quadrature, the nodes may be taken as equally spaced points in a suitable interval and the weights as the corresponding renormalized densities. For instance, under the assumption that the ability has a standard normal distribution, the uniform quadrature consists in taking the nodes as k points in an interval, say $[-4, 4]$, with weights computed as

$$\pi_v = \frac{\phi(\xi_v)}{\sum_{u=1}^{k} \phi(\xi_u)}, \tag{5.8}$$

with $\phi(\cdot)$ denoting the density function of this distribution. Bock and Lieberman (1970) also suggested using a Newton–Raphson algorithm to maximize the resulting log-likelihood.

A further development is due to Bock and Aitkin (1981) in a form that is more computationally feasible and, under the assumption that the latent distribution is correctly specified, it leads to a consistent estimation of the item parameters. In particular, the method of Bock and Aitkin (1981) is based on an expectation–maximization (EM) algorithm (Dempster et al., 1977). In the following, we provide a formulation of their approach, which is suitable for the class of IRT models here proposed and is based on the assumption that the ability has normal distribution with arbitrary parameters μ and σ^2. These parameters are collected in the vector ψ. We also show how it is possible to relax the assumption of normality of the latent distribution by assuming that the ability has a discrete distribution with arbitrary support points and probabilities. This approach is well known in the literature; see Thissen (1982), Lindsay et al. (1991), Pfanzagl (1993), Formann (1995), Schilling and Bock (2005), and Bartolucci (2007); and is related to the latent class (LC) model (Lazarsfeld and Henry, 1968; Goodman, 1974b).

5.4.1 Estimation with Normal Distribution for the Ability

First, consider that using a quadrature of type (5.7) is equivalent to assuming that the ability has a discrete distribution with support points equal to $\xi_v' = \mu + \sigma \xi_v$ and probabilities π_v computed as in (5.8), under a uniform quadrature, on the basis of the normal standard density function. Moreover, the EM paradigm exploits the log-likelihood of the *complete data*, which corresponds to the log-likelihood that we could compute if we knew the ability level of each subject in the sample. Upon an additive constant term, this function may be expressed as

$$\ell_M^*(\psi) = \sum_{i=1}^{n} \sum_{v=1}^{k} z_{iv} \log p(y_i | \xi_v')$$

$$= \sum_{i=1}^{n} \sum_{v=1}^{k} z_{iv} \sum_{j=1}^{J} \left[y_{ij} \log p_j(\xi_v') + (1 - y_{ij}) \log \left(1 - p_j(\xi_v')\right) \right],$$

where z_{iv} is a dummy variable equal to 1 when subject i has ability level ξ_v' and to 0 otherwise. In the previous expression, we exploit the fact that variables y_{ij} in y_i are conditionally independent given the ability level.

Since the ability level of each subject in the sample is unknown, the EM algorithm maximizes $\ell_M(\psi)$ by alternating the following steps until convergence:

- *E-step*: It consists in computing the expected value of $\ell_M^*(\psi)$ given the observed data and the current value of the parameters.
- *M-step*: It consists in maximizing the previous expected value with respect to the model parameters.

In practice, the E-step is performed by computing, for each i, the conditional expected value of z_{iv}, $v = 1, \ldots, k$, given the response configuration y_i. This expected value is equal to the posterior probability

$$\hat{z}_{iv} = p\left(\xi_v'|y_i\right) = \frac{p\left(y_i|\xi_v'\right)\pi_v}{p(y_i)} = \frac{p\left(y_i|\xi_v'\right)\pi_v}{\sum_{u=1}^{k}p\left(y_i|\xi_u'\right)\pi_u}, \tag{5.9}$$

which is defined on the basis of Bayes' theorem.

Then, the M-step consists in maximizing, with respect to ψ, function $\ell_M^*(\psi)$ once each z_{iv} has been substituted with $\hat{z}_{iv}$. The resulting function is denoted by $\hat{\ell}_M^*(\psi)$. This maximization may be performed by a standard Newton–Raphson or Fisher-scoring algorithm for each item parameter. This requires to compute derivatives of $\hat{\ell}_M^*(\psi)$ with respect to each element of ψ. In particular, for what concerns the 3PL model, we need to use the derivates reported in Appendix 5.A.3 of this chapter.

Convergence of the algorithm may be checked in the usual way on the basis of the difference in the log-likelihood and in the parameter estimates between two consecutive steps, as made for the JML method. We can also estimate the ability parameter of each subject i as

$$\hat{\theta}_i = \sum_{v=1}^{k}\hat{z}_{iv}\hat{\xi}_v', \quad i = 1, \ldots, n. \tag{5.10}$$

This is an expected a posteriori (EAP) estimate of the ability, as it is based on the a posteriori distribution of this latent trait given the observed responses.

Finally, the information matrix of the incomplete data can be obtained as minus the numerical derivative of the score vector. From this matrix, the standard errors for the parameter estimates can be obtained in the usual way. An alternative method consists in estimating the information matrix by an estimate of the variance–covariance matrix of the score function (McLachlan and Peel, 2000, pp. 64–66).

5.4.2 Estimation with Discrete Distribution for the Ability

In this case, we assume that the ability Θ_i has a discrete distribution with k support points $\xi_1, \ldots, \xi_k$ and probabilities $\pi_1, \ldots, \pi_k$. An equivalent way to express this assumption is by associating to each subject i a discrete latent variable V_i with k support points labeled from 1 to k. This latent variable defines k latent classes in the population of interest, with subjects in latent class v sharing the common ability level ξ_v. Moreover, the corresponding weight π_v is a measure of dimension of class v, being equal to the proportion of subjects in this class.

The marginal log-likelihood is still equal to (5.6), with ψ including the parameters of this distribution and $p(y_i)$ computed as in (5.7). On the other hand, the complete log-likelihood has a different expression with respect to the one previously defined; it is defined as

$$\ell_M^*(\psi) = \sum_{i=1}^n \sum_{v=1}^k z_{iv} \sum_{j=1}^J \left[y_{ij} \log p_j(\xi_v) + (1 - y_{ij}) \log \left(1 - p_j(\xi_v) \right) \right]$$

$$+ \sum_{i=1}^n \sum_{v=1}^k z_{iv} \log \pi_v.$$

In order to maximize $\ell_M(\psi)$, we can still use the EM algorithm, which has the same structure described previously. In particular, the E-step consists in computing the posterior probabilities $\hat{z}_{iv}$ defined in (5.9) with $\xi_v' = \xi_v$. The M-step is based on a Fisher-scoring algorithm in which the parameters in ψ are updated by maximizing the expected value of the complete log-likelihood. We can use the same Fisher-scoring algorithm previously described, which is based on the same derivatives for λ_j, β_j, and δ_j with $\xi_v = \xi_v'$; the same happens for the elements of the information matrix. A Fisher-scoring algorithm may also be applied to update the support points ξ_v on the basis of the derivatives

$$\frac{\partial \hat{\ell}_M^*(\psi)}{\partial \xi_v} = \sum_{i=1}^n \hat{z}_{iv} \sum_{j=1}^J \lambda_j \left(y_{ij} - p_j(\xi_v) \right) \frac{p_j^\dagger(\xi_v)}{p_j(\xi_v)},$$

with $p_j^\dagger(\xi_v)$ defined in Appendix 5.A.1 and information

$$-E\left(\frac{\partial^2 \hat{\ell}_M^*(\psi)}{\partial \xi_v^2} \right) = \sum_i \hat{z}_{iv} \sum_j \lambda_j^2 (1 - \delta_j) \frac{p_j^\dagger(\xi_v)^2}{p_j(\xi_v)} \left(1 - p_j^\dagger(\xi_v) \right).$$

Finally, the mass probabilities π_v are simply updated as

$$\pi_v = \frac{1}{n} \sum_{i=1}^n \hat{z}_{iv}, \quad v = 1, \ldots, k.$$

Two crucial points are how to select the number of support points, k, and how to initialize the EM algorithm. Concerning the first point, we can use selection criteria and, in this regard, we refer the reader to Section 5.7.1. Regarding the second point, we recall that a standard rule consists in taking as starting values those obtained from the EM algorithm run for the case of ability having a normal distribution. Trying different starting values is useful when the marginal likelihood of the observed data is multimodal. These starting values may be obtained by randomly perturbating those obtained as previously indicated.

After the end of the algorithm, we can also obtain EAP estimates of the abilities by expression (5.10). Alternatively, we can assign each subject i to a latent class on the basis of the maximum of the posterior probabilities $\hat{z}_{iv}$ and then predict his or her ability level by the ability estimate of this class. This is a maximum a posteriori (MAP) estimate of the ability.

5.5 Estimation of Models for Polytomous Items

Until now, all estimation methods have been illustrated with reference to models for dichotomous items (see Chapter 3) and, therefore, binary responses. Obviously, these methods may also be used for the models for polytomous items described in Chapter 4. In this section, we briefly discuss this case, referring the reader to specialized texts, such as Baker and Kim (2004), for a detailed description.

The first estimation method that may be used with polytomous items is the JML method. We recall that this method follows a fixed-effects approach based on the idea that each individual ability level is a parameter to be estimated along with the item parameters. The likelihood function on which this method is based may be expressed as

$$L_J(\psi) = \prod_{i=1}^{n} \prod_{j=1}^{J} p_{jy}(\theta_i), \tag{5.11}$$

where $p_{jy}(\theta_i) = p(Y_{ij} = y|\theta_i)$ is the probability that individual i with latent trait level θ_i answers by category y to item j. This probability is defined according to the model of interest, as for instance the graded response model (GRM) or PCM illustrated in Chapter 4; see Equations 4.4 and 4.10, respectively. Consequently, further to the ability parameters θ_i, ψ includes the set of item parameters defined according to the specific model of interest.

The log-likelihood function corresponding to (5.11) is simply

$$\ell_J(\psi) = \sum_{i=1}^{n} \sum_{j=1}^{J} \log p_{jy}(\theta_i).$$

This function may be maximized by a Newton–Raphson algorithm implemented along the same lines as in Section 5.2. In particular, the algorithm is still based on two steps that are alternated until convergence. At the first step each ability level is updated, while keeping the value of all other parameters fixed at the current value. Similarly, at the second step, the parameters of each item are updated while keeping all the other parameters fixed. Properties of this estimation method are the same holding for models for dichotomous items. In particular, the resulting estimates are not ensured to be consistent, but they may be considered as reliable when many test items are used. The same holds for the standard errors.

The second estimation method, still based on a fixed-effects approach, is the CML method that, on one hand, has the advantage of providing consistent estimates of the item parameters and, on the other hand, has the disadvantage of not being applicable to all IRT models for polytomous items.

The principle on which the CML method is based is that of eliminating the ability parameters θ_i by conditioning on suitable sufficient statistics. However, finding such statistics is possible only for certain models for polytomous items, that is, the models that may be seen as extensions of the Rasch model. This is the case of PCM and of models for nominal responses based on reference-category logits, described in Section 4.5, provided that these models do not contain discrimination parameters. In these cases, we can implement a likelihood function of type

$$L_C(\psi) = \prod_{i=1}^{n} p_C(y_i),$$

where $p_C(y_i)$ is the conditional probability of the sequence of responses y_i provided by subject i, given the value of a suitable sufficient statistic for the ability parameter. For the Rasch model for binary data, this statistic reduces to $y_{i\cdot}$, that is, the individual total score. The parameter vector ψ contains only the item parameters and the corresponding log-likelihood,

$$\ell_C(\psi) = \sum_{i=1}^{n} \log p_C(y_i),$$

may be maximized by a Newton–Raphson algorithm of the type described in Section 5.3. A similar algorithm may be used for GRM without discrimination parameters, case in which conditional inference may be implemented on the basis of a dichotomization of the responses provided by every subject to each item, according to the approach described in Baetschmann et al. (2011).

Finally, the MML method, which is based on the random-effects formulation in which the ability levels are considered as drawn from a certain distribution, remains a very general method as it may be applied to any model for polytomous items both in its formulation based on the normal distribution and in that based on a discrete distribution for the latent trait. In both cases, the likelihood has the same expression as in (5.5), that is,

$$L_M(\psi) = \prod_{i=1}^{n} p(y_i),$$

where

$$p(y_i) = \sum_{v=1}^{k} p(y_i|\xi_v)\pi_v = \sum_{v=1}^{k} \left(\prod_{j=1}^{J} p_{jy}(\xi_v)\right) \pi_v.$$

The marginal log-likelihood $\ell_M(\psi)$ has the same expression given in (5.6) and may be maximized by an EM algorithm of the type described in Section 5.4.1 for the case of normally distributed ability and in Section 5.4.2 for that of ability with discrete distribution. See Bacci et al. (2014) and Bartolucci et al. (2014) for a detailed description.

5.6 Graphical Diagnostic Tools

Among the most popular instruments to evaluate the appropriateness of a test and of its single items, graphical tools have a prominent position, because they permit an immediate evaluation of the structure of the test. In the following, we focus on the concepts of item information and test information. We also focus on the person–item map, the test characteristic function, and the receiver operation characteristic curve.

5.6.1 Item Information and Test Information Functions

A relevant difference between CTT and IRT consists in the different perspective from which the psychometric quality of a questionnaire is considered (Furr and Bacharach, 2008). As illustrated in Chapter 2, in CTT, the quality of a test is assessed through the reliability coefficient and a single reliability estimate is computed. On the other hand, IRT admits that the quality of a questionnaire can vary, so that a test may provide better information at some latent variable levels than at other levels. For instance, a test might be able to well discriminate between individuals with similar low levels of ability, but, at the same time, it might not be able to equally well discriminate between individuals with similar higher levels of ability.

In practice, the quality of a test in the IRT setting is related to the concept of *information* which concerns the precision of a parameter estimator. Since this precision is measured by the variability of the estimator around the value of the parameter itself, the information corresponds to the reciprocal of the variance of the parameter estimate.

In IRT, the interest is typically in estimating the value of the ability parameter θ_i for an individual. If the information is large, an individual whose true ability is at that level can be estimated with precision; that is, possible estimates will be reasonably close to the true value. If the amount of information is small, it means that the ability cannot be estimated with precision and these estimates will be widely scattered around the true ability.

Since in IRT each item of a test contributes to measuring the underlying latent trait, items also contribute to defining the precision of the measurement for each person. In particular, the *item information curve* (or *item information function*) expresses the information arising from a single item against the ability θ_i and is denoted as $I_j(\theta_i)$.

The information associated with a specific response category y of an item j is defined as (Samejima, 1969, 1972; Baker and Kim, 2004)

$$I_{jy}(\theta_i) = -\frac{\partial^2 \log p_{jy}(\theta_i)}{\partial \theta_i^2}$$

and it contributes to define the item information $I_j(\theta_i)$ proportionally to the corresponding conditional probability, according to the following formula:

$$I_j(\theta_i) = \sum_{y=0}^{l_j-1} I_{jy}(\theta_i) p_{jy}(\theta_i). \tag{5.12}$$

It should be noted that $I_j(\theta_i) \neq \sum_{y=0}^{l_j-1} I_{jy}(\theta_i)$ and that increasing the number of response categories will result in an increase of the item information; thus, a polytomous IRT model will estimate more precisely the ability levels than a dichotomous IRT model.

In the binary case (Birnbaum, 1968), Equation 5.12 reduces to

$$I_j(\theta_i) = \left(\frac{\partial p_j(\theta_i)}{\partial \theta_i}\right)^2 \frac{1}{p_j(\theta_i)\left(1 - p_j(\theta_i)\right)}; \tag{5.13}$$

see also Appendix 5.A.1. In more detail, the item information function of the 2PL model (Hambleton and Swaminathan, 1985; McDonald, 1999; Baker and Kim, 2004) is obtained by Equation 5.13 as

$$I_j(\theta_i) = \lambda_j^2 p_j(\theta_i)\left(1 - p_j(\theta_i)\right),$$

where $p_j(\theta_i)$ is the ICC given in Equation 3.8 for fixed β_j and λ_j parameters. By setting $\lambda_j = 1$, we obtain the item information function for a Rasch model, where $p_j(\theta_i)$ is given in Equation 3.4. The maximum value of $I_j(\theta_i)$ is proportional to the square of the discrimination parameter and it is equal to $0.25\lambda_j^2$. This maximum value is obtained at the ability level corresponding to the item difficulty parameter. Obviously, in the Rasch model, the maximum of $I_j(\theta_i)$ constantly equals 0.25.

For the 3PL model, a more complex formula exists (Hambleton and Swaminathan, 1985; McDonald, 1999; Baker and Kim, 2004), that is,

$$I_j(\theta_i) = \lambda_j^2 \frac{\left(1 - p_j(\theta_i)\right)}{p_j(\theta_i)} \frac{\left(p_j(\theta_i) - \delta_j\right)^2}{\left(1 - \delta_j\right)^2},$$

where $p_j(\theta_i)$ is given in (3.11) for fixed β_j, λ_j, and δ_j parameters. The maximum point for $I_j(\theta_i)$ also depends on the pseudo-guessing parameter; for the detailed formulas, see Hambleton and Swaminathan (1985) and Baker and Kim (2004). In fact, as δ_j decreases, the item information increases and it reaches its maximum when $\delta_j = 0$, that is, when the 3PL model specializes in the 2PL.

For GRM, the item information function (Samejima, 1969; Baker and Kim, 2004) is retrieved by Equation 5.12 as

$$I_j(\theta_i) = \sum_{y=0}^{l_j-1} \frac{\lambda_j^2\left[p_{j,y-1}^*(\theta_i)\left(1 - p_{j,y-1}^*(\theta_i)\right) - p_{jy}^*(\theta_i)\left(1 - p_{jy}^*(\theta_i)\right)\right]^2}{p_{j,y-1}^*(\theta_i) - p_{jy}^*(\theta_i)},$$

where $p_{jy}^*(\theta_i)$ is defined in Equation 4.5. Samejima (1969) shows that the item information increases if categories are split into more categories.

Finally, the item information functions of partial credit models may be obtained as special cases of those of Bock's nominal response model, whose detailed equations are provided in Baker and Kim (2004, Chapter 9).

With respect to the results previously discussed, the item information function is usually bell shaped and its shape depends on the item parameters. In more detail, it reaches its maximum value at the ability level corresponding to the item difficulty parameter (or close to it for the 3PL model). Therefore, the item information decreases as the ability level departs from the item difficulty and approaches zero at the extremes of the ability scale (Figure 5.1). The higher the item discrimination parameter, the more peaked the curve is (compare curves for $j = 1$ and $j = 2$ in Figure 5.1). Besides, the difficulty parameter influences the location of the curve along the latent continuum (compare curves for $j = 2$ and $j = 3$ in Figure 5.1), and, more precisely, low values of difficulty provide information for low latent variable values, whereas high values of difficulty provide information for high values of this variable.

The information produced by a single item is useful to evaluate the quality of this item. In general, we are usually more interested in evaluating the test as a whole: due to the local independence assumption, the overall information yielded by the test at any ability level is simply the sum of the item information functions at that level (Birnbaum, 1968; Timminga and Adema, 1995). Thus, the *test information curve* (or *test information function*) is defined as

$$I(\theta) = \sum_{j=1}^{J} I_j(\theta).$$

Since the test information function is calculated by summing all the single item information functions, then the order of the items does not influence

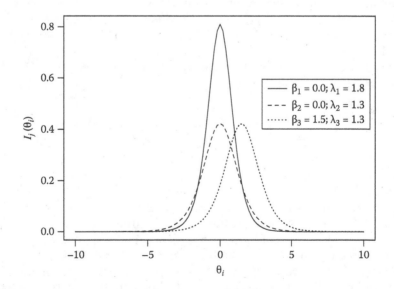

FIGURE 5.1
Item information curves for items $j = 1, 2, 3$ under the 2PL model (different levels of difficulty and discrimination parameters are assumed).

the total information. In addition, it is obvious that longer tests will usually measure an individual's ability with greater precision than shorter tests.

Figure 5.2 shows the test information curves in two hypothetical situations. Both tests consist of 10 binary items, but differ in the difficulty parameters. In the first test, for which the test information curve is represented by a solid line, item difficulties are assumed to be equally distributed in the range [−2.0; 2.0], whereas in the second test, for which the test information curve is represented by a dashed line, a prevalence of very easy and very difficult items is supposed. As a consequence, in both cases the test information curve peaks at a certain point on the ability scale, then it rapidly decreases. Therefore, both tests measure ability with unequal precision along the ability scale. However, the first test differentiates better than the second test between subjects whose latent trait is around an intermediate level and it is poorer at discriminating between those with high or low levels of the latent trait.

To conclude, a good test should be made up of items targeted near the ability person levels. Indeed, it is also necessary to take into account that the analysis of persons' fit considers the possibility of unexpected responses, so that when highly probable responses are not observed, or when very low probable responses are observed, a misfit of a person is implied. When all items are targeted near the ability levels, this kind of analysis would not be possible. Thus, a proper test design requires a compromise between these

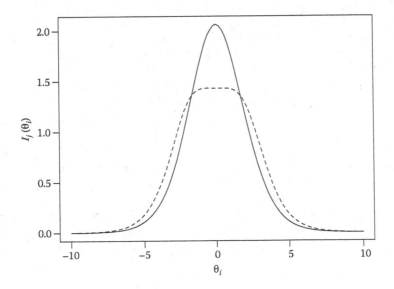

FIGURE 5.2
Examples of test information curves: Rasch model with $\beta_1 = -2.0$, $\beta_2 = -1.0$, $\beta_3 = -0.5$, $\beta_4 = -0.3$, $\beta_5 = \beta_6 = 0.0$, $\beta_7 = 0.3$, $\beta_8 = 0.5$, $\beta_9 = 1.0$, $\beta_{10} = 2.0$ (solid line); Rasch model with $\beta_1 = \beta_2 = \beta_3 = -2.0$, $\beta_4 = -1.5$, $\beta_5 = \beta_6 = 0.0$, $\beta_7 = 1.5$, $\beta_8 = \beta_9 = \beta_{10} = 2.0$ (dashed line).

extremes: on one side, items that provide maximum information allow minimum misfit detection and, on the other side, items that allow for maximum misfit detection provide minimum information. In conclusion, a good test should be made up of both items with high information and items with difficulty parameters far from the expected ability levels (Wright and Stone, 1999).

5.6.2 Person–Item Map and Test Characteristic Curve

The *person–item map* (Wright and Masters, 1982; Wright and Stone, 1999; Bond and Fox, 2007) is a very simple graphical tool, used especially for Rasch-type models (see Figures 5.5 and 5.9 for two examples). It displays the distribution of estimated item and threshold (in the polytomous case) difficulty parameters as well as the distribution of estimated person parameters along the latent dimension (both expressed on the logit scale) on the same plot. This tool is typically based on the CML estimates.

Items should ideally be located along the whole scale to meaningfully measure the latent trait levels of all subjects. Globally, bad calibrated tests or single problematic items can be discovered immediately. More precisely, a test might be made of too many difficult items. In such a case, subjects with low ability levels cannot be satisfactorily measured through this kind of test. Similarly, a test might be too simple for the ability level of the population. Finally, a test might be made of very similar items, which prove to be well calibrated only for an average ability person: individuals located at the extremes of the distribution (i.e., individuals with lower or higher ability levels than difficulties of any item) cannot be properly evaluated by this kind of test.

Special attention should also focus on items with the same difficulty parameters, which provide redundant information. In this case, because short tests are usually preferred to long tests, so as to avoid missing responses due to tiredness, one may wish to remove one or more of these redundant items.

In the case of polytomous responses, the difficulties of each item threshold are represented on the person–item map, so that one can detect items with two (or more) thresholds with the same difficulty or, more in general, items with disordered thresholds.

For more complex IRT models, where items also have other parameters beyond difficulties, the person–item map is not so useful. More complete information on each item and on the test as a whole can be obtained by IRCCCs (or ICCs in the binary case) and by the *test characteristic curve* (TCC) or test characteristic function; see McDonald (1999) and Raykov and Marcoulides (2011).

ICCs and IRCCCs have already been described in Chapters 3 and 4 in the case of binary and ordered polytomous items. We remind the reader that the IRCCCs of an item j describe the probability of observing the response categories of that item, given the latent variable. In the dichotomous case, only the curve referred to the response coded as 1 is usually represented. In addition, IRCCCs and ICCs are useful to detect items with differential functioning (for more details, see Section 5.9), and they contribute to building the characteristic curve of a test.

Other than in the characteristic functions of each item, there can be interest in a global graphical representation of the whole test, which is usually provided by TCC. First, the expected score for each item is obtained as

$$E(Y_{ij}) = \sum_{y=0}^{l_j-1} y p_{jy}(\theta_i)$$

and, then, TCC is obtained as a function of θ_i summing the expected scores across all items

$$\sum_{j=1}^{J} E(Y_{ij}) = \sum_{j=1}^{J} \sum_{y=0}^{l_j-1} y p_{jy}(\theta_i).$$

In the dichotomous case, the expected score of item j simplifies to $E(Y_{ij}) = p_j(\theta_i)$, so that the TCC is defined as the sum of the probabilities of endorsing the test items, that is,

$$\sum_{j=1}^{J} p_j(\theta_i),$$

and it describes the expected number of right responses as a function of the latent trait.

The shape of TCC depends on a number of elements, such as the number of items in the test (Figure 5.3a), the type of model, and the values of the item parameters (Figure 5.3b–d). Usually, it looks like an IRCCC (or an ICC in the dichotomous case), but it can also have regions with varying steepness and substantially flat regions. In general, the steepness of TCC increases with the reduction of the item difficulties range (Figure 5.3c) and with the increasing of the average value of the item discrimination parameters (Figure 5.3d).

5.6.3 Received Operation Characteristic Curve

Another graphical instrument that can be used to evaluate the goodness-of-fit of an IRT model is borrowed from the context of logistic regression and is based on the predictive capabilities of the assumed IRT model.

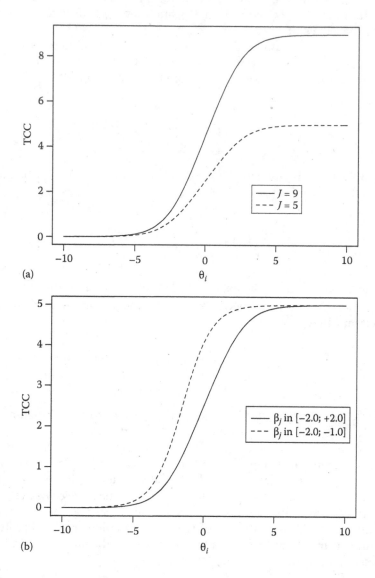

FIGURE 5.3

Examples of TCC: Rasch model with $J = 9$ items versus Rasch model with $J = 5$ items (a); Rasch model with β_j belonging to the range $[-2.0; +2.0]$ versus Rasch model with β_j belonging to the range $[-2.0; -1.0]$, $j = 1, \ldots, 5$ (b).   (*Continued*)

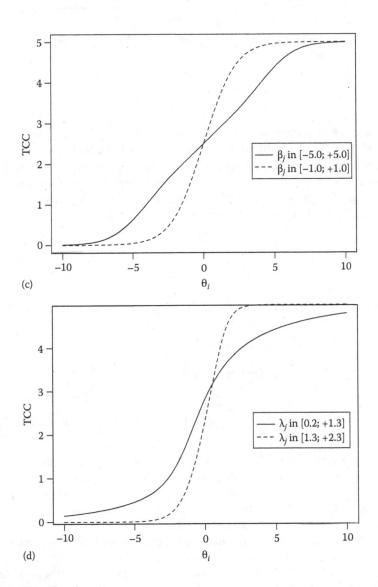

(c)

(d)

FIGURE 5.3 (*Continued*)
Examples of TCC: Rasch model with β_j belonging to the range $[-5.0; 5.0]$ versus Rasch model with β_j belonging to the range $[-1.0; +1.0]$, $j = 1, \ldots, 5$ (c); 2PL model with λ_j belonging to the range $[0.2; 1.3]$ versus 2PL model with λ_j belonging to the range $[+1.3; +2.3]$, $j = 1, \ldots, 5$ (d).

TABLE 5.1

Confusion Matrix for a Set of Binary Items

	Observed Response		
Predicted Response	0	1	Total
0	m_{00}	m_{01}	$m_{0.}$
1	m_{10}	m_{11}	$m_{1.}$
Total	$m_{.0}$	$m_{.1}$	Jn

In the case of binary items, predicted responses to a set of items can be obtained by assuming a given threshold, say, c in $[0, 1]$, and by dichotomizing the estimated response probabilities $\hat{p}_j(\hat{\theta}_i)$ into 1, when $\hat{p}_j(\hat{\theta}_i) > c$, and into 0 otherwise, for $i = 1, \ldots, n$ and $j = 1, \ldots, J$. Observed and predicted responses can then be organized in a 2×2 contingency table, known as the *confusion matrix* (see Table 5.1), with frequencies

$$m_{00} = \sum_{i=1}^{n}\sum_{j=1}^{J} 1\left\{\hat{p}_j\left(\hat{\theta}_i\right) \leq c, y_{ij} = 0\right\},$$

$$m_{01} = \sum_{i=1}^{n}\sum_{j=1}^{J} 1\left\{\hat{p}_j\left(\hat{\theta}_i\right) \leq c, y_{ij} = 1\right\},$$

$$m_{10} = \sum_{i=1}^{n}\sum_{j=1}^{J} 1\left\{\hat{p}_j\left(\hat{\theta}_i\right) > c, y_{ij} = 0\right\},$$

$$m_{11} = \sum_{i=1}^{n}\sum_{j=1}^{J} 1\left\{\hat{p}_j\left(\hat{\theta}_i\right) > c, y_{ij} = 1\right\}.$$

On the basis of these data, we can calculate the *true-positive rate* (or *sensitivity*) $m_{11}/m_{.1}$, the *true-negative rate* (or *specificity*) $m_{00}/m_{.0}$, its complementary value known as the *false-positive rate* (or *false alarm rate*) $m_{10}/m_{.0}$, and the *false-negative rate* $m_{01}/m_{.1}$, where $m_{.1} = m_{01} + m_{11}$, $m_{.0} = m_{00} + m_{10}$, and so on.

Other than these indices, a more sophisticated measure of the predictive capacity of a given set of items can be achieved through the so-called *received operation characteristic* (ROC) curve (Metz, 1978; Haley and McNeil, 1982). The ROC curve is obtained by varying the threshold point c within $[0, 1]$ and by plotting the corresponding false-positive rate against the corresponding true-positive rate on a Cartesian axes system (for instance, see Figures 5.7 and 5.8): one starts with $c = 1$, which results in point $(0, 0)$, and ends with $c = 0$, which results in point $(1, 1)$. All the other points lie on the ROC curve.

There are several interesting points on the ROC space (Fawcett, 2006; Mair et al., 2008): the lower left point $(0, 0)$ refers to situations with no false positives, but also with no true positives; the upper right point $(1, 1)$, instead, returns true positives with probability 0.5. The point (0, 1) denotes the ideal situation, with no classification errors, that is, no false positives and only true positives. If the observed point lies on the upper left-hand side of the ROC plot, then one is in a conservative situation, where a point is classified as 1 only with strong evidence, which can result in a low true-positive rate; at the same time, false-positive errors are uncommon. On the other hand, if the observed point is on the upper right-hand side, it means that it is classified as 1 also with weak evidence, which results not only in a high true-positive rate but also in a high false-positive rate. In general, a good ROC curve should pass as closer as possible to point (0, 1). A model with no predictive ability will provide a curve along the diagonal line, which denotes a strategy based on chance (similar to a coin toss). A model with ROC curve below the diagonal line is also theoretically possible, denoting a performance worse than a random guessing. However, note that, in this last case, a performance better than random guessing is obtained by inverting the labels of predictions (i.e., inverting "0" with "1"; Hand and Till, 2001).

Another interesting way to evaluate the predictive goodness of an IRT model is based on calculating the *area under the ROC curve* (AUC); see Haley and McNeil (1982) and Swets and Pickett (1982). From a theoretical point of view, AUC varies in [0; 1]; however, a realistic model usually has an AUC greater than 0.5: the closer to 1 is the AUC value, the better is the classification capacity of the model. According to the rule of thumb suggested by Hosmer and Lemeshow (2000), the predictive capacity of a model may be considered acceptable for $0.7 \leq AUC < 0.8$, excellent for $0.8 \leq AUC < 0.9$, and outstanding in the case of AUC values greater than 0.9. Therefore, one can compare two IRT models on the basis of AUC and prefer the model with the highest AUC value. Some attention is necessary when the ROC curves of two models cross each other, that is, one model being better than the other one in some region of their ROC space and worse in the remaining space; see Fawcett (2006) and Hand and Till (2001) and the references therein.

Until now, the discussion has dealt with the case of binary items, coherently with most of the literature concerning ROC curve and AUC. However, it should also be interesting to extend the mentioned criteria to models for polytomous items. In general, it is not simple to deal with a confusion matrix having dimension $l \times l$, where l is the number of response categories of the items (assuming that $l_j = l$ for $j = 1, \ldots, J$). Fawcett (2006) proposes to plot the ROC curve for each possible dichotomization of category y ($y = 0, \ldots, l - 1$) versus the remaining ones, so that y corresponds to category "1" and all the others correspond to category "0." Consequently, AUC can be computed for each ROC curve and a total value is obtained as follows (Provost and Domingos, 2001):

$$\sum_{y=0}^{l-1} AUC_y \frac{m_{.1}}{Jn},$$

where AUC_y is the area under the ROC curve of category y and $m_{.1}/Jn$ is the proportion of observed responses in category y. A different approach is proposed by Hand and Till (2001), based on the comparison of all the possible pairs of distinct categories. In practice, an observation is classified in category y_1 rather than in category y_2 ($y_1, y_2 = 0, \dots, l-1, y_1 < y_2$) if $\hat{p}_{jy_1}(\theta_i) > \hat{p}_{jy_2}(\theta_i)$; otherwise, it is classified in category y_2, being the other categories ignored. A synthetic measure of the predictive capacity of the model is then defined as the average of the AUC values obtained for each pair (y_1, y_2), that is,

$$\frac{2}{l(l-1)} \sum_{y_1=0}^{l-2} \sum_{y_2=1}^{l-1} AUC_{y_1,y_2},$$

where AUC_{y_1,y_2} is the area under the ROC curve involving categories y_1 and y_2 and $l(l-1)/2$ is the number of all the possible pairs of distinct categories.

5.7 Goodness-of-Fit

To thoroughly delve into the diagnostics for IRT models, graphical analyses should be assisted by more objective statistical instruments, such as synthetic indices (Section 5.7.1) and hypothesis tests (Section 5.7.2).

5.7.1 Information Criteria and Pseudo-R^2

Before describing the most popular IRT tests, it is useful to remind the reader that, sometimes, it is desirable to measure the goodness-of-fit of a model through synthetic indices. Similar to other statistical models, a common approach used for model selection is represented by information criteria. The information criteria are based on indices that are, essentially, penalized versions of the maximum log-likelihood. The most common criteria of this type are the *Akaike information criterion* (AIC) (Akaike, 1973) and the *Bayesian information criterion* (BIC) (Schwarz, 1978).

The first criterion is a measure of the goodness-of-fit of a model, which describes the trade-off between the accuracy and complexity of the model. In particular, AIC is based on estimating the Kullback–Leibler distance between the true density and the estimated density, which focuses on the expected log-likelihood, and is defined on the basis of the following index:

$$AIC = -2\hat{\ell} + 2\,\#\mathrm{par},$$

where $\hat{\ell}$ is the maximum log-likelihood of the model of interest and $\#$par denotes the number of free parameters. The criterion of Schwarz (1978) is, instead, based on the following index:

$$BIC = -2\,\hat{\ell} + \log(n)\,\#\text{par},$$

where $\hat{\ell}$ and $\#$par are previously defined. This criterion relies on a larger penalty for the model complexity (measured in terms of number of parameters) than AIC index.

According to both criteria mentioned earlier, one should select the model with the minimum value of *AIC* or *BIC*. In practice, the information criteria are useful for the choice between nonnested models, that is, models that cannot be obtained one from the other through suitable constraints on the parameters. A typical situation that requires the use of information criteria is the selection of LC-IRT models, which differ in the number of mixture components, in which case $\hat{\ell} = \ell_M\left(\hat{\psi}_M\right)$; see Section 5.4.

Alternatively to information criteria, whose value cannot be interpreted in an absolute way, we can use goodness-of-fit measures that are bounded between 0 and 1. These measures are usually adaptations to logistic regression models of the coefficient of determination R^2 used for linear regression models, which describes the rate of variance explained by the model. One of the most well-known examples of such measures (for other criteria, see Mair et al., 2008) is the *pseudo-R^2* due to McFadden (1974), which can be expressed as

$$R^2_{\text{MF}} = \frac{\hat{\ell}_0 - \hat{\ell}_1}{\hat{\ell}_0},$$

where
$\hat{\ell}_1$ is the maximum log-likelihood of the IRT model of interest
$\hat{\ell}_0$ is the corresponding quantity of a reduced model

These log-likelihoods are obtained from the MML method under the assumption that the ability has a normal distribution. Typically, the latter is the model without individual and item effects. Values of R^2_{MF} at least equal to 0.2 are usually considered satisfactory.

5.7.2 Parametric and Nonparametric Hypothesis Tests

Several tests have been proposed for the evaluation of the model fit in the IRT framework and, especially, in the context of Rasch-type models. To present an overview of the various approaches, a distinction between parametric and nonparametric tests is considered. Moreover, another important taxonomy of model tests concerns the assumptions of the model to be tested. Generally speaking, model tests are usually constructed for specific alternatives, but it is not always possible to completely separate the assumptions in order to test each of them separately. Therefore, each test can also be considered as a global fit test, where the null hypothesis concerns the validity of the model of interest.

The great majority of the tests has been developed to evaluate the goodness-of-fit of the Rasch paradigm: the rejection of the null hypothesis implies that a Rasch-type model does not fit the observed data; therefore, a more general IRT model (e.g., a 2PL model) has to be chosen or a different set of items has to be selected and taken into account.

5.7.2.1 Parametric Tests

Among the parametric tests, one can find a group directly borrowed from the logistic regression setting (for an in-depth review, see Mair et al., 2008); these tests are suitable for evaluating the global goodness-of-fit of any IRT model. Besides, a further group of tests was elaborated in connection with the Rasch paradigm.

Logistic Regression–Based Tests for IRT Models

- *Likelihood ratio (LR) test*: The most well-known goodness-of-fit test used in the logistic regression literature is represented by the standard LR test, which allows for a comparison between nested IRT models. More precisely, the LR test is useful to evaluate the goodness-of-fit of a model formulated by suitable constraints on the parameters, which correspond to the null hypothesis denoted by H_0.

  For instance, the LR test can be used to compare models that differ in the number of latent traits (i.e., H_0 corresponds to the unidimensional model in comparison to a bidimensional one) or in the constraints on the discrimination parameters, such as $\lambda_j = 1$ for $j = 1, \ldots, J$ (i.e., in the case of dichotomously scored items, H_0 corresponds to the Rasch model in comparison to the 2PL model), all other elements being constant.

  Denoting by $\hat{\ell}_0$ and $\hat{\ell}_1$ the maximum log-likelihood of the constrained model and the general model, respectively, the LR test is based on the statistic

$$LR = -2\left(\hat{\ell}_0 - \hat{\ell}_1\right), \tag{5.14}$$

  which, under the usual regularity conditions, has null asymptotic distribution of χ^2 type with a number of degrees of freedom equal to the number of constraints imposed by H_0 on the unconstrained model.

  In addition to the comparison between two nested models, the LR test can also be used to formulate a global judgment about a specific IRT model. For this, the overall sample is divided into H

groups and, for each group, a suitable frequency is considered, such as the number of responses equal to 1 to a certain item. Then, the frequencies predicted by the model of interest are compared with the observed ones. Note that the observed frequencies may be seen as corresponding to the *saturated model*, which represents the largest model one can fit. However, given that there are different ways to define groups and statistics of interest, there does not exist a unique definition of a saturated model and this ambiguity leads to a set of alternative LR tests based on different test statistics; see also Simonoff (1998). What follows is explicitly referred to as dichotomously scored items, as the software implementation in the IRT setting is limited to binary data (Mair et al., 2008, 2013).

- *Collapsed deviance test*: It is based on grouping subjects in H subgroups according to the total raw score, or to more sophisticated strategies to collapse response configurations (Simonoff, 1998), and in considering the number of responses equal to 1 to any item provided by the subjects in each group. The resulting test statistic has an asymptotic null distribution χ^2 with $(H-1)J$ degrees of freedom.
- *Rost's deviance test*: In this case, each group corresponds to a certain response configuration, so that $H = 2^J$ and the frequency of interest corresponds to the number of subjects in each group. The resulting test statistic has an asymptotic null distribution χ^2 with $2^J - J$ degrees of freedom for the Rasch model. Consequently, the test is reasonable with a reduced number of test items.
- *Hosmer–Lemeshow test*: This test is inspired by the test of Hosmer and Lemeshow (1980) and is based on H groups defined by percentiles of the total score and frequencies of interest similar to those used for the collapsed deviance test. The test statistic has asymptotic χ^2 distribution with $H - 2$ degrees of freedom. Typically, $H = 10$ so that this asymptotic distribution has 8 degrees of freedom.

Global and Item-Specific Tests for Rasch-Type Models

Other than the aforementioned tests borrowed from the logistic regression setting, several parametric tests have been elaborated to verify the null hypothesis of validity of the Rasch paradigm. A taxonomy and detailed description of these tests is due to Glas and Verhelst (1995a) for the binary case, which is considered in the following; and to Glas and Verhelst (1995b) for the polytomous case, to which we refer the reader for details.

- *Andersen's test* (Andersen, 1973): This test is based on the property that, in a Rasch-type model, the invariance of the parameter estimates holds also for subgroups of subjects. Based on this property, the sample is split into H subgroups and, for each subgroup, the CML estimates are calculated. The resulting LR-type test statistic is given by

$$-2\left(\hat{\ell}_C - \sum_{h=1}^{H} \hat{\ell}_{Ch}\right),$$

where $\hat{\ell}_C$ is the conditional maximum log-likelihood for the Rasch model of interest, whereas $\hat{\ell}_{Ch}$ is the conditional maximum log-likelihood for group h ($h = 1, \ldots, H$). It can be verified that Andersen's LR-type test statistic has an asymptotic χ^2 distribution with a number of degrees of freedom equal to the number of parameters estimated in the subgroups minus the number of parameters estimated in the total dataset. Depending on the way subgroups are defined, different violations can be verified. For example, if the subgroups correspond to different modalities of a covariate (e.g., gender, geographic area), then the Andersen's test is sensitive to DIF (for more details, see Section 5.9). Moreover, if the subgroups are defined according to raw score levels, the Andersen's test is sensitive to differences in item discrimination parameters, and it is substantially equivalent to tests Q_1 and R_{1c} described in the following.

- R_{1c} test (Glas, 1988) and Q_1 test (Van den Wollenberg, 1982): Both tests are based on Pearson-type statistics, which compare observed sample values and expected frequencies under the null hypothesis. The sample of individuals is partitioned into H subgroups according to the raw score level, and both tests are based on the following differences, calculated for each item j:

$$d_h^{(j)} = o_{1h}^{(j)} - \hat{o}_{1h}^{(j)}, \tag{5.15}$$

where $o_{yh}^{(j)}$ is the number of persons belonging to group h and giving response y to item j and $\hat{o}_{yh}^{(j)}$ is the corresponding expected value under the Rasch model. In both tests, R_{1c} and Q_1, $\hat{o}_{1h}^{(j)}$ are calculated as

$$\hat{o}_{1h}^{(j)} = \sum_{r=0}^{J} n_r \hat{q}_j(r),$$

where n_r is the number of subjects who scored $y_{i.} = r$ ($r = 0, \ldots, J$) and $q_j(r) = p(Y_{ij} = 1 | Y_{i.} = r)$ is the probability of endorsing item j for a subject with a raw score equal to r, and its value is estimated through the CML method.

The general idea of the tests based on quantities $d_h^{(j)}$ is that, if the observed number of responses equal to 1 is too small at low score levels and too large at high score levels, then the corresponding ICC will have a greater or smaller slope than that predicted by the Rasch model. To properly evaluate quantities $d_h^{(j)}$, their estimated standard deviation $\hat{se}\left(d_h^{(j)}\right)$ has to be taken into account. As a consequence, both tests at issue are based on

$$T_h = \sum_{j=1}^{J} \frac{\left(d_h^{(j)}\right)^2}{\hat{se}\left(d_h^{(j)}\right)^2}, \quad h = 1, \ldots, H,$$

where $\hat{se}\left(d_h^{(j)}\right)$ differs between R_{1c} test and Q_1 test (for detailed formulas, see Van den Wollenberg, 1982; Glas, 1988; Hardouin, 2007), so that Q_1 test may be considered as a good approximation of R_{1c} test.

More precisely, R_{1c} test verifies the assumptions of sufficiency of raw scores and parallel ICCs through the following test statistic:

$$R_{1c} = \sum_{h=1}^{H} T_h,$$

having an asymptotic χ^2 distribution with $(H-1)(J-1)$ degrees of freedom. This test is adopted in the context of the CML estimation, but it has an analogue that can be used in the framework of MML estimation, when the latent variable is assumed to be normally distributed (for details, see Glas and Verhelst, 1995a). Alternatively, the Q_1 test is based on the test statistic

$$Q_1 = \frac{J-1}{J} \sum_{h=1}^{H} T_h,$$

which, under the null hypothesis, has the same asymptotic distribution of χ^2 type with $(H-1)(J-1)$ degrees of freedom.

- S_j *test of Verhelst and Eggen (1989) and U_j test of Molenaar (1983b)*: R_{1c} and Q_1 tests evaluate the global fit of the model, but do not indicate which items show specific problems. Alternatively, tests S_j (see, also, Glas and Verhelst, 1995a) and U_j are explicitly item oriented and allow us to detect which items violate the Rasch paradigm. Both of them are based on CML estimates, but they have analogues in the MML framework.

Test S_j is based on the same components as the R_{1c} test, that is, on $d_h^{(j)}$ as defined in Equation 5.15. The test statistic is given by

$$S_j = \sum_{h=1}^{H} \frac{\left(d_h^{(j)}\right)^2}{\hat{se}\left(d_h^{(j)}\right)^2},$$

which has an asymptotic χ^2 distribution with $H-1$ degrees of freedom under the null hypothesis that the Rasch model holds.

On the other hand, in the U_j test, the sample is divided into three subsamples indexed by $h = 1, 2, 3$. The first subsample includes all individuals with a raw score smaller than a given threshold c_1, the second sample includes all individuals with a raw score in the range (c_1, c_2), with $c_2 > c_1$, and the third sample includes all individuals with a raw score greater than c_2. The two thresholds c_1 and c_2 are usually put equal to the first and the third quartile of the observed distribution of the raw scores. Then, the statistic on which the U_j test is based has the expression

$$U_j = \frac{1}{\sqrt{c_1 + J - c_2}} \left(\sum_{h=1}^{c_1} \frac{d_h^{(j)}}{\hat{se}\left(d_h^{(j)}\right)} - \sum_{h=c_2}^{J-1} \frac{d_h^{(j)}}{\hat{se}\left(d_h^{(j)}\right)} \right),$$

having an asymptotic standard normal distribution. It can be verified that if the discriminating power of item j is significantly higher than the mean of the discriminating parameters of the other items, then U_j tends to be negative, whereas if the discriminating power of item j is significantly smaller than expected, then U_j tends to be positive.

- *Martin-Löf test* (Martin-Löf, 1973): This is perhaps the most well-known test to verify the unidimensionality assumption of the Rasch model. It is an LR test, which only works in the case of binary items and compares a general model characterized by all items partitioned into two subsets, $\mathcal{J}_1$ and $\mathcal{J}_2$, with $\mathcal{J}_1$ containing J_1 items and $\mathcal{J}_2$ containing J_2 items, versus a restricted version defined by considering all $J = J_1 + J_2$ items into the same dimension. An extension of this test that also works in the presence of polytomous items and allows us to test unidimensionality against an alternative with at least two dimensions, is provided by Christensen et al. (2002); see also Bartolucci (2007). Similar to Andersen's, R_{1c}, and Q_1 tests, the Martin-Löf test is based on parameter estimates obtained from the CML approach.

Denoting the maximum log-likelihood of the general model with $\tilde{\ell}_1$, and that of the constrained model with $\tilde{\ell}_0$, the statistic on which the test is based is defined as

$$ML = -2\left(\tilde{\ell}_0 - \tilde{\ell}_1\right).$$

In particular, log-likelihood $\tilde{\ell}_1$ is obtained as

$$\tilde{\ell}_1 = \hat{\ell}_{C1}^{(1)} + \hat{\ell}_{C1}^{(2)} + \hat{\ell}_{M1}, \tag{5.16}$$

where
$\hat{\ell}_{C1}^{(1)}$ is the maximum conditional log-likelihood for the items in $\mathcal{J}_1$
$\hat{\ell}_{C1}^{(2)}$ is the maximum conditional log-likelihood for the items in $\mathcal{J}_2$
$\hat{\ell}_{M1}$ is the maximum marginal log-likelihood of the saturated multinomial model for the distribution of the scores

The last component may be expressed as

$$\hat{\ell}_{M1} = \sum_{r_1=0}^{J_1} \sum_{r_2=0}^{J_2} n_{r_1 r_2} \log \frac{n_{r_1 r_2}}{n},$$

where
r_1 is the test score for the items in the subset $\mathcal{J}_1$
r_2 is the test score for the items in the subset $\mathcal{J}_2$
$n_{r_1 r_2}$ is the frequency of subjects with scores r_1 and r_2 at the two subsets

A decomposition similar to that in (5.16) also holds for $\tilde{\ell}_0$. We have

$$\tilde{\ell}_0 = \hat{\ell}_{C0} + \hat{\ell}_{M0},$$

where $\hat{\ell}_{C0}$ is the maximum conditional log-likelihood for the items in the full set of items $\mathcal{J}_1 \cup \mathcal{J}_2$, and

$$\hat{\ell}_{M0} = \sum_{r=0}^{J} n_r \log \frac{n_r}{n}.$$

In the previous expression, r denotes the test score achieved on the items of both subsets, the frequency of which is denoted by n_r. Test statistic ML has an asymptotic χ^2 distribution with $J_1 J_2 - 1$ degrees of freedom under the null hypothesis of unidimensionality. If this null hypothesis is rejected, then the sample contains sufficient evidence against unidimensionality and, possibly, a multidimensional model must be adopted (see Section 6.4).

Other than the Martin-Löf test, an empirical method that can be used in an explorative way to verify the unidimensionality assumption is the *modified parallel analysis* (MPA) (Drasgow and Lissak, 1983), which has the relevant advantage of working for general dichotomous and polytomous IRT models, not necessarily of Rasch type. MPA consists of performing a factor analysis on the tetrachoric correlations between test items and in calculating the eigenvalues of the unrotated solution. Then, a given number of new datasets is generated, which are parallel to the observed one in the sense that each of them has the same number of items with the same parameters and the same number of subjects. For each subject, the ability is simulated from a normal distribution with a mean equal to the ability estimate in the original dataset and standard deviation given by the corresponding standard error. For each new simulated dataset, a factor analysis is performed on the tetrachoric correlations, and the corresponding eigenvalues are calculated, retaining the second one. Therefore, unidimensionality is verified by computing the proportion of simulated second eigenvalues that are at least equal to the observed second eigenvalue. The general idea is that, if such proportion is sufficiently high, then the set of items is considered unidimensional.

The most relevant drawbacks of MPA are the following: (1) it lacks a measure for the formal assessment of the closeness of eigenvalues, from the observed and the parallel datasets, and therefore, the judgment on the unidimensionality is strongly subjective and (2) the comparison between two eigenvalues only lacks a theoretical justification, and the differences between the remaining eigenvalues are ignored. Despite these drawbacks, MPA is useful for an explorative analysis of the dimensionality of the dataset. For other tests of unidimensionality, see Christensen et al. (2002) and Bartolucci (2007).

5.7.2.2 Nonparametric Tests

Most goodness-of-fit tests illustrated in the previous section are influenced by the subjectivity of the number of groups and of the adopted partitioning criterion. Besides, all of them are asymptotic; therefore, they cannot be reliably used with small samples. An alternative approach to asymptotic parametric tests that allows us to deal with small samples is represented by nonparametric tests. In particular, Ponocny (2001) proposed a family of nonparametric tests for Rasch-type models, which allows us to test several violations of the basic assumptions.

All the tests belonging to the family proposed by Ponocny (2001) rely on statistics computed on the basis of the observed sample and measuring a certain type of deviance from the Rasch model. Then, a p-value for this statistic is computed by a Monte Carlo procedure that draws matrices of data of the same dimension as the observed one, which satisfy suitable constraints about their margins. The main test statistics that are used within this approach are listed in the following:

- $T_1 = \sum_{i=1}^{n} 1\{y_{ij_1} = y_{ij_2}\}$, where j_1, j_2 is a pair of items of interest. For these items, cases are counted with equal responses on both items. If more pairs of items are of interest, the test statistic is obtained by summing this function over all these pairs. The resulting test allows to check for local independence violation via increased inter-item correlation.

- $T_{11} = \sum_{j_1=1}^{J} \sum_{j_2=1, j_2 \neq j_1}^{J} |\text{cor}_{j_1 j_2} - \overline{\text{cor}}_{j_1 j_2}|$, where $\text{cor}_{j_1 j_2}$ is the observed pairwise item correlation and $\overline{\text{cor}}_{j_1 j_2}$ is the corresponding average value computed on all the simulated data matrices. This statistic allows us to globally test the violation of local independence and may be used together with statistic T_1 to detect pairs of problematic items.

- T_2 is equal to the variance of the score attained at a certain subset of items or to the range between minimum and maximum scores or the mean (or median) absolute deviation. With this test, violations of local independence are verified via increased dispersion of the individual raw scores to a given subset of items. A specific subset of items is chosen because they are suspected to represent some kind of subgroup, which means items with higher correlation than expected. Similarly, with T_2 equal to minus the variance, test $-T_2$, we can presumably detect heterogeneous items in a given subset, via a smaller dispersion of individual raw scores than expected.

- $T_5 = \sum_{i=1}^{n} 1\{y_{ij} = 0\} y_{i.}$ denotes the sum of raw scores of all individuals who scored 0 at item j. As statistic T_5 is a monotonic function of the point-biserial correlation between the score on item j and the total score, then it can be used to test for the hypothesis of equal discrimination item parameters or, in other words, for the adequacy of the Rasch model instead of the 2PL model: if the correlation is too low or too high, item j shows different discriminations compared to the other items of the test (or of a given subset of items).

5.8 Infit and Outfit Statistics

Other than the aforementioned statistics, the most popular statistics measuring fit, which are specific for Rasch models, are based on the comparison between observed and expected responses (Wright and Masters, 1982); for the binary case, see also Linacre and Wright (1994) and Wright and Stone (1999). The following relationships can be obtained, where $\hat{p}_{jy}(\hat{\theta}_i)$ denotes the IRCC with parameters estimated under a Rasch-type model through the CML approach:

- $\hat{y}_{ij} = \sum_{y=0}^{l_j-1} y \hat{p}_{jy}(\hat{\theta}_i)$ is the response expected value, which, in the binary case, coincides with $\hat{p}_j(\hat{\theta}_i)$.
- $e_{ij} = y_{ij} - \hat{y}_{ij}$ is the residual score.
- $\hat{se}(\hat{y}_{ij}) = \sqrt{\sum_{y=0}^{l_j-1} (y - \hat{y}_{ij})^2 \hat{p}_{jy}(\hat{\theta}_i)}$ is the estimated standard deviation for the predicted response $\hat{y}_{ij}$, which is largest when item j is well calibrated for person i, whereas it decreases when the ith person ability becomes further apart from the jth item difficulty parameters, β_{jy}.
- $e_{ij}/\hat{se}(\hat{y}_{ij})$ is the estimated standardized residual.

In order to evaluate the fit of an item to the Rasch model, a simple or weighted average of the squared standardized residuals is calculated. In particular, the *outfit* or *unweighted mean square statistics* out$_j$ is defined as the simple average of the squared standardized residuals and it is computed as

$$\text{out}_j = \frac{1}{n} \sum_{i=1}^{n} \frac{e_{ij}^2}{\hat{se}(\hat{y}_{ij})^2}.$$

Statistic out$_j$ has a mean equal to 1 and an estimated standard error equal to

$$\hat{se}(\text{out}_j) = \sqrt{\frac{1}{n^2} \sum_{i=1}^{n} \frac{\sum_{y=0}^{l_j-1} (y - \hat{y}_{ij})^4 \hat{p}_{jy}(\hat{\theta}_i)}{\hat{se}(\hat{y}_{ij})^4} - \frac{1}{n}}.$$

Since $\hat{se}(\text{out}_j)$ depends on the number of subjects and $\hat{se}(\hat{y}_{ij})$ is different between items and samples, it is not simple to define a cutoff level to evaluate the goodness-of-fit of an item. It is common practice to consider items with values of out$_j$ greater than 1.3 or less than 0.75 as having an unsatisfactory fit to the assumed Rasch-type model (Bond and Fox, 2007). Alternatively, it is convenient to apply to out$_j$ the Wilson–Hilferty transformation (Wilson and Hilferty, 1931), obtaining the so-called *standardized outfit* or *standardized unweighted mean square statistics* out$_j^*$, which has an approximately standard normal distribution and it is computed as

$$\text{out}_j^* = \left(\text{out}_j^{1/3} - 1\right) \frac{3}{\hat{se}(\text{out}_j)} + \frac{\hat{se}(\text{out}_j)}{3}. \tag{5.17}$$

Since an outfit statistic is simply an unweighted average, unexpected responses far from item difficulty parameters have relatively more influence;

thus, one could reject an item as misfitting just because of a few unexpected responses made by individuals for whom the item is badly calibrated (i.e., it is too difficult or too easy).

To overcome this problem, the outfit statistic is usually substituted by the *infit* or *weighted mean square statistic* in$_j$. In such a case, residuals are weighted by their individual variance, which, as outlined earlier, is as greater as difficulty and ability estimates are similar. Therefore, infit statistic in$_j$ gives a greater weight to responses made by subjects for whom the item has a good calibration, whereas responses made by subjects for whom the item is inappropriate have less impact. The infit statistic in$_j$ is computed through the following formula:

$$\text{in}_j = \frac{\sum_{i=1}^{n} e_{ij}^2}{\sum_{i=1}^{n} \hat{se}(\hat{y}_{ij})^2}$$

and has a mean equal to 1 and an estimated standard deviation equal to

$$\hat{se}(\text{in}_j) = \sqrt{\frac{1}{\sum_{i=1}^{n} \hat{se}(\hat{y}_{ij})^4} \sum_{i=1}^{n} \left[\sum_{y=0}^{l_j-1} (y - \hat{y}_{ij})^4 \hat{p}_{jy}(\hat{\theta}_i) - \hat{se}(\hat{y}_{ij})^4 \right]}.$$

Also in this case, the *standardized infit statistic* or *standardized weighted mean square statistic* in$_j^*$ is more convenient. It is obtained by applying the Wilson–Hilferty transformation to in$_j$, that is,

$$\text{in}_j^* = \left(\text{in}_j^{1/3} - 1 \right) \frac{3}{\hat{se}(\text{in}_j)} + \frac{\hat{se}(\text{in}_j)}{3}. \tag{5.18}$$

The same kind of statistics can be defined to evaluate the fit of each individual *i*. In such a case, the person's outfit statistic is denoted as out$_{(i)}$ and it is calculated as

$$\text{out}_{(i)} = \frac{1}{J} \sum_{j=1}^{J} \frac{e_{ij}^2}{\hat{se}(\hat{y}_{ij})^2},$$

whereas the person's infit statistics is denoted as in$_{(i)}$ and is obtained as

$$\text{in}_{(i)} = \frac{\sum_{j=1}^{J} e_{ij}^2}{\sum_{j=1}^{J} \hat{se}(\hat{y}_{ij})^2}.$$

The Wilson–Hilferty transformation is then performed as in Equations 5.17 and 5.18 to define the person's standardized outfit and infit statistics, denoted by out$_{(i)}^*$ and in$_{(i)}^*$, respectively.

When observed responses conform to the model, the standardized infit and outfit statistics have mean 0 and standard deviation 1. Values of out_j^* and in_j^* (or $out_{(i)}^*$ and $in_{(i)}^*$) less than -2 or greater than $+2$ usually indicates that the item (person) has a poorer compatibility with the model than expected at the 5% level (Bond and Fox, 2007). In more detail, negative values less than -2 denote less variation than modeled, that is, the response string of the jth item (ith person) is closer to a perfect Guttman scale than the Rasch paradigm leads one to suppose. This means that, in the binary case, all individuals with a low ability level fail item j and all individuals with a high ability level endorse item j (all easy items are endorsed by person i and all of the most difficult items are failed by person i). Note that according to the Rasch assumptions, a perfect Guttman response pattern (e.g., 11111000) is unrealistic and unexpected, because of its deterministic nature: rather, a zone of uncertainty around the item level of difficulty (the person's level of ability) is intrinsic in the probabilistic nature of the Rasch-type models. On the other hand, positive values of out_j^* and in_j^* (or $out_{(i)}^*$ and $in_{(i)}^*$) greater than $+2$ indicate more haphazard responses than expected or higher tendency to guess. In both cases, items and persons with standardized infit and outfit statistics denoting misfitting should be eliminated from the analysis. Alternatively, one can leave the Rasch paradigm for another IRT model (e.g., 2PL, 3PL); however, in this case, the specific objectivity property is lost.

5.9 Differential Item Functioning

DIF, also known as *item bias*, occurs when subjects from different groups (e.g., clustered on the basis of gender or geographic area) with the same level of the latent trait have a different probability of giving a certain response to a given item. Tests containing such items may have a reduced validity for between-group comparisons, because their scores may be indicative of a variety of attributes other than those the scale is intended to measure (Thissen et al., 1988). Detection of DIF is useful to validate a questionnaire and provides aid to interpreting the physiological process underlying group differences in answering the given items. More precisely, since in the framework of IRT models, item parameters are assumed to be invariant to group membership, differences in the item parameters (i.e., differences in ICCs or IRCCCs) estimated separately for each group indicate the presence of a differential functioning for that item.

IRT models allow for a different number of item parameters to be estimated from the data and, thus, allow for the evaluation of DIF for different item properties. For binary data, analyses based on the Rasch model investigate DIF in the difficulty parameters: differential functioning for a given item is observed if the ICCs estimated separately for two or more groups are shifted, so that for a group the conditional probability of endorsing the item is systematically higher or lower than that for other groups, for all latent trait levels. In this case, the DIF effect is said to be *uniform*, because the differences between groups in the conditional probabilities are independent of the common latent trait value.

Furthermore, the 2PL model also allows for investigating DIF in the discriminant parameters. This type of DIF corresponds to an interaction between the latent variable and the group membership, so that ICCs estimated separately for two or more groups have different slopes and, as a consequence, cross one other. This means that the conditional probability of response 1 to a given item for one group is higher than that for another group for some values of the latent trait and it is smaller for the remaining values. In this case, the DIF effect is said to be *nonuniform*.

The distinction between uniform and nonuniform DIF is not immediately generalizable to polytomous items, because of the presence of a number of characteristic curves $p_{jy}(\theta_i)$, with $y = 0, \ldots, l_j - 1$, which are not monotone as in the dichotomous case. Glas and Verhelst (1995b) state that a polytomous item is affected by uniform DIF if the expected score on that item, given the latent trait level, is systematically higher or smaller for one group of subjects with respect to the others.

It is important to note that, under certain conditions, the notion of DIF technically coincides with that of multidimensionality (de Gruijter and van der Kamp, 2008). For instance, with $H = 2$ groups, let the conditional probability of correct answer to item j by person i coming from group h be described by the Rasch model

$$p_j(\theta_{hi}) = \frac{e^{\theta_{hi} - \beta_j^{(h)}}}{1 + e^{\theta_{hi} - \beta_j^{(h)}}},$$

where
$\beta_j^{(h)}$ is the item difficulty parameter in the case person i belongs to group h
θ_{hi} is the ability level of the subject

The fact that a given item is affected by uniform DIF may be interpreted in two perfectly equivalent ways. On the one hand, one can think that the difficulty of the biased item is shifted to the right or to the left for one group, that is, $\beta_j^{(1)} = \beta_j^{(2)} + c$, where c is a constant. On the other hand, this is equivalent to assuming that $\beta_j^{(1)} = \beta_j^{(2)} = \beta_j$ and introducing two latent traits that are shifted of an amount equal to $-c$ for subjects in group $h = 2$ with respect to those in group $h = 1$. In other words, uniform DIF may be interpreted in terms of multidimensionality, in the case that each ability is perfectly associated with a given group.

The analysis of ICCs and IRCCCs can be strengthened by another simple analysis (Crocker and Algina, 1986; Wright and Stone, 1999) that consists in estimating item parameters separately for each group and, then, plotting them against each other. An identity line is drawn through the origin of this plot with slope one and the statistical control lines are constructed around this identity line: items that fall outside the control lines are statistically identified as possibly differential functioning.

In addition to the graphical analyses of ICCs and IRCCCs and statistical control lines of item parameters estimated separately for each group, other diagnostic methods have been developed to identify items affected by DIF, depending on whether they rely on an item response model or not (Crocker and Algina, 1986; de Gruijter and van der Kamp, 2008; Magis et al., 2010). In the first case, the estimation of an IRT model is required, and a statistical testing procedure is followed based on the asymptotic properties of statistics derived from the estimation results. In the second case, the detection of DIF items is usually based on statistical methods for categorical data, with the total test score as a matching criterion.

In addition, the usual setting of DIF detection methods consists of comparing the responses from two groups (e.g., males and females), and when there are more than two groups, the common approach is to perform pairwise comparisons between all of the groups or between one reference group and each of the other groups. Only some methods have also been generalized to simultaneous testing.

Here, some of the main testing procedures for DIF detection are outlined, with special attention for those that are implemented in the statistical software of interest for this book. Furthermore, a promising procedure to detect DIF is based on using latent class IRT (LC-IRT) models with item parameters that differ among the latent classes. In such a way, the analysis could reveal classes of individuals who adopt different response strategies, resulting in different item parameters (de Gruijter and van der Kamp, 2008). Moreover, the generalization of IRT models to include individual covariates (Section 6.2 for details) represents another alternative approach for the DIF detection (Zwinderman, 1997; De Boeck and Wilson, 2004a).

5.9.1 Non-IRT-Based Methods for DIF Detection

- *Mantel–Haenszel test* (Mantel and Haenszel, 1959; Holland and Thayer, 1988): The goal of this method is to test whether there is an association between a group membership and item response, conditionally on the total test score. It is based on the analysis of the contingency tables of correct/incorrect (1/0) responses to a given item by two different groups of subjects, for the various levels of the total test score.

    Let $o_{yhr}^{(j)}$ be the number of individuals attaining test score $r = 0, \ldots, J$, belonging to group $h = 1, 2$ and having response $y = 0, 1$

to item $j = 1, \ldots, J$; also, let n_r be the number of subjects from both groups attaining raw test score r.

Then, the Mantel–Haenszel test is based on the following statistic:

$$MH_j = \frac{\left[\left|\sum_{r=0}^{J}\left(o_{11r}^{(j)} - o_{\cdot1r}o_{1\cdot r}^{(j)}/n_r\right)\right| - 0.5\right]^2}{\sum_{r=0}^{J} o_{\cdot1r}o_{\cdot2r}o_{1\cdot r}^{(j)}o_{0\cdot r}^{(j)}/\left[n_r^2(n_r - 1)\right]},$$

where

$o_{\cdot hr}$ is the number of subjects from group h having test score r

$o_{y\cdot r}^{(j)}$ is the number of responses equal to y from both groups given test score r

It can be verified that, under the null hypothesis of absence of DIF, the MH_j statistic has an asymptotic χ^2 distribution with one degree of freedom. A generalization of the Mantel–Haenszel test to multiple groups is due to Penfield (2001), whereas the extension to the case of polytomous items is proposed by Wang and Su (2004).

As an alternative to the MH_j statistic, we can also use the common odds ratio across all j values (Mantel and Haenszel, 1959), given by

$$\alpha_{MHj} = \frac{\sum_{r=0}^{J} o_{11r}^{(j)}o_{02r}^{(j)}/n_r}{\sum_{r=0}^{J} o_{01r}^{(j)}o_{12r}^{(j)}/n_r},$$

whose logarithm is asymptotically normally distributed under the assumption of absence of DIF.

- *Standardized p difference test* (Dorans and Kulick, 1986): The standardized p difference method is based on the comparison of the proportions of success rates between two groups for all levels of the test score. The related statistic is defined as a weighted average of the differences between the proportions of responses 1 to a given item j among groups 1 and 2:

$$\text{ST}-p-\text{DIF} = \frac{\sum_{r=0}^{J} w_r \left(o_{11r}^{(j)}/o_{\cdot1r} - o_{12r}^{(j)}/o_{\cdot2r}\right)}{\sum_{r=0}^{J} w_r},$$

where the w_r are weights, usually chosen as the proportion of subjects from group $h = 1$ with test score r.

Since the distribution of the $\text{ST}-p-\text{DIF}$ statistic is unknown, the usual classification rule consists of fixing a threshold, such as 0.05 or 0.10, such that an item is classified as affected by DIF if the $\text{ST}-p-\text{DIF}$ is larger than this threshold. More precisely, it is common practice to classify DIF as negligible for $|ST-p-DIF| \leq 0.05$, moderate for $0.05 < |ST-p-DIF| \leq 0.10$, and large for $|ST-p-DIF| > 0.10$ (Magis et al., 2010). Dorans and Kulick (1986) also propose a generalization of the standardized p difference test to account for polytomous items.

- *Logistic regression analysis* (Swaminathan and Rogers, 1990): A different approach for DIF evaluation is provided by the estimation of a suitable logistic regression model. This is a model for the probability of answering 1 (or 0) to the tested binary item, using as covariates the test score, the group membership, and the interaction between these two variables. A uniform DIF effect is detected by a significant effect of group membership, whereas a nonuniform DIF effect is attributable to a significant interaction effect. More formally, the regression coefficients of the model can be estimated and tested through the usual statistical test procedures, such as a Wald test and an LR test. The null hypothesis of no uniform DIF is rejected when the regression coefficients for group membership are not significantly different from zero. Moreover, the null hypothesis of no nonuniform DIF is rejected when the regression coefficients for the interaction effect are not significantly different from zero.

  The logistic regression approach may be easily generalized to account for two important elements. First, it can be adopted in the case of polytomous items, using a logistic regression model for ordered responses (usually, a proportional-odds model is adopted) instead of a logistic model for binary responses. Moreover, the logistic regression can also account for multiple grouping, introducing in the regression equation a suitable set of dummies among the covariates.

5.9.2 IRT-Based Methods for DIF Detection

- *IRT-based LR test* (Thissen et al., 1988): This is an LR-type test comparing two nested IRT models; the null hypothesis is that of equal item parameters in two groups. First, an IRT model is estimated with identical item parameters for both groups of subjects and the maximum marginal log-likelihood is obtained. Then, an augmented model is specified, where the parameters of the item suspected to be biased are allowed to take different values in the two groups. The maximum marginal log-likelihood of this augmented model is also computed. The LR statistic is then obtained in the usual way, see Equation 5.14, and it has an asymptotic null χ^2 distribution with a number of degrees of freedom equal to the number of unconstrained item parameters in the augmented model with respect to the constrained one. This test is particularly appealing because it can be used in the case of both binary and polytomous items and with any

IRT model. According to the assumed IRT model, item difficulties, discrimination indices, and pseudo-guessing parameters can vary between the groups and can be the object of investigation for DIF.

- *Lord's test* (Lord, 1980): This is a test for the null hypothesis of no DIF, and is based on the comparison between the item parameters estimated via the MML approach for the two groups, under a given IRT model. The resulting test statistic has a χ^2 distribution with as many degrees of freedom as the number of item parameters (i.e., 1 for the Rasch model, 2 for the 2PL model, and 3 for the 3PL model), under the null hypothesis. Note that the item parameters from the two groups must be scaled with a common metric prior to statistical testing. For instance, for the Rasch model for binary responses, the Lord's test is based on the following test statistic:

$$Q_j = \frac{\left(\hat{\beta}_j^{(1)} - \hat{\beta}_j^{(2)}\right)^2}{\hat{\text{se}}\left(\hat{\beta}_j^{(1)}\right)^2 + \hat{\text{se}}\left(\hat{\beta}_j^{(2)}\right)^2},$$

having a χ^2 distribution with just one degree of freedom. For the generalization of Lord's test to multiple groups, see Kim et al. (1995).

- *Raju's test* (Raju, 1988): Raju's method aims to test the null hypothesis of no DIF through a normally distributed statistic that computes the area included between the ICC of group 1 and the ICC of group 2 for a given item. Under the null hypothesis, this area is zero and the larger this area, the more likely the rejection of the null hypothesis is. Raju's method may be applied to any IRT model, with the restriction that under the 3PL model, the pseudo-guessing parameters are constrained to be equal across the two groups. Besides, under the Rasch model, Raju's statistic coincides with Q_j. For an alternative DIF detection test based on the geometric difference between ICCs, see also Rudner et al. (1980).

- *Nonparametric test* (Ponocny, 2001): This test is formulated for DIF detection within the Rasch model. Relying on the framework described in Section 5.7.2, the test is based on the following statistic:

$$T_4 = \sum_{r=0}^{J} o_{1hr}^{(j)},$$

which counts how often an item is responded positively within the specified group. It may be used to detect that item j works differently between groups of persons as this count is compared with the expected value of positive responses as predicted by the Rasch model. If more than one item is suspected of differential functioning, then the sum of T_4 on all the specified items provides a total test statistic.

Finally, an interesting extension of the logistic regression approach in the IRT framework is due to Crane et al. (2006); see Choi et al. (2011) for the software implementation. This approach substitutes the test score values in the linear predictor of the logistic model with the ability estimates, obtained on the basis of an IRT model. In addition, an iterative process is employed to account for DIF in the latent trait estimation by using group-specific IRT item parameters for items identified to be affected by DIF. We also remind the proposal of Gnaldi and Bacci (2015) for DIF detection, characterized by the discreteness of the latent trait and an extended IRT parameterization to account for items affected by DIF.

5.10 Examples

In this section, an implementation of the diagnostic instruments previously described is illustrated using Stata and R software. The RLMS and HADS datasets are used in the first case, whereas the INVALSI dataset is considered in the second case.

5.10.1 RLMS Binary Data and Depression Dimension of HADS: Diagnostics in Stata

Stata does not offer many possibilities for a thorough diagnostic analysis of IRT models, with the only exception of modules raschtest (Hardouin, 2007) and pcmtest that implement some of the earlier described instruments for the binary Rasch model and for PCM and RSM, respectively. In the following, we first refer to the binary version of the RLMS dataset regarding Russian job satisfaction (see also the Stata analysis in Section 3.8.1), and to the items of the HADS questionnaire that measures depression (see also the Stata analysis in Section 4.6.1).

Concerning the case of binary items, raschtest provides some item-oriented statistics, such as the U_j test of Molenaar (1983b) and the infit and outfit statistics, and some tests to evaluate the global fit of the model, such as the R_{1c} test and the Andersen's test. The function raschtest is invoked, as already described in Section 3.8.1, by specifying cml instead of mml for the estimation method and with two more options:

```
.  * Load data
. use "RLMS_bin.dta", clear
.  * Define an identificative for each subject
. gen ind=_n
.  * Rasch model
. raschtest Y1-Y4, id(ind) method(cml) information graph
```

```
Estimation method: Conditional maximum likelihood (CML)
Number of items: 4
Number of groups: 5 (3 of them are used to compute the statistics of test)
Number of individuals: 1418 (0 individuals removed for missing values)
Number of individuals with null or perfect score: 699
Conditional log-likelihood: -762.2636
Log-likelihood: -1597.4755
```

	Difficulty					Standardized		
Items	Parameters	Std. Err.	R1c	df	p-value	Outfit	Infit	U
Y1	-1.90674	0.12500	7.091	2	0.0289	-1.862	-2.756	-2.385
Y2	-1.69294	0.12156	2.728	2	0.2556	-1.538	-2.269	-1.906
Y3	0.56648	0.10882	15.888	2	0.0004	2.559	1.069	3.100
Y4*	0.00000	.	44.638	2	0.0000	3.581	1.137	4.170
R1c test		R1c=	65.567	6	0.0000			
Andersen LR test		Z=	54.262	6	0.0000			

```
*: The difficulty parameter of this item had been fixed to 0.

[...] Output omitted
```

Option `information` allows us to plot the information curve (Figure 5.4), whereas `graph` displays the person–item map (Figure 5.5). For each item, the outfit and infit statistics and the U_j statistic are provided, other than the contribution of each item to the R_{1c} statistic. In addition, R_{1c} and Andersen's global test statistics are displayed. All the performed analyses agree on a negative judgment about the goodness-of-fit of a Rasch model for the data at issue.

Finally, Figure 5.5 shows the way individual abilities and item difficulties are distributed along the same metric. As outlined in Section 5.6.2, a

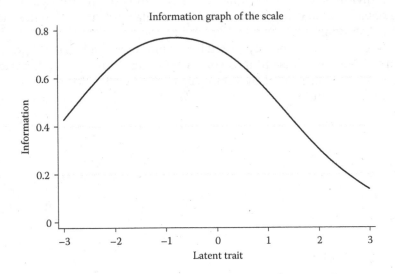

FIGURE 5.4
Information curve for the RLMS dataset.

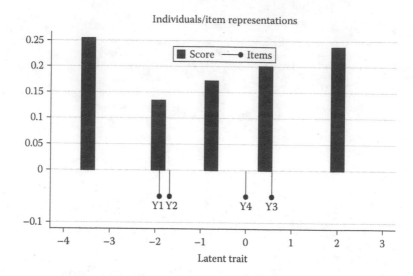

FIGURE 5.5
Person–item map for the RLMS dataset.

well-calibrated questionnaire generally should show the following characteristics: (1) the item difficulty range covers adequately the ability range of
the individuals and (2) the item difficulties are well spaced one another. As
shown in this figure, there are too many individuals for whom the test is too
easy (individuals with score equal to −3.5) or too difficult (individuals with
score equal to 2) and items 1 and 2 show similar difficulties, so that one of
them is redundant.

In the case of polytomously scored items, modules `pcmodel` and `pcmtest`
perform analyses similar to those of `raschtest`. In more detail, `pcmodel`
estimates PCM (and RSM by option `rsm`) through the MML method and
provides the pseudo-R^2 of McFadden (1974), which assumes a very small
value:

```
. * Load data
. use "HADS.dta", clear
. * Select only items related to depression and rename them
. keep Y1 Y3-Y5 Y9 Y13-Y14
. rename Y3 Y2
. rename Y4 Y3
. rename Y5 Y4
. rename Y9 Y5
. rename Y13 Y6
. rename Y14 Y7
. * Install pcmodel
. ssc install pcmodel
. * PCM through function pcmodel
. pcmodel Y1-Y7
```

```
Model : Partial Credit Model

    log likelihood: -1441.917409422232
                    Marginal McFadden's pseudo R2: 10.5%
```

```
[…] Output omitted
```

Using estimates from `pcmodel`, function `pcmtest` allows us to test the fit between the observed data and PCM (or RSM) through the R_{1m} test and the item-specific S_j test (option `sitest`). The R_{1m} test corresponds to the R_{1c} test when the estimation is performed through the MML method. We observe that in all cases, the goodness-of-fit of PCM and of single items is bad.

```
. * Install pcmtest
. ssc install pcmtest
. * Run pcmtest
. pcmtest, sitest

Performing R1m test

[…] Output omitted

Global tests of the fit : test R1m
                   groups : 4 6 9 21
                   Number of individuals with missing data : 0 (0%)

---------------------------------------------
                             N =     201
              df       R1m   p-val  Power
---------------------------------------------
  R1m         79      114.9  0.0052 1.0000
---------------------------------------------

Items specific tests of the fit : tests Si

---------------------------------------------
                             N =     201
  Item        df       Si    p-val  Power
---------------------------------------------
  Y1 :         9      44.6   0.0000 0.9996
---------------------------------------------
  Y2 :         9      28.0   0.0009 0.9794
---------------------------------------------
  Y3 :         9      28.2   0.0009 0.9801
---------------------------------------------
  Y4 :         9      27.8   0.0010 0.9783
---------------------------------------------
  Y5 :         9      55.9   0.0000 1.0000
---------------------------------------------
  Y6 :         9      24.9   0.0031 0.9609
---------------------------------------------
  Y7 :         9      31.7   0.0002 0.9907
---------------------------------------------
```

5.10.2 INVALSI Mathematics Data: Diagnostics in R

In the following, the main diagnostics tools available in R are illustrated with reference to the INVALSI reduced dataset, analyzed in Section 3.8.2.

```
> load("Invalsi_reduced.RData")
```

We first treat the graphical analyses, and then, illustrations of the main hypothesis tests and of the outfit and infit statistics are provided. We conclude with the DIF analysis.

5.10.2.1 Graphical Diagnostic Tools

The graphical analyses described in Section 5.6 may be performed through the R packages ltm, eRm, and mirt, other than package ROCR for the ROC and AUC analyses.

After estimating a given IRT model (say the 2PL, as in Section 3.8.2), the item information may be calculated through function information of package ltm (also package mirt allows for item and test information analyses), which allows us to specify one or more items of interest, through option items. For instance, for item Y19 we have

```
> require(ltm)
> out3 = ltm(Y ~ z1)
> information(out3, range=c(-5,5), items = 19)

Call:
ltm(formula = Y ~ z1)

Total Information = 1.77
Information in (-5, 5) = 1.77 (99.96%)
Based on items 19
```

The test information is obtained in a similar way without specifying option items:

```
> information(out3, range=c(-5,5), items = NULL)

Call:
ltm(formula = Y ~ z1)

Total Information = 27.55
Information in (-5, 5) = 25.84 (93.8%)
Based on all the items
```

For a better diagnosis, we suggest to graphically represent the item and test information curves, through function plot, which requires the following main inputs: the output of the estimated 2PL model (out3), the specification of the type of curve (i.e., IIC, which means item information curve), the list of items of interest (put 0 in the case of test information curve), and the range of the latent trait level:

```
> plot(out3, type="IIC" , items = c(7, 19, 23), zrange = c(-5.0, 5.0),
+    labels = c("Y7","Y19","Y23"), legend=TRUE, lty=c(1,2,4))
> plot(out3, type="IIC" , items = 0, zrange = c(-5.0, 5.0))
```

The resulting plots are shown in Figure 5.6. We observe that the information of the three analyzed items Y7, Y19, and Y23 (Figure 5.6a) is coherent with the item parameter estimates reported in Section 3.8.2. Item Y7, which presents a very low discriminating power (0.238), has a flat item information curve with a very small maximum value. Item Y19, which discriminates very well (1.766), has a very peaked information curve. Item Y23, which is characterized by a greater difficulty than item Y19 (0.581 versus −0.380), is shifted on the right. Both curves for items Y19 and Y23 are bell shaped, so that they provide more accurate estimates of latent trait levels near the corresponding item difficulties. Overall, the test information curve is bell shaped and it reaches a peak near to the average ability; furthermore, the information is greater for negative ability levels rather than for positive levels (Figure 5.6b).

The ROC curve and the AUC value may be obtained through function `performance` of package ROCR, based on function `prediction`, which requires in input the original observations (input `labels`), that is, one value (0/1 in the case of binary items) for each item-response pattern combination, and the corresponding estimated posterior probabilities of answering 1 (input `predictions`). Since the ROCR package is not specifically tailored to deal with item responses, these inputs have to be obtained in a suitable way: the detailed code that defines the two vectors at issue, denoted as `obspattern_vec` and `postprob_vec`, is reported in the following.

First, the coefficients of the 2PL model and the ability values are extrapolated (the code is already illustrated in Section 3.8.2):

```
> out3$beta = -out3$coefficients[,1]/out3$coefficients[,2]
> out3$lambda = out3$coefficients[,2]
> coeff = cbind(beta=out3$beta,lambda=out3$lambda)
> out3scores = factor.scores(out3)
> ability = as.matrix(out3scores$score[,30])
```

and, then, the response patterns and the posterior probabilities are computed on the basis of the estimated abilities:

```
> # Response patterns
> obspattern = as.matrix(out3scores$score[,1:27])

> n = nrow(ability)
> J = ncol(obspattern)
> # Posterior probabilities of endorsing each item
> postprob = matrix(rep(0), nrow = n, ncol=J)
> for (i in 1: n){
+ for (j in 1 : J) {
+ postprob[i,j] =
+ exp(coeff[j,2]*(ability[i]-coeff[j,1]))/(1+exp(coeff[j,2]*(ability[i]-coeff[j,1]) ))
+ }
+ }
> postprob_vec = as.vector(t(postprob))
> obspattern_vec = as.vector(t(obspattern))
```

After these preliminary analyses, the ROC curve (Figure 5.7) and the related AUC value are obtained through the `performance` function, by declaring

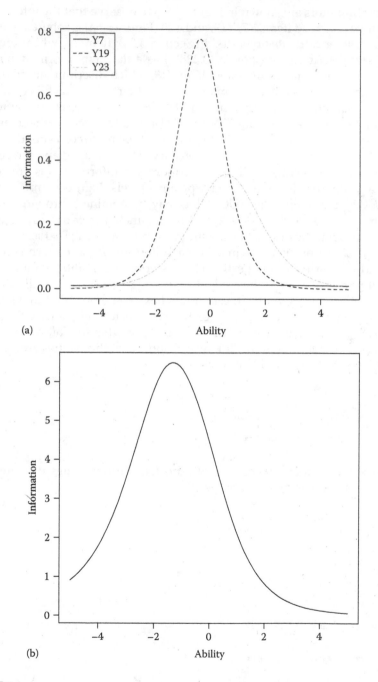

(a)

(b)

FIGURE 5.6
Item information curves for items Y7, Y19, Y23 (a) and test information curve (b).

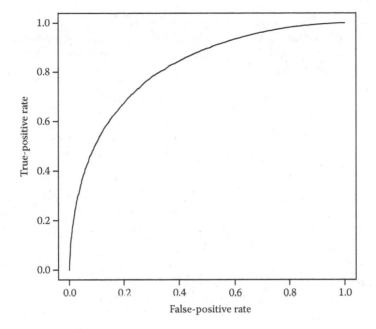

FIGURE 5.7
ROC curve for the 2PL model.

inputs `tpr`, `fpr` (standing for true-positive rates and false-positive rates), and `auc`, respectively:

```
> require(ROCR)
> pred = prediction(predictions=postprob_vec, labels=obspattern_vec)
> # ROC curve
> perf = performance( pred, "tpr", "fpr" )
> plot( perf )
> # AUC value
> performance( pred, "auc")@y.values
 0.8017324
```

For the Rasch model, the ROC curve and the AUC value may be computed in an easier way, through function `gofIRT` of package eRm. Function `gofIRT` relies on package ROCR and it requires the person parameter values in input. These are estimated through function `person.parameter`, which is applied to the output of function RM of the eRm package (see Section 3.8.2 for an alternative estimation of the Rasch model with package `ltm`):

```
> require(eRm)
> out = RM(Y)
> # Ability estimates
> est_theta = person.parameter(out)
> gof = gofIRT(est_theta)
```

The output provided by `gofIRT` resumes several interesting elements, such as the confusion matrix and the related rates of accuracy, sensitivity, and specificity described in Section 5.6.3, which give a general idea of the predictive capacities of the model:

```
> # Confusion matrix and rates of accuracy, sensitivity, and specificity
> gof$classifier
 $confmat
         observed
predicted    0     1
        0  4257  2127
        1  6929 33343

$accuracy
[1] 0.8058985

$sensitivity
[1] 0.9400338

$specificity
[1] 0.380565
```

In addition, commands `plot(gof$ROC)` and `gof$AUC` give the graph of the ROC curve (Figure 5.8) and the AUC value, respectively:

```
> # ROC curve
> plot(gof$ROC)
> # Alternative way to represent the ROC curve
```

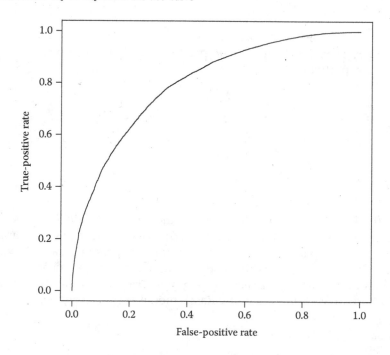

FIGURE 5.8
ROC curve for the Rasch model.

```
> TP = gof$ROC@y.values
> FP = gof$ROC@x.values
> plot(FP[[1]], TP[[1]], xlab = "False positive rate" , ylab = "True positive rate")
> # AUC value
> gof$AUC
[1] 0.8188212
```

The comparison with the analysis performed previously for the 2PL model leads to the conclusion of a very similar predictive power of the two models: in both cases, the AUC value is greater than 0.80, denoting an excellent predictive capability.

As outlined in Section 5.6.2, a graphical tool specific for Rasch-type models is the person–item map, which may be plotted through function plotPImap of package eRm (this function is also available for polytomous Rasch models). It requires as main argument, the output of a Rasch model, obtained by function RM; the option sorted=TRUE allows us to display the item from the easiest to the most difficult:

```
> # Person-item map
> plotPImap(out, sorted=TRUE)
```

The output of function plotPImap is shown in Figure 5.9 (also compare with Figure 3.7). We conclude that the INVALSI mathematics test is generally well calibrated for the set of students at issue, with the evident exception of item 5, whose level of difficulty is definitely too low compared with the ability distribution. Besides, the skewed shape of the ability distribution shows that individuals with an extremely high level of ability in mathematics cannot be adequately measured by the test at issue.

5.10.2.2 Hypothesis Tests for the Adequacy of the Rasch Model

The goodness-of-fit of the Rasch model is more deeply evaluated through the tests described in Section 5.7.2. First, we investigate about the null hypothesis of unidimensionality, by means of MPA, Martin-Löf test, and the nonparametric $-T_2$ test of Ponocny (2001).

We begin with the explorative analysis based on MPA, which is implemented in package ltm, through function unidimTest. The procedure can also be applied to the 2PL model, but it is not yet extended to polytomous items. Function unidimTest requires the output from the Rasch model estimation in input (estimated through the appropriate function implemented in ltm) and provides as output the second eigenvalue in the observed data, the average of the second eigenvalues in a given number of Monte Carlo–simulated samples (100 is the default value), and the p-value that allows us to compare the two types of eigenvalues. In our case, the p-value equals 0.0198, so that the unidimensionality assumption cannot be rejected at the 1% significance level:

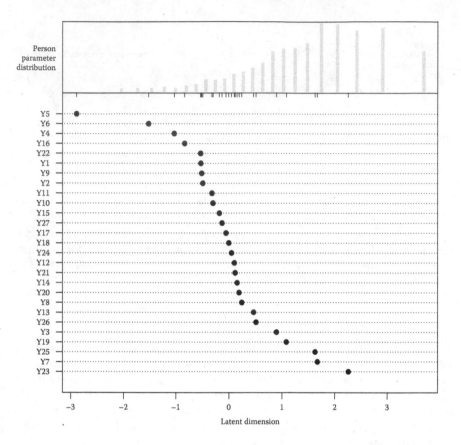

FIGURE 5.9
Person–item map for the Rasch model.

```
> (mpa <- unidimTest(rasch(Y)))
[...] Output omitted

Alternative hypothesis: the second eigenvalue of the observed data is substantially larger
than the second eigenvalue of data under the assumed IRT model

Second eigenvalue in the observed data: 1.463
Average of second eigenvalues in Monte Carlo samples: 0.717
Monte Carlo samples: 100
p-value: 0.0198
```

The MPA analysis may be completed with the plot of all the eigenvalues (Figure 5.10), to empirically verify the differences between all pairs of eigenvalues, other than the second ones:

```
> plot(mpa, type="b",pch=1:2)
> legend("topright", c("Observed data", "Average simulated data"), lty=1, pch=1:2, col=1:2,
+ bty="n")
```

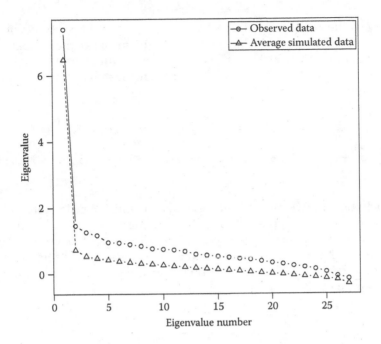

FIGURE 5.10
Eigenvalues of observed data and average eigenvalues of parallel simulated data.

After MPA, two formal tests are performed. First, the Martin-Löf test, which is typical of the Rasch model, is implemented in function MLoef of package eRm. It may be declared by specifying two given subsets of items, which are suspected to detect two separate dimensions, or by specifying a split criterion (such as the mean test score or the median test score) so that items are divided into two plausible groups, as follows:

```
> MLtest = MLoef(out, "mean")
> summary(MLtest)

Martin-Loef-Test (split criterion: mean)

Group 1:
Items: 1, 2, 4, 5, 6, 9, 10, 11, 12, 14, 15, 16, 17, 18, 21, 22, 24, 27
Log-Likelihood: -8802.841

Group 2:
Items: 3, 7, 8, 13, 19, 20, 23, 25, 26
Log-Likelihood: -6028.075

Overall Rasch-Model:
Log-Likelihood: -17010.36

LR-value: 121.635
Chi-square df: 161
p-value: 0.991
```

With a *p*-value equal to 0.991, we do not reject the null hypothesis of unidimensionality. The same conclusion is reached by means of the nonparametric $-T_2$ test of Ponocny (2001), which is performed by function NPtest of package eRm, choosing option method="T2m". This test may also be applied to specific subsets of items.

```
> NPtest(as.matrix(Y), method="T2m",stat="var",idx = 1:27)
Nonparametric RM model test: T2m (multidimensionality - model deviating subscales)
    (decreased dispersion of subscale person rawscores)
Number of sampled matrices: 500
Items in subscale: 1 2 3 4 5 6 7 8 9 10 11 12 13 14 15 16 17 18 19 20 21 22 23 24 25 26 27
Statistic: variance
one-sided p-value: 1
```

To conclude, all three methods agree about the unidimensionality of the set of 27 items, which compose the INVALSI mathematics test. The analysis of the adequacy of the Rasch model continues with some global tests described in Section 5.7.2.1. The logistic regression–based tests and Andersen's test are implemented in the package eRm through functions gofIRT (already applied earlier) and LRtest, respectively:

```
> gof$test.table
                          value        df      p-value
Collapsed Deviance  1066.29813        621 0.00000000
Hosmer-Lemeshow       17.39393          8 0.02625873
Rost Deviance      18704.72736 134217701 1.00000000
Casewise Deviance  39340.49886      46607 1.00000000

> LRtest(out, splitcr = "median")

Andersen LR-test:
LR-value: 235.009
Chi-square df: 26
p-value:  0
```

Function LRtest allows us to specify the split criterion for the subgroups of individuals, and it tests the null hypothesis of constant item parameters in the different groups. In all cases, the performed global tests agree on rejecting the null hypothesis of adequacy of the Rasch model.

In addition to the aforementioned global parametric tests, the estimated Rasch model is also analyzed through the nonparametric tests for the null hypothesis of local independence. For this, the global test T_{11} is first performed through function NPtest:

```
> # Test T11
> (t11=NPtest(as.matrix(Y), method="T11"))

Nonparametric RM model test: T11 (global test - local dependence)
(sum of deviations between observed and expected inter-item correlations)
Number of sampled matrices: 500
one-sided p-value: 0

> # Test statistic of test T11
> (sum(abs(t11$T11r-t11$T11rho)))
32.84712
```

On the basis of test T_{11} (the test statistic equals 32.812 with p-value < 0.001), we conclude that the null hypothesis of local independence is violated for at least one pair of items. The specific pairs of items that concur with this conclusion are detected through test T_1:

```
> # Pairs of items that contribute to test statistic of test T11
> (NPtest(as.matrix(Y), method="T1"))

Nonparametric RM model test: T1 (local dependence - increased inter-item correlations)
    (counting cases with equal responses on both items)
Number of sampled matrices: 500
Number of Item-Pairs tested: 351
Item-Pairs with one-sided p < 0.05
(2,11)  (2,18)  (2,19)  (2,23)  (2,24)  (3,4)  (4,9)  (4,10)  (4,11)  (4,19)  (5,6)
0.000   0.010   0.004   0.008   0.034   0.048  0.016  0.042   0.008   0.000   0.000
    (5,8)   (6,7)
    0.012   0.016
[...] Output omitted
```

The local independence assumption can also be tested by test T_2 for specific subsets of (two or more) items. For instance, let us consider the first pair of items (items 2 and 11) detected by the T_1 test and let us now perform test T_2 as follows:

```
> # Test T2
> (NPtest(as.matrix(Y), method="T2",stat="var",idx = c(2,11)))

Nonparametric RM model test: T2 (local dependence - model deviating subscales)
    (increased dispersion of subscale person rawscores)
Number of sampled matrices: 500
Items in subscale: 2 11
Statistic: variance
one-sided p-value: 0
```

The output of test T_2 confirms the deviation of items 2 and 11 from the local independence assumption.

On the basis of the described analyses, we conclude for an unsatisfactorily goodness-of-fit of the Rasch model to the data at issue. At this point, a first option consists of choosing a less restrictive model, such as the 2PL one. The two models can be easily compared through function anova of package 1tm, which provides values of AIC and BIC and performs the LR test:

```
> # AIC, BIC, and LR test for Rasch and 2PL models
> (anova(rasch(Y), out3))

   Likelihood Ratio Table
            AIC       BIC    log.Lik    LRT df p.value
rasch(Y) 43913.21 44066.87 -21928.61
out3     43487.30 43783.64 -21689.65 477.92 26  <0.001
```

Note that anova requires as inputs two nested models, estimated through package 1tm. In our case, we recall that rasch(Y) provides the output from the Rasch model estimation and out3 contains that for the 2PL model. The null hypothesis of equivalence between the Rasch and the 2PL model is rejected in favor of the latter.

The same goodness-of-fit analyses proposed for the Rasch model can be repeated for the 2PL model, using essentially the same functions. Besides, for the 2PL model, we can also perform an item-by-item analysis to test the different discriminating power of a group of items versus another group through a nonparametric test based, similar to test T_5, on the point-biserial correlations and implemented in function NPtest through option Tpbis. For instance, following what was suggested by the estimated discriminating parameters obtained for the 2PL model (see Section 3.8.2), we compare item Y7 versus items Y2, Y14, Y15, Y20, and Y27 and conclude for a significantly different discriminating power for the two sets of items:

```
> # Test T5
> (NPtest(as.matrix(Y), method = "Tpbis", idxt = 7, idxs = c(2, 14, 15, 20, 27)))

Nonparametric RM model test: Tpbis (discrimination)
    (pointbiserial correlation of test item vs. subscale)
Number of sampled matrices: 500
Test Item: 7
Subscale  - Items: 2 14 15 20 27
one-sided p-value (rpbis too low): 1
```

5.10.2.3 Outfit and Infit Statistics

Rather than choosing a less restrictive model, such as the 2PL, we can also remain in the Rasch framework and eliminate possible problematic items. For this aim, we base our analysis on the outfit and infit statistics, described in Section 5.8. The statistics at issue are computed for each item (also polytomous items are accepted) with function itemfit of package eRm. Note that another function exists under the mirt package, which is denominated in the same way but has different characteristics.

The function itemfit requires in input the predicted values of the latent trait, previously estimated through function person.parameter, and its output provides the outfit and infit statistics (Outfit MSQ and Infit MSQ, respectively), together with their standardized versions (Outfit t and Infit t, respectively):

```
> est_theta = person.parameter(out)
> itemfit(est_theta)

Itemfit Statistics:
        Chisq   df p-value Outfit MSQ Infit MSQ Outfit t Infit t
Y1  2175.731 1727   0.000      1.259     1.120     2.83    2.56
Y2  1677.002 1727   0.802      0.970     0.968    -0.33   -0.71
Y3  1933.999 1727   0.000      1.119     1.102     3.11    4.24
Y4  1581.189 1727   0.995      0.915     0.933    -0.74   -1.13
Y5  1732.948 1727   0.455      1.003     0.958     0.11   -0.21
Y6  1888.668 1727   0.004      1.093     0.995     0.67   -0.04
Y7  2369.835 1727   0.000      1.371     1.268    11.13   12.00
Y8  2049.970 1727   0.000      1.186     1.062     3.24    2.02
Y9  1534.807 1727   1.000      0.888     0.929    -1.34   -1.60
Y10 1505.270 1727   1.000      0.871     0.933    -1.77   -1.70
```

```
Y11 1237.687 1727    1.000      0.716    0.848    -4.13    -3.94
Y12 1717.304 1727    0.561      0.994    1.006    -0.08     0.18
Y13 1681.594 1727    0.779      0.973    0.942    -0.55    -2.16
Y14 1600.276 1727    0.986      0.926    0.970    -1.30    -0.94
Y15 1636.281 1727    0.941      0.947    0.970    -0.75    -0.78
Y16 1351.397 1727    1.000      0.782    0.893    -2.29    -2.07
Y17 1734.183 1727    0.447      1.004    1.022     0.08     0.62
Y18 1376.523 1727    1.000      0.797    0.905    -3.45    -2.83
Y19 1408.956 1727    1.000      0.815    0.876    -5.84    -5.91
Y20 1665.788 1727    0.851      0.964    0.997    -0.63    -0.10
Y21 1743.724 1727    0.384      1.009    1.021     0.17     0.64
Y22 1947.534 1727    0.000      1.127    1.021     1.45     0.47
Y23 1498.195 1727    1.000      0.867    0.927    -3.73    -3.22
Y24 1464.094 1727    1.000      0.847    0.926    -2.61    -2.24
Y25 1475.585 1727    1.000      0.854    0.896    -5.15    -5.25
Y26 1954.774 1727    0.000      1.131    1.084     2.73     3.03
Y27 1564.489 1727    0.998      0.905    0.981    -1.41    -0.51
```

From the above output, we note several problematic items. On the basis of values of `infit t`, the worse fitting is observed for those items that, under the 2PL model, show discriminating parameters much different from 1, such as items Y1, Y3, Y7, Y24, Y25, Y26 (see also the output in Section 3.8.2 about estimates of discriminating parameters under the 2PL model). After having eliminated all items whose infit statistic is greater than 2.00 or smaller than −2.00, the infit and outfit statistics are again computed:

```
# Estimate Rasch model after eliminating items with |infit t| > 2.00
> Y2 = Y[,c(2,4,5,6,9,10,12,14,15,17,20,21,22,27)]
> out2 = RM(Y2)
> est_theta2 = person.parameter(out2)
> itemfit(est_theta2)

Itemfit Statistics:
       Chisq    df p-value Outfit MSQ Infit MSQ Outfit t Infit t
Y2  1342.641 1372   0.709      0.978     0.983    -0.32   -0.37
Y4  1218.311 1372   0.999      0.887     0.918    -1.27   -1.44
Y5  1285.332 1372   0.953      0.936     0.934    -0.19   -0.38
Y6  1600.944 1372   0.000      1.166     1.011     1.37    0.17
Y9  1201.101 1372   1.000      0.875     0.924    -1.96   -1.78
Y10 1226.341 1372   0.998      0.893     0.934    -1.90   -1.73
Y12 1407.965 1372   0.244      1.025     1.017     0.59    0.56
Y14 1309.462 1372   0.885      0.954     0.976    -1.09   -0.79
Y15 1364.945 1372   0.549      0.994     0.980    -0.09   -0.54
Y17 1429.706 1372   0.136      1.041     1.039     0.85    1.14
Y20 1363.639 1372   0.559      0.993     1.005    -0.15    0.19
Y21 1472.678 1372   0.029      1.073     1.053     1.63    1.71
Y22 1578.137 1372   0.000      1.149     1.038     2.16    0.86
Y27 1368.703 1372   0.520      0.997     0.995    -0.04   -0.13
```

As shown by the newly estimated outfit and infit statistics, the selected subset of items presents now a satisfactory goodness-of-fit (specifically, note that all the `infit t` are between −2 and 2). Note that the same type of analysis may be performed on the sample of individuals, through function `personfit`, which is similar to `itemfit`:

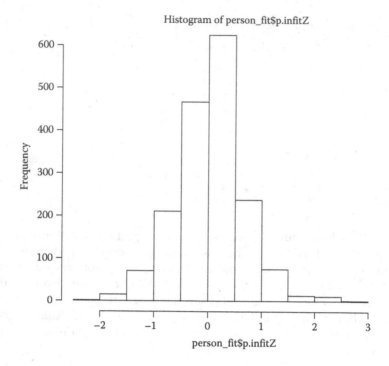

FIGURE 5.11

Distribution of estimated person's standardized infit statistics.

```
> person_fit = personfit(est_theta)
> hist(person_fit$p.infitZ)
```

Generally, students have a behavior in agreement with the Rasch model, although for some of them, we observe high levels of the latent trait, as shown by the skewed shape of the histogram in Figure 5.11.

5.10.2.4 DIF Analysis

The proposed diagnostics analysis of INVALSI mathematics data concludes with an investigation about DIF. For this aim, we use the complete dataset and select both male and female students, which come from the third geographic area:

```
> load("Invalsi_full.RData")
> # Select area_3=1
> data = data[data[,4]==1, ]
> # Select variable gender and mathematics items
> Y = data[ , c(2, 47:73)]
> # Rename items to compare with previous analyses
> names(Y)[2:28] = c("Y1", "Y2", "Y3", "Y4", "Y5", "Y6", "Y7", "Y8",
+ "Y9", "Y10", "Y11", "Y12", "Y13", "Y14", "Y15", "Y16", "Y17", "Y18",
+ "Y19", "Y20", "Y21", "Y22", "Y23", "Y24" ,"Y25", "Y26", "Y27")
```

The DIF analysis is performed through functions of the following packages: difR, which implements the main part of DIF detection methods described in Section 5.9, lordif, which is specialized in the logistic regression–modified method proposed by Crane et al. (2006), and eRm for the nonparametric test T_4.

Function dichoDIF of package difR allows us to perform most DIF tests for binary test items (in the case of polytomous items, function genDichoDif is used), that is, the Mantel–Haenszel test, the standardized p difference test, the logistic regression analysis, Lord's test, and Raju's test. Inputs required by dichoDIF are the matrix of item responses (Data), a vector denoting the group membership (group), the category indicating the group of interest (focal.name), and the specific DIF detection method (method). Moreover, several other options are available according to the flagged method: for instance, option correct=TRUE refers to the continuity correction included in the *MH* statistics, type refers to the type of DIF (i.e., uniform, nonuniform, or both) detected by the regression logistic analysis, and model denotes the IRT model, on which Lord's and Raju's tests are based (alternatively, a vector of estimated item parameters can be provided through option irt.param). The output of dichoDIF consists in a table, where the presence or absence of DIF is declared for each item at the significance level chosen by option alpha:

```
> require(difR)
> (dichoDif(Data=Y[,2:28], group=Y[,1], focal.name=0, method=c("MH","Std","Logistic","Lord",
+ "Raju"), alpha=0.05, correct=T, type="both", model="2PL"))

[...] Output omitted
```

	M-H	Stand.	Logistic	Lord	Raju	#DIF
Y1	NoDIF	NoDIF	NoDIF	NoDIF	NoDIF	0/5
Y2	NoDIF	NoDIF	NoDIF	DIF	DIF	2/5
Y3	NoDIF	NoDIF	NoDIF	DIF	DIF	2/5
Y4	NoDIF	NoDIF	NoDIF	DIF	NoDIF	1/5
Y5	NoDIF	NoDIF	NoDIF	NoDIF	NoDIF	0/5
Y6	NoDIF	NoDIF	NoDIF	NoDIF	NoDIF	0/5
Y7	NoDIF	NoDIF	NoDIF	DIF	NoDIF	1/5
Y8	NoDIF	NoDIF	NoDIF	DIF	DIF	2/5
Y9	NoDIF	NoDIF	NoDIF	DIF	DIF	2/5
Y10	NoDIF	NoDIF	NoDIF	DIF	DIF	2/5
Y11	NoDIF	NoDIF	NoDIF	DIF	DIF	2/5
Y12	NoDIF	NoDIF	NoDIF	DIF	DIF	2/5
Y13	NoDIF	NoDIF	NoDIF	DIF	DIF	2/5
Y14	NoDIF	NoDIF	NoDIF	DIF	DIF	2/5
Y15	NoDIF	NoDIF	NoDIF	DIF	DIF	2/5
Y16	DIF	NoDIF	DIF	DIF	DIF	4/5
Y17	NoDIF	NoDIF	NoDIF	DIF	DIF	2/5
Y18	NoDIF	NoDIF	NoDIF	DIF	DIF	2/5
Y19	NoDIF	NoDIF	NoDIF	DIF	DIF	2/5
Y20	NoDIF	NoDIF	NoDIF	DIF	DIF	2/5
Y21	DIF	NoDIF	DIF	DIF	DIF	4/5
Y22	NoDIF	NoDIF	NoDIF	NoDIF	NoDIF	0/5
Y23	NoDIF	NoDIF	NoDIF	DIF	DIF	2/5
Y24	DIF	NoDIF	NoDIF	DIF	DIF	3/5
Y25	NoDIF	NoDIF	DIF	DIF	DIF	3/5
Y26	NoDIF	NoDIF	NoDIF	DIF	NoDIF	1/5
Y27	NoDIF	NoDIF	NoDIF	DIF	DIF	2/5

A substantial agreement is observed between the non-IRT-based methods (see columns M-H, Stand., Logistic), which flag just a few items as biased, that is, items Y16, Y21, Y24, and Y25. On the other hand, the IRT-based tests (columns Lord and Raju) agree on detecting several problematic items. In general, items Y16 and Y21 are flagged as biased in 4 cases out of 5 and items Y24 and Y25 are flagged in 3 cases out of 5.

In addition, the regression logistic method of Crane et al. (2006), which is based on the latent trait estimates rather than on the raw test scores, is performed through function lordif of the homonymous package. Arguments in input are the matrix of the item responses, the vector of group membership (group), and the criterion to compare the nested logistic models (criterion). Three nested logistic models are assessed, which differ in the regressors taken into account: model 1, characterized only by the abilities as regressor; model 2, characterized by abilities and group membership as regressors; and model 3, which includes an interaction effect between the abilities and the group membership. In the case of criterion = "Chisqr", an LR test is performed on model 1 versus model 2 (null hypothesis of no uniform DIF) and on model 2 versus model 3 (null hypothesis of no nonuniform DIF); the corresponding *p*-values are denoted in output as chi12 and chi23, respectively. Note that if one or more items have less than five observations for each response category and each group, then the LR test is not performed. Therefore, after having verified that two males answered incorrectly to item 5, we proceed with the DIF analysis without such problematic item:

```
> table (Y$Y5, Y$gender)
    0   1
  0   5   2
  1 309 310
> require(lordif)
> DIF = lordif(Y[,c(2:5,7:28)], group=Y[,1], selection = NULL, criterion = "Chisqr",
+ alpha = 0.05)
> DIF$stats[c(1:3, 5)]
   item ncat  chi12  chi23
1     1    2 0.0437 0.6374
2     2    2 0.4799 0.2933
3     3    2 0.8522 0.2220
4     4    2 0.4613 0.0009
5     5    2 0.7114 0.6494
6     6    2 0.2053 0.0923
7     7    2 0.6767 0.6979
8     8    2 0.2185 0.5458
9     9    2 0.9866 0.6715
10   10    2 0.9129 0.6493
11   11    2 0.3999 0.5309
12   12    2 0.6944 0.1666
13   13    2 0.1816 0.0043
14   14    2 0.4297 0.1645
15   15    2 0.0056 0.8718
16   16    2 0.1231 0.9004
17   17    2 0.1050 0.1325
18   18    2 0.3485 0.2116
19   19    2 0.3737 0.6859
```

```
20   20    2 0.0032 0.1539
21   21    2 0.9047 0.8181
22   22    2 0.4118 0.2130
23   23    2 0.0245 0.3608
24   24    2 0.0563 0.0005
25   25    2 0.8555 0.0532
26   26    2 0.3451 0.4454
```

Note that items are now renumbered from 1 to 26, because of the elimination of item Y5, so that output for items Y1–Y4 is displayed in rows 1–4, whereas output for items Y6–Y27 is displayed in rows 5–26. Results of the LR test show that, at the 5% level, four items (Y1, Y16, Y21, Y24) are affected by uniform DIF. However, at significance level of 1%, only items Y16 and Y21 have a uniform differential functioning. Besides, clear evidence of nonuniform DIF (chi23 < 0.01) emerges for items Y4, Y14, and Y25.

To conclude the DIF analysis, it is useful to investigate about the direction of the DIF effect for each biased item. Therefore, we can plot the ICC of each item separately for males and females:

```
> plot(DIF,labels=c("Males", "Females"))
```

ICCs show that items Y1, Y16, and Y21 are more difficult for females than for males, whereas item Y24 is easier for females than for males. Concerning the three items suffering from nonuniform DIF (i.e., Y4, Y14, Y25), all of them are more difficult for females than for males for high values of ability, whereas the relation is inverted for small abilities.

We can also perform the nonparametric test T_4 (displayed here for item Y1, as specified through option idx):

```
> require(eRm)
> NPtest(as.matrix(Y[,2:28]), method="T4",idx = 1, group=Y[,1]==1, alternative="low")

Nonparametric RM model test: T4 (Group anomalies - DIF)
(counting low raw scores on item(s) for specified group)
Number of sampled matrices: 500
Items in Subscale: 1
Group: Y[, 1] == 1    n = 312
one-sided p-value: 0.036
```

As shown by the *p*-value, which is less than 0.05, the T_4 test agrees in detecting item Y1 as more difficult for females than for males. Indeed, by options group=Y[,1]==1 and alternative="low", we are evaluating if the number of positive responses on item Y1 is lower than expected for females (gender=1). The same piece of information is obtained by specifying group=Y[,1]==0 and alternative="high".

Exercises

1. Using the dataset `lsat.dta` (downloadable from `http://www.gllamm.org/docum.html`), perform the following analyses through `Stata`:

   (a) Estimate a Rasch model through module `raschtest` and compare the estimates of item difficulties with respect to those obtained with `gllamm` (see Exercise 1 of Chapter 3).

   (b) Discuss the values of outfit and infit statistics.

   (c) Discuss the results of the R_{1c} test and Andersen's test.

   (d) Plot the person–item map and compare the empirical distribution of abilities with that of the item difficulties.

   (e) Plot the test information curve and discuss its shape.

2. Consider the dataset `naep` available in the R package `MultiLCIRT` and estimate a Rasch model through the CML method. Then, perform the following analyses in R:

   (a) Evaluate the predictive capacity of the estimated Rasch model through the ROC curve and the AUC value, using function `performance` of package `ROCR` and function `gofIRT` of package `eRm`.

   (b) Plot the item information curves for each item.

   (c) Plot the person–item map.

   (d) Calculate the standardized outfit and infit statistics.

   (e) On the basis of the previous analyses, discuss the quality of the questionnaire, with special attention to the possible presence of problematic items (i.e., bad calibrated items with respect to the ability distribution, redundant items, misfitted items).

3. Consider the dataset `verbal` available in the R package `difR` and perform the following analyses through R:

   (a) For each item, compute the percentage of responses 1 (individual agrees with that item) for females (variable `Gender` equals 0) and males (variable `Gender` equals 1).

   (b) Consider the subset of males' responses and investigate the unidimensionality assumption under the Rasch model, by means of the Martin-Löf test and a suitable nonparametric test.

   (c) Consider the subset of males' responses and compare the goodness-of-fit of a Rasch model with that of a 2PL model, by means of information criteria, LR test, and a suitable nonparametric test.

(d) Estimate a 2PL model separately for males and females and plot the corresponding ICCs for each item: discuss the possible presence of items affected by DIF.

(e) On the basis of the 2PL model, investigate the possible presence of DIF by means of the Mantel–Haenszel test, the standardized p difference test, Lord's test, and Raju's test and compare these results with the graphical results obtained at point (d).

5.A Appendix

5.A.1 Derivatives of the Joint Log-Likelihood

5.A.1.1 Rasch Model

The first and second derivatives of $\ell_J(\psi)$ with respect to θ_i may be expressed as

$$\frac{\partial \ell_J(\psi)}{\partial \theta_i} = y_{i.} - \sum_{j=1}^{J} p_j(\theta_i)$$

and

$$\frac{\partial^2 \ell_J(\psi)}{\partial \theta_i^2} = -\sum_{j=1}^{J} p_j(\theta_i)\left(1 - p_j(\theta_i)\right),$$

for $i = 1, \ldots, n$, where we recall that

$$p_j(\theta_i) = \frac{e^{\theta_i - \beta_j}}{1 + e^{\theta_i - \beta_j}}.$$

Similarly, for $j = 2, \ldots, J$, the derivatives with respect to β_j may be expressed as

$$\frac{\partial \ell_J(\psi)}{\partial \beta_j} = -\left(y_{.j} - \sum_{i=1}^{n} p_j(\theta_i)\right),$$

and

$$\frac{\partial^2 \ell_J(\psi)}{\partial \beta_j} = -\sum_{i=1}^{n} p_j(\theta_i)\left(1 - p_j(\theta_i)\right).$$

In order to obtain the standard errors for these estimates, we need to compute the inverse of the observed information matrix

$$I(\psi) = \begin{pmatrix} I_{11} & I_{12} \\ I_{21} & I_{22} \end{pmatrix},$$

where

$$I_{11} = \text{diag}\left(\left\{ \sum_{j=1}^{J} p_j(\theta_i)\left(1 - p_j(\theta_i)\right), \ i = 1, \ldots, n \right\}\right),$$

$$I_{22} = \text{diag}\left(\left\{ \sum_{i=1}^{n} p_j(\theta_i)\left(1 - p_j(\theta_i)\right), \ j = 2, \ldots, J \right\}\right),$$

$$I_{12} = \left(\left\{ -p_j(\theta_i)\left(1 - p_j(\theta_i)\right), \ i = 1, \ldots, n, \ j = 2, \ldots, J \right\}\right),$$

$$I_{21} = I_{12}',$$

with diag($\cdot$) being an operator that transforms vector into a diagonal matrix. However, a direct calculation becomes soon unfeasible due to the large number of parameters. An efficient calculation exploits the block structure of the aforementioned matrix. Thus, the inverse has blocks

$$I_{11}^{(-1)} = \left(I_{11} - I_{12}I_{22}^{-1}I_{21}\right)^{-1},$$

$$I_{22}^{(-1)} = \left(I_{22} - I_{21}I_{11}^{-1}I_{12}\right)^{-1},$$

and the standard errors for $\hat{\psi}_J$ are obtained as the square root of the diagonal elements of these matrices; the first matrix is used for the ability parameter estimates collected in $\hat{\theta}_J$, and the second for the difficulty parameter estimates collected in $\hat{\beta}_J^*$.

5.A.1.2 3PL Model

Here we report the derivatives of $\ell_J(\psi)$ with respect to each single parameter of the 3PL model. For this aim, it is convenient to introduce the function

$$p_j^{\dagger}(\theta_i) = \frac{e^{\lambda_j(\theta_i - \beta_j)}}{1 + e^{\lambda_j(\theta_i - \beta_j)}},$$

so that

$$p_j(\theta_i) = \delta_j + (1 - \delta_j)p_j^{\dagger}(\theta_i).$$

Also let

$$a_{ij} = p_j(\theta_i)\left(1 - p_j(\theta_i)\right),$$

$$a_{ij}^\dagger = p_j^\dagger(\theta_i)\left(1 - p_j^\dagger(\theta_i)\right).$$

Then, the derivative with respect to each parameter θ_i is given by

$$\frac{\partial \ell_J(\psi)}{\partial \theta_i} = \sum_{j=1}^{J} \left(y_{ij} - p_j(\theta_i)\right)\frac{a_{ij}^\dagger}{a_{ij}}(1 - \delta_j)\lambda_j,$$

whereas the derivatives with respect to each item parameter λ_j, β_j, and δ_j are given by

$$\frac{\partial \ell_J(\psi)}{\partial \lambda_j} = \sum_{i=1}^{n} \left(y_{ij} - p_j(\theta_i)\right)\frac{a_{ij}^\dagger}{a_{ij}}(1 - \delta_j)(\theta_i - \beta_j),$$

$$\frac{\partial \ell_J(\psi)}{\partial \beta_j} = -\sum_{i=1}^{n} \left(y_{ij} - p_j(\theta_i)\right)\frac{a_{ij}^\dagger}{a_{ij}}(1 - \delta_j)\lambda_j,$$

$$\frac{\partial \ell_J(\psi)}{\partial \delta_j} = \sum_{i=1}^{n} \left(y_{ij} - p_j(\theta_i)\right)\frac{1 - p_j^\dagger(\theta_i)}{a_{ij}}.$$

The Fisher information matrix may be simply constructed by exploiting the fact that $E(y_{ij}) - p_j(\theta_i) = 0$ and then its element corresponding to the second derivative with respect to θ_i may be obtained as

$$-E\left(\frac{\partial^2 \ell_J(\psi)}{\partial \theta_i^2}\right) = \sum_{j=1}^{J} \frac{\partial p_j(\theta_i)}{\partial \theta_i}\frac{a_{ij}^\dagger}{a_{ij}}(1 - \delta_j)\lambda_j = \sum_{j=1}^{J} \frac{\left(a_{ij}^\dagger\right)^2}{a_{ij}}(1 - \delta_j)^2\lambda_j^2.$$

Similarly, concerning the elements corresponding to the second derivatives with respect to the item parameters, we have

$$-E\left(\frac{\partial^2 \ell_J(\psi)}{\partial \lambda_j^2}\right) = \sum_{i=1}^{n} \frac{\left(a_{ij}^\dagger\right)^2}{a_{ij}}(1 - \delta_j)^2(\theta_i - \beta_j)^2,$$

$$-E\left(\frac{\partial^2 \ell_J(\psi)}{\partial \beta_j^2}\right) = \sum_{i=1}^{n} \frac{\left(a_{ij}^\dagger\right)^2}{a_{ij}}(1 - \delta_j)^2\lambda_j^2,$$

$$-E\left(\frac{\partial^2 \ell_J(\psi)}{\partial \delta_j^2}\right) = \sum_{i=1}^{n} \frac{\left(1 - p_j^\dagger(\theta_i)\right)^2}{a_{ij}}.$$

We also have

$$-E\left(\frac{\partial^2 \ell_J(\psi)}{\partial\theta_i\partial\lambda_j}\right) = \frac{\left(a_{ij}^\dagger\right)^2}{a_{ij}}(1-\delta_j)^2\lambda_j(\theta_i-\beta_j),$$

$$-E\left(\frac{\partial^2 \ell_J(\psi)}{\partial\theta_i\partial\beta_j}\right) = -\frac{\left(a_{ij}^\dagger\right)^2}{a_{ij}}(1-\delta_j)^2\lambda_j^2,$$

$$-E\left(\frac{\partial^2 \ell_J(\psi)}{\partial\theta_i\partial\delta_j}\right) = \frac{a_{ij}^\dagger}{a_{ij}}(1-\delta_j)\lambda_j\left(1-p_j^\dagger(\theta_i)\right),$$

$$-E\left(\frac{\partial^2 \ell_J(\psi)}{\partial\lambda_j\partial\beta_j}\right) = -\sum_{i=1}^n\frac{\left(a_{ij}^\dagger\right)^2}{a_{ij}}(1-\delta_j)^2\lambda_j(\theta_i-\beta_j),$$

$$-E\left(\frac{\partial^2 \ell_J(\psi)}{\partial\lambda_j\partial\delta_j}\right) = \sum_{i=1}^n\frac{a_{ij}^\dagger}{a_{ij}}(1-\delta_j)\lambda_j(\theta_i-\beta_j)\left(1-p_j^\dagger(\theta_i)\right),$$

$$-E\left(\frac{\partial^2 \ell_J(\psi)}{\partial\beta_j\partial\delta_j}\right) = -\sum_{i=1}^n\frac{a_{ij}^\dagger}{a_{ij}}(1-\delta_j)\lambda_j\left(1-p_j^\dagger(\theta_i)\right).$$

All the other elements of the information matrix are equal to 0.

Taking into account the identifiability constraints, the information matrix has dimension $[n + 2(J-1) + J] \times [n + 2(J-1) + J]$ under the 3PL model, which reduces to $[n + 2(J-1)] \times [n + 2(J-1)]$ under the 2PL model, when $\delta_j = 0$ for all j. Its diagonal elements are used within the estimation algorithm to update each parameter. At the end, the overall matrix is used to compute the standard errors as usual.

5.A.2 Conditional Log-Likelihood and Its Derivatives

First consider that, since each term ω_i depends on the subject i only through the score $y_{i\cdot}$, we can express the conditional log-likelihood in (5.4) as

$$\ell_C(\psi) = -(y^*)'\beta^* - \sum_{r=1}^{J-1} n_r \log \omega_r(\exp(-\beta)),$$

where $y^* = \left(y_{\cdot 2}^*, \ldots, y_{\cdot J}^*\right)'$, n_r is the frequency of subjects with score r, and

$$\omega_r(e) = \sum_{y\in\mathcal{Y}(r)} e_j^{y_j}, \tag{5.19}$$

which is known as an *elementary symmetric function*. Note that vector β has the first element equal to 0 and all the other elements are equal to the corresponding elements of β^*. Then, the first and second derivatives of $\ell_C(\psi)$ may be expressed as

$$\frac{\partial \ell_C(\psi)}{\partial \psi} = -y^* - \sum_{r=1}^{J-1} \frac{n_r}{w_r(\exp(-\beta))} \frac{\partial w_r(\exp(-\beta))}{\partial \beta^*},$$

$$\frac{\partial^2 \ell_C(\psi)}{\partial \psi \partial \psi'} = \sum_{r=1}^{J-1} n_r \left[\frac{1}{(\psi_r^*)^2} \frac{\partial w_r(\exp(-\beta))}{\partial \beta^*} \frac{\partial w_r(\exp(-\beta))}{\partial \beta^{*\prime}} \right.$$
$$\left. - \frac{1}{w_r(\exp(-\beta))} \frac{\partial^2 w_r(\exp(-\beta))}{\partial \beta^* \partial \beta^{*\prime}} \right],$$

where the derivatives of $w_r(\exp(-\beta))$ with respect to β^* are obtained by removing the appropriate elements from the first derivative vector

$$\frac{\partial w_r(\exp(-\beta))}{\partial \beta} = -\text{diag}(e) \frac{\partial w_r(e)}{\partial e}$$

and the second derivative matrix

$$\frac{\partial^2 w_r(\exp(-\beta))}{\partial \beta \partial \beta'} = \text{diag}(e) \frac{\partial^2 w_r(\exp(-\beta))}{\partial e \partial e'} \text{diag}(e) + \text{diag}(e) \text{diag}\left(\frac{\partial w_r(e)}{\partial e}\right),$$

both evaluated at $e = \exp(-\beta)$.

The main advantage of the present formulation is that the elementary symmetric function and its derivatives may be computed through a simple recursion even for very large values of J; for further details, see Formann (1986), Gustafson (1980), and Liou (1994). In fact, for $j = 1, \ldots, J$ and $r = 1, \ldots, J-1$, we have

$$w_r(e) = w_r(e_{-j}) + e_j w_{r-1}(e_{-j}), \tag{5.20}$$

where e_{-j} is the vector e without its jth element, so that $w_r(e_{-j})$ is computed by a sum of type (5.19) extended to vectors y of dimension $J-1$, with $w_r(e_{-j})$ similarly defined. We also have

$$\frac{\partial w_r(e)}{\partial e_j} = w_{r-1}(e_{-j}) \tag{5.21}$$

and

$$\frac{\partial^2 w_r(e)}{\partial e_{j_1} \partial e_{j_2}} = \begin{cases} w_{r-2}(e_{-(j_1,j_2)}) & \text{if } j_1 \neq j_2 \text{ and } r = 2,\ldots,J, \\ 0 & \text{otherwise,} \end{cases}$$

where $e_{-(j_1,j_2)}$ is the vector e without the elements j_1 and j_2. A more convenient expression for the second derivative is based on the following result:

$$w_{r-2}\left(e_{-(j_1,j_2)}\right) = \frac{w_{r-1}\left(e_{-j_1}\right) - w_{r-1}\left(x_{-j_2}\right)}{e_{j_2} - e_{j_1}}. \tag{5.22}$$

On the basis of (5.20), Gustafson (1980) proposed the *difference algorithm* to compute $w_r(e)$. It consists in computing $w(e_{-j},0) = 1$, for $j = 1,\ldots,J$, and $w(e,1) = \sum_{j=1}^J e_j$ and then performing the following operations for $j = 1,\ldots,J-1$:

$$w_r(e_{-j}) = w_r(e) - e_j w_{r-1}(e_{-j}), \quad j = 1,\ldots,J,$$

$$w_{r+1}(e) = \frac{1}{r+1} \sum_{j=1}^J e_j w_r(e_{-j}).$$

The first and second derivatives may then be obtained on the basis of (5.21) and (5.22), paying attention to applying the latter to the cases in which $e_{j_1} = e_{j_2}$.

5.A.3 Derivatives for the MML Method

We have

$$\frac{\partial \hat{\ell}_M^*(\psi)}{\partial \lambda_j} = \sum_{i=1}^n \sum_{v=1}^k \hat{z}_{iv}\left(y_{ij} - p_j(\xi_v')\right)\frac{p_j^\dagger(\xi_v')}{p_j(\xi_v')}(\xi_v' - \beta_j),$$

$$\frac{\partial \hat{\ell}_M^*(\psi)}{\partial \beta_j} = -\lambda_j \sum_{i=1}^n \sum_{v=1}^k \hat{z}_{iv}(y_{ij} - p_j(\xi_v'))\frac{p_j^\dagger(\xi_v')}{p_j(\xi_v')},$$

$$\frac{\partial \hat{\ell}_M^*(\psi)}{\partial \delta_j} = \frac{1}{1-\delta_j} \sum_{i=1}^n \sum_{v=1}^k \hat{z}_{iv}(y_{ij} - p_j(\xi_v'))\frac{1}{p_j^\dagger(\xi_v')}.$$

The corresponding elements of the information matrix are

$$-E\left(\frac{\partial^2 \hat{\ell}_M^*(\psi)}{\partial \lambda_j^2}\right) = (1 - \delta_j) \sum_{i=1}^{n} \sum_{v=1}^{k} \hat{z}_{iv} \frac{p_j^\dagger(\xi_v')^2}{p_j(\xi_v')} \left(1 - p_j^\dagger(\xi_v')\right)(\xi_v' - \beta_j)^2,$$

$$-E\left(\frac{\partial^2 \hat{\ell}_M^*(\psi)}{\partial \beta_j^2}\right) = (1 - \delta_j)\lambda_j^2 \sum_{i=1}^{n} \sum_{v=1}^{k} \hat{z}_{iv} \frac{p_j^\dagger(\xi_v')^2}{p_j(\xi_v')} \left(1 - p_j^\dagger(\xi_v')\right),$$

$$-E\left(\frac{\partial^2 \hat{\ell}_M^*(\psi)}{\partial \delta_j^2}\right) = \frac{1}{1 - \delta_j} \sum_{i=1}^{n} \sum_{v=1}^{k} \hat{z}_{iv} \frac{1 - p_j^\dagger(\xi_v')}{p_j(\xi_v')}.$$

Concerning the parameters μ and σ defining the latent distribution, we have

$$\frac{\partial \hat{\ell}_M^*(\psi)}{\partial \mu} = \sum_{i=1}^{n} \sum_{v=1}^{k} \hat{z}_{iv} \sum_{j=1}^{J} \lambda_j \left(y_{ij} - p_j(\xi_v')\right) \frac{p_j^\dagger(\xi_v')}{p_j(\xi_v')},$$

$$\frac{\partial \hat{\ell}_M^*(\psi)}{\partial \sigma} = \sum_{i=1}^{n} \sum_{v=1}^{k} \hat{z}_{iv}\xi_v \sum_{j=1}^{J} \lambda_j \left(y_{ij} - p_j(\xi_v')\right) \frac{p_j^\dagger(\xi_v')}{p_j(\xi_v')},$$

and

$$-E\left(\frac{\partial^2 \hat{\ell}_M^*(\psi)}{\partial \mu^2}\right) = \sum_{i=1}^{n} \sum_{v=1}^{k} \hat{z}_{iv} \sum_{j=1}^{J} \lambda_j^2 (1 - \delta_j) \frac{p_j^\dagger(\xi_v')^2}{p_j(\xi_v')} \left(1 - p_j^\dagger(\xi_v')\right),$$

$$-E\left(\frac{\partial^2 \hat{\ell}_M^*(\psi)}{\partial \sigma^2}\right) = \sum_{i=1}^{n} \sum_{v=1}^{k} \hat{z}_{iv}\xi_v^2 \sum_{j=1}^{J} \lambda_j^2 (1 - \delta_j) \frac{p_j^\dagger(\xi_v')^2}{p_j(\xi_v')} \left(1 - p_j^\dagger(\xi_v')\right).$$

The maximization of the expected value of the complete data at the M-step has to be made under the usual identifiability constraints, which are $\beta_1 = 0$, for the Rasch model, and $\lambda_1 = 1$ and $\beta_1 = 0$ for the 2PL and 3PL models.

6

Some Extensions of Traditional Item Response Theory Models

6.1 Introduction

In this chapter, we illustrate several extensions of the models discussed in the previous chapters. We refer in particular to item response theory (IRT) models with covariates, to models for multilevel data in which units are collected in separate clusters, to multidimensional IRT models, and to IRT models in the structural equation modeling setting.

As shown in the previous chapters, traditional IRT models assume that the distribution of individual abilities, either continuous or discrete, is the same for all subjects. A first way to extend classical IRT models is by enclosing in the model individual covariates so that the distribution of the ability becomes individual specific. In the presence of individual covariates, a random-effects formulation of IRT models is adopted. Moreover, when the latent trait is assumed to have a continuous distribution, these covariates are typically included by a simple linear model. On the other hand, when the latent trait is assumed to have a discrete distribution, a more complex approach is necessary, which is based on the specification of multinomial or global logit models.

Another way to extend traditional IRT models is by allowing a dependence between the item responses of individuals belonging to the same group or cluster. This latter case is rather common in educational settings where students belong to the same schools. In these situations, one can expect that the latent trait levels of individuals within the same cluster are correlated to one another due to factors related to the schools.

In addition, according to the IRT models presented in previous chapters, the dependence between individual responses is accounted for by only one latent trait. However, questionnaires are often made up of several subsets of items measuring different constructs. When this happens, the assumption of unidimensionality of traditional IRT models may be too restrictive, and it is necessary to explicitly account for the multidimensional structure of the questionnaire. In this way, it is also possible to estimate the correlation between each pair of latent traits.

Traditional IRT models as well as their extensions may be embedded in the common framework of structural equation models (SEMs). Indeed, SEMs represent a wide class of models characterized by a system of multiple equations that accommodate relations among both latent and manifest variables. Traditional IRT models are examples of SEMs involving just one latent variable and a set of manifest variables, which in the present case correspond to test items. Starting from this type of SEMs, the extensions at issue are implemented by adding latent variables (e.g., for multidimensional and multilevel models) and manifest variables (e.g., covariates) and by specifying the reciprocal relations in a suitable way.

This chapter is organized as follows. The following section illustrates extensions of IRT models to account for the presence of covariates, in the case of both continuous and discrete latent distributions. Section 6.3 discusses IRT extensions for clustered and longitudinal data, and Section 6.4 illustrates multidimensional IRT models. Section 6.5 deals with IRT models in the structural equation modeling setting. Finally, examples in `Stata` and `R` are described in Section 6.6.

6.2 Models with Covariates

Under the random-effects formulation, the IRT models illustrated in the previous chapters assume that the distribution of the individual ability, represented by the random variable Θ_i, is the same for all subjects. This distribution may be either continuous or discrete. The random variables $\Theta_1, \ldots, \Theta_n$ are also assumed to be independent and to represent the only explanatory factors of the response variables Y_{ij}. A possible way of relaxing the latter assumption is by including in the model individual covariates, if available, so that the distribution of each Θ_i becomes individual specific, but these latent variables remain independent.

Essentially, there are two ways to enclose individual covariates. In the first formulation, the covariates affect the distribution of the latent variables; whereas in the second, the covariates affect the response variables in addition to the latent variables. Such a direct effect on the item responses represents a possible strategy to analyze differential item functioning, as an alternative to what is described in Section 5.9, to which we refer the reader for details. Here we focus, in particular, on the first extension, which is of interest when, as typically happens in IRT applications, the study aims at understanding how the covariates affect the ability level. Such an extension, based on enclosing individual characteristics to explain the latent variable, is also called the *person explanatory approach*, as opposed to the second extension, also known as the *item explanatory approach* (Wilson and De Boeck, 2004).

In both the cases described earlier, the traditional point of view, based on considering an IRT model as a measurement instrument, is replaced by a more general point of view, which consists in attributing to IRT models a measurement and an explanatory function at the same time. Such a global approach, which directly relates observed item responses with a set of explanatory variables, allows us to solve the problems that arise with a two-step approach, that is, when latent variable estimates, resulting from a measurement model, are used—as if they were known values instead of estimated values—in a second step regression analysis (Goldstein, 1980; Lord, 1984; Zwinderman, 1991; Hoijtink, 1995; Mesbah, 2004). In particular, estimates of regression coefficients and item parameters are more precise when based on such a global approach (Mislevy, 1987; Zwinderman, 1991; Adams et al., 1997a). The advantages of using a global rather than a two-step approach are especially evident when the number of items is small (5–10 items), whereas in the presence of longer tests, the gain in efficiency may be negligible.

In the following, we consider both the case in which the ability is assumed to have a continuous (usually normal) distribution and the case in which it is assumed to have a discrete distribution. In the first case, it is equivalent to assuming that the covariates have a direct effect on the latent variable or on the responses. In the second case, we assume that the effect of the covariates is on the weights of the latent classes, so that a multinomial logit parameterization or similar parameterizations are used.

6.2.1 Continuous Latent Distribution

As an extension of the assumption that the random latent variables Θ_i follow a normal distribution, we can assume the linear model

$$\Theta_i = \kappa_0 + x_i'\kappa_1 + \varepsilon_i, \quad i = 1, \ldots, n, \tag{6.1}$$

where
- x_i denotes the column vector of individual covariates that are considered as fixed and given
- κ_0 is the constant term
- κ_1 is a vector of the regression parameters
- ε_i are independent and identically normal distributed error terms with mean zero and constant variance σ_ε^2, for $i = 1, \ldots, n$

Model (6.1) has to be combined with a suitable measurement model, which relates Θ_i to the observed item responses y_{ij}, $j = 1, \ldots, J$. More precisely, Equation 6.1 may be directly substituted in the item response function of a given IRT model, so that the item responses are directly related to the covariates. For instance, in the case of binary items, if one substitutes Equation 6.1 in Equation 3.4, we obtain the *latent regression Rasch model* (Zwinderman, 1991), under which the manifest distribution of a vector of responses y_i given x_i has the following expression:

$$p_i\left(y_i\right) = \int_{\mathbb{R}} \left\{ \prod_{j=1}^{J} \frac{\exp\left[y_{ij}\left(\kappa_0 + x_i'\kappa_1 + \varepsilon_i - \beta_j\right)\right]}{1 + \exp\left(\kappa_0 + x_i'\kappa_1 + \varepsilon_i - \beta_j\right)} \right\} \phi\left(\varepsilon_i; 0, \sigma_\varepsilon^2\right) d\varepsilon_i, \quad (6.2)$$

where $\phi\left(\varepsilon_i; 0, \sigma_\varepsilon^2\right)$ stands for the density function of the latent distribution with mean 0 and variance σ_ε^2; since the covariates are considered as fixed and given, we omit explicit reference to x_i as conditioning argument in the manifest distribution. The same convention is typically used throughout the chapter. On the basis of this distribution, we can implement a marginal maximum likelihood estimator. In this regard, some restrictions are needed to make the model identifiable. As usual, we can fix $\beta_1 = 0$ or $\sum_{j=1}^{J} \beta_j = 0$, so that the intercept parameter κ_0 is free. It is also worth noting that, for model (6.2), assumptions of unidimensionality, local independence, and monotonicity, illustrated in Section 3.2, are still valid.

The other IRT models described in Chapters 3 and 4 may be extended in a similar way (for specific examples, see, among others, Mislevy, 1987; Zwinderman, 1997; Christensen et al., 2004), being the identifiability constraints on the item discrimination parameters and/or the threshold difficulties still necessary (see Sections 5.2.1 and 5.2.2).

6.2.2 Discrete Latent Distribution

In order to formulate the extended IRT model with covariates affecting the discrete distribution of the ability, we use the notation already introduced in Sections 3.6.2 and 5.4.2. It is important to remember, in particular, that k is the number of support points of the distribution of each latent variable Θ_i, which corresponds to the latent classes of subjects with the same ability level. These support points are denoted by $\xi_1, \ldots, \xi_k$. However, the corresponding weights are now individual specific, as they depend on the available covariates. Similar to the continuous case, we denote by x_i the column vector of covariates corresponding to sample unit i, which are considered as fixed and known. Moreover, we denote by $\pi_{iv} = p(\Theta_i = \xi_v)$ the probability that the ability level of unit i is equal to ξ_v, $v = 1, \ldots, k$. We recall that assuming a discrete distribution for the ability amounts to associate a discrete latent variable V_i to each subject i with support points labeled from 1 to k. This variable defines k latent classes in the population, with subjects in class v having ability level ξ_v.

Different from the continuous case, the effect of the covariates in x_i on the discrete latent trait Θ_i cannot be expressed through a linear model as in Equation 6.1, but a different type of model is needed. A natural approach is based on modeling the dependence of the probabilities π_{iv} on the covariates in x_i

through a logit model (Smit et al., 1999, 2000; Maij-de Meij et al., 2008; Tay et al., 2011). The resulting types of models belong to the class of *concomitant variable latent class models* (Dayton and Macready, 1988; Huang and Bandeen-Roche, 2004; Formann, 2007b). The most general parameterization is based on reference-category logits, as follows:

$$\log \frac{\pi_{iv}}{\pi_{i1}} = \kappa_{0v} + x_i' \kappa_{1v}, \quad i = 1, \dots, n, \ v = 2, \dots, k, \tag{6.3}$$

where

κ_{0v} are intercepts

κ_{1v} are vectors of regression parameters to be estimated as shown in the following

The parameters involved in the multinomial parameterization may be of difficult interpretation when several latent classes are used because there is a set of regression parameters referred to the comparison between each latent class (apart from the first) and the first. In particular, it may be difficult to understand if the ability level tends to increase or to decrease with each covariate in x_i. An alternative and more parsimonious parameterization that makes sense when the ability levels are increasingly ordered, that is,

$$\xi_1 < \cdots < \xi_k,$$

is based on global logits that are formulated as

$$\log \frac{\pi_{iv} + \cdots + \pi_{ik}}{\pi_{i1} + \cdots + \pi_{i,v-1}} = \kappa_{0v} + x_i' \kappa_1, \quad i = 1, \dots, n, \ v = 2, \dots, k. \tag{6.4}$$

In this way, the regression parameters, collected in κ_1, are common to all latent classes and are of simple interpretation. In particular, if one coefficient in κ_1 is positive, this implies that the corresponding covariate in x_i has a positive effect on the ability level: as the covariate increases, the ability level also increases. On the contrary, if this element is negative, then the corresponding covariate has a negative effect on the ability level. Finally, note that the intercepts κ_{0v} are decreasingly ordered in v, as the probability at numerator, corresponding to $p(\Theta_i \geq \xi_v)$, decreases with v.

Regardless of the adopted parameterization of the class probabilities, the manifest distribution of the response variables is now individual specific. The corresponding probability function is

$$p_i(y_i) = \sum_{v=1}^{k} p(y_i | \xi_v) \pi_{iv},$$

where $p(y_i|\xi_v)$ is defined as in Equation 3.1. Similarly, the posterior distribution of Θ_i is now defined as

$$p_i\left(\xi_v|y_i\right) = \frac{p\left(y_i|\xi_v\right)\pi_{iv}}{p_i\left(y_i\right)}, \tag{6.5}$$

and may be used to assign each subject i to the latent classes in the usual way.

Based on the manifest distribution derived earlier, the marginal likelihood function has expression

$$L_M(\psi) = \prod_{i=1}^{n} p_i\left(y_i\right),$$

where ψ includes the support points $\xi_1, \ldots, \xi_k$ in addition to the parameters in (6.3) or (6.4) depending on the adopted parameterization. We also recall that the vector ψ includes the subvector of the item parameters, which has the same structure as outlined in the previous chapters according to the model specification. The corresponding log-likelihood function is

$$\ell_M(\psi) = \sum_{i=1}^{n} \log p_i\left(y_i\right).$$

In order to maximize the likelihood function mentioned earlier, or equivalently its logarithm, we can use an expectation-maximization (EM) algorithm implemented in a similar way as in Section 5.4.2. As usual, the algorithm alternates two steps until convergence and is based on the complete-data log-likelihood function that, for binary response variables, is defined as follows:

$$\ell_M^*(\psi) = \sum_{i=1}^{n}\sum_{v=1}^{k} z_{iv} \sum_{j=1}^{J} \left[y_{ij}\log p_j\left(\xi_v\right) + \left(1-y_{ij}\right)\log\left(1-p_j(\xi_v)\right)\right]$$

$$+ \sum_{i=1}^{n}\sum_{v=1}^{k} z_{iv}\log\pi_{iv}.$$

We recall that z_{iv} is an indicator variable equal to 1 if unit i belongs to latent class v and to 0 otherwise.

At the E-step of the EM algorithm, the expected value of $\ell_M^*(\psi)$ is obtained. This expected value is conditional on the observed data, that is, x_i and y_i for $i = 1, \ldots, n$, and the current value of the parameters. This step amounts to computing the posterior probabilities

$$\hat{z}_{iv} = p_i\left(\xi_v|y_i\right), \quad i = 1, \ldots, n, \ v = 1, \ldots, k,$$

according to (6.5).

At the M-step, the function obtained by substituting each indicator variable z_{iv} with its conditional expected value $\hat{z}_{iv}$ is maximized with respect to the model parameters. For the item parameters and the support points $\xi_1, \ldots, \xi_k$, this maximization is performed as for models without covariates. On the other hand, the maximization with respect to the parameters in (6.3) or (6.4), depending on the adopted parameterization, is more complex.

6.3 Models for Clustered and Longitudinal Data

In Section 3.6.1, the Rasch and two parameter logistic (2PL) models were formulated in terms of nonlinear mixed (or multilevel/hierarchical) models, and in Chapter 4 it was pointed out that polytomous models may be included in the same framework. The unifying framework of nonlinear mixed models is especially useful as it allows us to extend and generalize IRT models to more complex data structures, such as multilevel and longitudinal structures.

Indeed, it is far from unusual to deal with item response data collected in a certain number of clusters of individuals (e.g., an ability test submitted to students coming from different schools) or in a certain number of time occasions (e.g., a psychological trait measured repeatedly during a follow-up study). In all these situations, one can expect that latent trait levels of persons within the same cluster (or measured at different time occasions within the same individual) are correlated to one another.

Some naive approaches are often used in the situations described earlier. A first solution consists in ignoring the hierarchical structure of the data and then fitting a pooled IRT model. However, in this case, we cannot disentangle the effect of the different hierarchical levels on the item responses and an inaccurate inference may result, especially in terms of standard errors for the parameter estimates. A second approach, which is sometimes used, consists in formulating a distinct IRT model for each cluster, thus limiting the interest in the relations within such clusters. Clearly, this solution precludes the possibility to compare clusters and ignores the *intraclass correlation* (Snijders and Bosker, 2012). Third, item responses may be summarized at cluster level through suitable statistics, so formulating a model for the summary statistics allows us to study the relations between groups. However, interactions between hierarchical levels are ignored and an *aggregation bias* (also known as an *ecological fallacy*, Robinson, 1950) typically results. A further alternative consists in a two-step approach, which disentangles the measurement phase and the explanatory phase of the analysis. In the measurement phase, the individual levels of the latent trait are predicted through an IRT model, whereas the explanatory phase consists in using these estimated values in a further analysis (e.g., based on a multilevel linear regression model) as fixed and known quantities. In this way, one accounts for the hierarchical structure of the data (units in clusters or repeated measures in individuals), but the variability of the latent variable estimates is completely ignored, so that several problems may arise, such as the bias of regression coefficient estimates (Maier, 2001) and the underestimation of their standard errors (Smit et al., 1999, 2000; Fox, 2005), other than the underestimation of the association between latent traits and covariates (Bolck et al., 2004; Mesbah, 2004).

All the mentioned problems can be overcome by adding hierarchical levels in the IRT models. In such a case, we are dealing with *multilevel (or mixed or hierarchical) IRT models*, where the term multilevel (or mixed or hierarchical) refers to the presence of (at least) one supplemental aggregation level compared to standard IRT models, which are organized on the basis of just two hierarchical levels (item responses within individuals; see Section 3.6.1). As will be clarified in the following sections, the resulting model may be formulated either as a two-level random intercept logit model for multivariate responses or as a three-level random intercept logit model for univariate responses (Vermunt, 2008). In the first case, item responses are considered as a multivariate dependent variable and the hierarchical structure is represented by individuals nested within clusters. In the second case, item responses represent a further hierarchical level (item responses within individuals within groups). In both cases, latent traits involved at individual and cluster levels represent the random intercepts.

In the following, we first illustrate the multilevel IRT models under the assumption of normality of the latent trait and, then, we focus on abilities with discrete distribution. We suppose that every individual belongs to one of H clusters, with each cluster h having dimension n_h. The overall sample size is $n = \sum_{h=1}^{H} n_h$. We also denote by Y_{hij} the response provided by subject i in cluster h to item j. We also have covariates at individual level in the column vectors $x_{hi}^{(1)}$ and covariates at cluster level in the column vectors $x_h^{(2)}$. Consider that the adopted notation is something different from that usually adopted in the multilevel models setting (Goldstein, 2003), where the first subscript denotes the individual units and the second subscript denotes the group units. However, the reversed order of subscripts allows us to accommodate the third subscript j, which refers to the items, coherently with the notation adopted throughout this book.

6.3.1 Continuous Latent Distribution

The approach followed here is due to Kamata (2001) (see also Pastor, 2003), but many other authors have discussed the topic at issue, such as Mislevy and Bock (1989), Adams et al. (1997a), Fox and Glas (2001), Maier (2001), and Fox (2005); see also the general approaches due to Skrondal and Rabe-Hesketh (2004) and De Boeck and Wilson (2004a), which include the hierarchical one as a special case. For some illustrative examples and applications, it is also worth considering, among others, the works of Raudenbush and Sampson

(1999), Cheong and Raudenbush (2000), Maier (2002), Fox (2004), Pastor and Beretvas (2006), Kamata and Cheong (2007), Bacci and Caviezel (2011), and Bacci and Gnaldi (2015).

Relying on Equations 3.12 through 3.14 concerning the Rasch model, an extension to the multilevel case without covariates is obtained as described in the following. As an example, one can assume a hierarchical structure organized on three levels with individuals in clusters, so that individuals represent the second-level units and clusters are the third-level units (and item responses are the first-level units); it is also assumed that a single continuous latent trait is involved.

First, the item model and the person model of Equations 3.12 and 3.13, respectively, are generalized through the new index h, indicating the cluster:

$$\text{logit}\left[p_j\left(\theta_{hi}^{(1)}, \theta_h^{(2)}\right)\right] = \gamma_{hi0} + \gamma_{hi1}1\{j=1\} + \cdots + \gamma_{hiJ}1\{j=J\},$$

$$h = 1, \ldots, H, \ i = 1, \ldots, n, \qquad (6.6)$$

and

$$\begin{cases} \gamma_{hi0} = \gamma_{h00} + \Theta_{hi}^{(1)}, & \Theta_{hi}^{(1)} \sim N\left(0, \sigma_{\theta^{(1)}}^2\right), \\ \gamma_{hi1} = \gamma_{h01}, \\ \vdots \qquad \vdots \\ \gamma_{hiJ} = \gamma_{h0J}, \end{cases} \qquad (6.7)$$

where $\Theta_{hi}^{(1)}$ is the person-level latent variable corresponding to the deviation of latent trait for person i belonging to cluster h from the average value in cluster h, which is $\Theta_h^{(2)}$. Therefore, it is different from Θ_i in a Rasch model, which denotes the deviation of the latent trait for person i from the population's mean ability.

Second, a third-level model (or cluster model) has to be introduced to accommodate parameters $\gamma_{h00}, \gamma_{h01}, \ldots, \gamma_{h0J}$

$$\begin{cases} \gamma_{h00} = \beta_0 + \Theta_h^{(2)}, & \Theta_h^{(2)} \sim N(0, \sigma_{\theta^{(2)}}^2), \ \text{cor}\left(\Theta_{hi}^{(1)}, \Theta_h^{(2)}\right) = 0, \\ \gamma_{h01} = -\beta_1, \\ \vdots \qquad \vdots \\ \gamma_{h0J} = -\beta_J, \end{cases} \qquad (6.8)$$

where $\Theta_h^{(2)}$ corresponds to the deviation of the mean latent trait for cluster h from the population mean latent trait. Moreover, β_0 is the average mean ability and β_j, for $j = 1, \ldots, J$, are interpretable in terms of difficulty of item j.

Alternative to a formulation based on separate models, the complete model results by substituting Equations 6.8 and 6.7 in Equation 6.6, obtaining:

$$\text{logit}\left[p_j\left(\theta_{hi}^{(1)}, \theta_h^{(2)}\right)\right] = \beta_0 - \beta_1 1\{j = 1\} \cdots - \beta_J 1\{j = J\} + \theta_{hi}^{(1)} + \theta_h^{(2)},$$

$$i = 1, \ldots, n,$$

which, under the constraint $\beta_0 = 0$, is equivalent to

$$p_j\left(\theta_{hi}^{(1)}, \theta_h^{(2)}\right) = \frac{e^{\theta_{hi}^{(1)} + \theta_h^{(2)} - \beta_j}}{1 + e^{\theta_{hi}^{(1)} + \theta_h^{(2)} - \beta_j}} \qquad (6.9)$$

for $h = 1, \ldots, H, i = 1, \ldots, n_h$, and $j = 1, \ldots, J$.

The multilevel Rasch model introduced here takes into account the hierarchical structure of the data, and it can be further extended in order to account for person and cluster characteristics (Van den Noortgate and Paek, 2004). Let $x_{hi}^{(1)}$ denote a vector of covariates observed for subject i in cluster h and let $x_h^{(2)}$ be a vector of covariates observed for cluster h. Assuming that $x_{hi}^{(1)}$ and $x_h^{(2)}$ directly contribute to explain $\Theta_{hi}^{(1)}$ and $\Theta_h^{(2)}$, respectively, two latent regression models are introduced for $h = 1, \ldots, H$ and $i = 1, \ldots, n$

$$\begin{cases} \Theta_{hi}^{(1)} = \kappa_0^{(1)} + \left(x_{hi}^{(1)}\right)' \kappa_1^{(1)} + \varepsilon_{hi}^{(1)}, & \varepsilon_{hi}^{(1)} \sim N\left(0, \sigma_{\varepsilon^{(1)}}^2\right), \\ \Theta_h^{(2)} = \kappa_0^{(2)} + \left(x_h^{(2)}\right)' \kappa_1^{(2)} + \varepsilon_h^{(2)}, & \varepsilon_h^{(2)} \sim N\left(0, \sigma_{\varepsilon^{(2)}}^2\right), \end{cases} \qquad (6.10)$$

where

$\kappa_1^{(1)}$ and $\kappa_1^{(2)}$ are vectors of the regression coefficients for $x_{hi}^{(1)}$ and $x_h^{(2)}$, respectively

$\varepsilon_{hi}^{(1)}$ and $\varepsilon_h^{(2)}$ are the corresponding individual- and cluster-level residuals obtained after controlling for the covariates

$\kappa_0^{(1)}$ and $\kappa_0^{(2)}$ are the intercepts, which may be constrained to 0 as in Equations 6.7 and 6.8

Note that random effects $\Theta_{hi}^{(1)}$ and $\Theta_h^{(2)}$ in Equation 6.9 are now substituted by $\varepsilon_{hi}^{(1)}$ and $\varepsilon_h^{(2)}$, respectively, and the corresponding variance components are denoted by $\sigma_{\varepsilon^{(1)}}^2$ and $\sigma_{\varepsilon^{(2)}}^2$ instead of $\sigma_{\theta^{(1)}}^2$ and $\sigma_{\theta^{(2)}}^2$.

The multilevel Rasch model (without and with covariates) described previously can be modified in a suitable way to account for the presence of discriminant parameters and/or polytomously scored items, according to what was described in Sections 3.6.1 and 4.2, respectively. Here, we do not dwell on these issues, rather we discuss some peculiarities that arise when the hierarchical structure of the data is due to repeated measures of the latent trait of the same individuals along the time.

Two main elements have to be taken into account. First, as time occasions are aggregated in individuals (rather than individuals aggregated in clusters), the hierarchical data structure is composed by item responses $j = 1, \ldots, J$ at the first level, time occasions $h = 1, \ldots, H$ at the second level, and individuals $i = 1, \ldots, n$ at the third level. Moreover, it can often be of some interest to add a second-level covariate describing the time effect on the latent trait and/or on the item difficulties and to consider the possibility that the time trend of the latent trait differs among individuals. The inclusion of all these elements in Equations 6.6 through 6.8 allows us to define a multilevel Rasch model for longitudinal data, as shown in detail in the following (for more details and for the extension to polytomous data, see Pastor and Beretvas, 2006). The resulting model is as follows, where subscripts i and h are in reversed order with respect to the case of Equations 6.6 through 6.10:

- First-level model (or item model)

$$\text{logit}\left[p_j\left(\theta_{ih}^{(1)}, \theta_i^{(2)}\right)\right] = \gamma_{ih0} + \gamma_{ih1}1\{j = 1\} + \cdots + \gamma_{ihJ}1\{j = J\},$$
$$h = 1, \ldots, H, \; i = 1, \ldots, n,$$

- Second-level model (or time model) based on the linear trend

$$\begin{cases} \gamma_{ih0} = \gamma_{i00} + \gamma_{i10}h + \varepsilon_{ih}^{(1)}, & \varepsilon_{ih}^{(1)} \sim N\left(0, \sigma_{\varepsilon^{(1)}}^2\right), \\ \gamma_{ih1} = \gamma_{i01} + \gamma_{i11}h, \\ \vdots \qquad \vdots \\ \gamma_{ihJ} = \gamma_{i0J} + \gamma_{i1J}h. \end{cases}$$

- Third-level model (or person model)

$$\begin{cases} \gamma_{i00} = \beta_{00} + \varepsilon_{i0}^{(2)}, \\ \gamma_{i10} = \beta_{10} + \varepsilon_{i1}^{(2)}, \\ \gamma_{i01} = \beta_{01}, \\ \gamma_{i11} = \beta_{11}, \\ \vdots \qquad \vdots \\ \gamma_{i0J} = \beta_{0J}, \\ \gamma_{i1J} = \beta_{1J}, \end{cases}$$

where $\varepsilon_{i0}^{(2)}$ and $\varepsilon_{i1}^{(2)}$ are independent of $\varepsilon_{ih}^{(1)}$ and are bivariate normally distributed with mean $\mathbf{0}$ and variance and covariance matrix

$$
\begin{pmatrix}
\sigma^2_{\varepsilon_0^{(2)}} & \sigma_{\varepsilon_0^{(2)}\varepsilon_1^{(2)}} \\
\sigma_{\varepsilon_1^{(2)}\varepsilon_0^{(2)}} & \sigma^2_{\varepsilon_1^{(2)}}
\end{pmatrix}.
$$

In comparison with the multilevel Rasch model for clustered data, based on Equations 6.6 through 6.8, a higher number of parameters are now involved. The time-level error terms $\varepsilon_{hi}^{(1)}$ correspond to the deviation of the latent trait at time h for person i from the average value of the latent trait for person i, whereas the person-level error terms $\varepsilon_{i0}^{(2)}$ denote the deviation of the latent trait for person i from the population mean latent trait. Besides, the error terms $\varepsilon_{i1}^{(2)}$, which appear in the person model, describe the individual effect of the time on the latent trait: when $\sigma^2_{\varepsilon_1^{(2)}} = 0$, then the time effect on the latent trait is constant among all individuals. The two fixed terms β_{00} and β_{01} represent the average overall latent trait at the initial time occasion (i.e., when the h-th covariate equals 0) and the overall (linear) time effect on the latent trait, respectively.

Concerning the interpretation of the item difficulties, we can distinguish two types of parameter. The first ones are $\beta_{01}, \ldots, \beta_{0J}$ that describe the item difficulties at the first time occasion. The second type of item parameters is given by $\beta_{11}, \ldots, \beta_{1J}$ that refer to the variation of item difficulties along the time. It should be noted that a linear trend is assumed (but other functions of the time may be introduced) and that the time effect is constant among individuals. If all parameters β_{1j}, for $j = 1, \ldots, J$, equal 0, then item difficulties are constant over time.

For details about the estimation of the class of models described in this section (i.e., three-level logit model with random intercepts or with random intercepts and slopes), see, among others, McCulloch and Searle (2001), Goldstein (2003), Skrondal and Rabe-Hesketh (2004), and Snijders and Bosker (2012). For a general discussion about similar models, see Bollen and Curran (2006), and for alternative models based on a hidden Markov formulation, see Bartolucci et al. (2013).

6.3.2 Discrete Latent Distribution

The multilevel IRT model under the assumption of discreteness of latent trait Θ_i is defined in a similar way as in the previous section, with some adjustments due to the different nature of Θ_i.

In agreement with the notation introduced in Section 5.4.2, to each subject i within cluster h, we associate the latent variable $V_{hi}^{(1)}$ having a discrete distribution with k_1 support points, from 1 to k_1. In addition to each cluster h we associate the latent variable $V_h^{(2)}$ having a discrete distribution with

k_2 support points, from 1 to k_2. These support points define k_1 and k_2 latent classes of individuals and clusters, respectively. In particular, individuals in latent class $v^{(1)}$ have an ability level denoted by $\xi_{v^{(1)}}$.

With respect to the weights, we have to consider that the distribution of each latent variable $V_{hi}^{(1)}$ now depends on $V_h^{(2)}$ and on the covariates $x_{hi}^{(1)}$, and then, we introduce the notation

$$\pi_{hi,v^{(1)}|v^{(2)}}^{(1)} = p\left(V_{hi}^{(1)} = v^{(1)}|V_h^{(2)} = v^{(2)}, x_{hi}^{(1)}\right).$$

Such dependence can be formulated according to a multinomial logit parameterization, as follows:

$$\log \frac{\pi_{hi,v^{(1)}|v^{(2)}}^{(1)}}{\pi_{hi,1|v^{(2)}}^{(1)}} = \kappa_{0v^{(1)}v^{(2)}}^{(1)} + \left(x_{hi}^{(1)}\right)' \kappa_{1v^{(1)}v^{(2)}}^{(1)}, \quad v^{(1)} = 2, \ldots, k_1, \ v^{(2)} = 1, \ldots, k^{(2)},$$

$$(6.11)$$

where the elements of vector $\kappa_{1v^{(1)}v^{(2)}}^{(1)}$ denote the effect of individual covariates $x_{hi}^{(1)}$ on the logit of $\pi_{hi,v^{(1)}|v^{(2)}}^{(1)}$ with respect to $\pi_{hi,1|v^{(2)}}^{(1)}$ and $\kappa_{0v^{(1)}v^{(2)}}^{(1)}$ is the intercept specific for subjects of class $v^{(1)}$ that belong to a cluster in class $v^{(2)}$. In a similar way, we can formulate a global logit parameterization instead of the multinomial logit one, which is more parsimonious and easier to interpret (see Section 6.2.2 for details).

The distribution of each latent variable $V_h^{(2)}$ depends on the cluster covariates $x_h^{(2)}$ and, then, we introduce a further weight denoted by

$$\pi_{hv^{(2)}}^{(2)} = p\left(V_h^{(2)} = v^{(2)}|x_h^{(2)}\right).$$

Then, a multinomial logit parameterization similar to (6.11) is adopted

$$\log \frac{\pi_{hv^{(2)}}^{(2)}}{\pi_{h1}^{(2)}} = \kappa_{0v^{(2)}}^{(2)} + \left(x_h^{(2)}\right)' \kappa_{1v^{(2)}}^{(2)}, \quad v^{(2)} = 2, \ldots, k_2,$$

where the elements of vector $\kappa_{1v^{(2)}}^{(2)}$ measure the effect of cluster covariates $x_h^{(2)}$ on the logit of $\pi_{hv^{(2)}}^{(2)}$ with respect to $\pi_{h1}^{(2)}$ and $\kappa_{0v^{(2)}}^{(2)}$ is the intercept specific for clusters within class $v^{(2)}$.

For k_1 and k_2, the parameters of the resulting multilevel latent class IRT (LC-IRT) model may be estimated by maximizing the log-likelihood:

$$\ell_M(\psi) = \sum_{h=1}^{H} \log \sum_{v^{(2)}=1}^{k_2} \pi_{hv^{(2)}}^{(2)} p_h\left(v^{(2)}\right),$$

where ψ includes all the fixed regression parameters $\kappa^{(1)}_{0v^{(1)}v^{(2)}}$, $\kappa^{(1)}_{1v^{(1)}v^{(2)}}$, $\kappa^{(2)}_{0v^{(2)}}$, and $\kappa^{(2)}_{1u}$, other than the support points $\xi_{v^{(1)}}$, $v^{(1)} = 1, \ldots, k_1$, and to the item parameters coherently with the IRT-adopted parameterization. Moreover, we have

$$
\rho_h\left(v^{(2)}\right) = \prod_{i=1}^{n_h} \sum_{v^{(1)}=1}^{k_1} \pi^{(1)}_{hi,v^{(1)}|v^{(2)}} \prod_{j=1}^{J} p\left(y_{hij}|V^{(1)}_{hi} = v^{(1)}\right),
$$

with $p(y_{hij}|V^{(1)}_{hi} = v^{(1)})$ defined according to one of the IRT models described in Chapters 3 and 4.

In order to maximize the log-likelihood, we make use of an EM algorithm developed along the same lines as described in Sections 5.4.2 and 6.2.2; for a detailed description see Gnaldi et al. (2015).

6.4 Multidimensional Models

According to the IRT approach presented in the previous chapters, the associations between the responses of a subject are entirely accounted for by only one latent trait Θ_i. The characterization of the latent person space in terms of a single unidimensional ability means that all items are located on the same scale, contributing to measure the same latent trait. However, a questionnaire is often composed by some subsets of items measuring different but potentially related constructs. In such a case, the traditional IRT assumption of only one underlying latent variable may be too restrictive. In fact, the unidimensional approach has the clear disadvantage of ignoring the differential information about individual ability levels relative to several latent traits, which are *confused* in the same measurement. For a thorough analysis on the consequences of applying unidimensional models to multidimensional data, see, among others, Embretson (1991), Camilli (1992), and Luecht and Miller (1992).

In order to avoid problems related to the violation of unidimensionality, a *consecutive approach* (Briggs and Wilson, 2003) is commonly adopted, which consists of modeling each latent trait independently of the others. Therefore, the consecutive approach consists in formulating a specific unidimensional IRT model for each ability. It has the advantage of providing person ability estimates and standard errors for each latent variable; however, the possibility that these latent variables are related is ignored. On the other hand, this aspect is accounted for in the *multidimensional approach*, which encloses the correlation between the latent variables.

In a multidimensional IRT model, each latent trait is assumed to have a direct influence on the responses to a certain set of items and also an indirect influence on the responses to other items. Hence, the main advantage of using the multidimensional approach is that the structure of the questionnaire is explicitly taken into account, so that estimates for the correlation between the latent traits are provided and more accurate parameter estimates are obtained. Several authors have dealt with multidimensional extensions of traditional IRT models. Among the main contributions, we remind the reader the works of Duncan and Stenbeck (1987), Agresti (1993), Kelderman and Rijkes (1994), and Kelderman (1997), who proposed a number of examples of loglinear multidimensional IRT models, Muraki and Carlson (1995) for a multidimensional normal ogive graded response model (GRM), Kelderman (1996) for a multidimensional version of the partial credit model, Rijmen and Briggs (2004) for a multidimensional Rasch model, and Yao and Schwarz (2006) for a multidimensional generalized partial credit model. In addition, for a wide and thorough overview of the topic at issue, we refer the reader to Reckase (2009). Finally, for an in-depth illustration of models based on a discrete latent variable formulation, see Bartolucci (2007) and Bacci et al. (2014).

Let D denote the number of abilities of the model or, equivalently, the *dimension* of the model. The random-effect formulation assumes the existence of a vector of latent variables of dimension D for each subject i, denoted by $\Theta_i = (\Theta_{i1}, \ldots, \Theta_{iD})'$. The main issue is how to relate the vector of abilities Θ_i to the distribution of the response variables Y_{ij}. Here, we focus on the so-called *between-item multidimensional* approach (Adams et al., 1997b; Wang et al., 1997), in which each item contributes to measure a single latent trait. An alternative and more general approach, which is here ignored, is the *within-item multidimensional* one (Adams et al., 1997b; Wang et al., 1997), in which each item can depend on more than one dimension.

For any set or subset $\mathcal{J}_d$ of items measuring a distinct ability ($d = 1, \ldots, D$), a standard IRT parameterization of the type illustrated in the previous chapters is assumed, which now depends on a specific ability. With binary responses, the item characteristic curve is indicated by

$$p_j(\theta_i) = p\left(Y_{ij} = 1 | \theta_i\right),$$

where θ_i denotes a realization of Θ_i with elements θ_{id}, $d = 1, \ldots, D$. For instance, under a two-parameter logistic (2PL) parameterization, we extend model (3.8) as follows:

$$\log \frac{p_j(\theta_i)}{1 - p_j(\theta_i)} = \lambda_j \left(\sum_{d=1}^{D} 1\{j \in \mathcal{J}_d\}\theta_{id} - \beta_j \right), \quad j = 1, \ldots, J.$$

Similarly, with polytomous items, the multidimensional extension is based on a category boundary curve (Section 4.2) denoted by

$$p_{jy}^{*}(\Theta_i) = p\left(Y_{ij} \geq y | \Theta_i\right), \quad y = 1, \ldots, l_j - 1.$$

For instance, under a GRM parameterization we have

$$\log \frac{p_{jy}^{*}(\Theta_i)}{1 - p_{jy}^{*}(\Theta_i)} = \lambda_j \left(\sum_{d=1}^{D} 1\{j \in \mathcal{J}_d\}\theta_{id} - \beta_{jy} \right), \quad j = 1, \ldots, J, \ y = 1, \ldots, l_j - 1,$$

which extends the traditional GRM of Equation 4.6 to the multidimensional setting. We refer the reader to Bacci et al. (2014) for details about other possible parameterizations.

The local independence assumption implies that the joint distribution of the response vector y_i, conditional on Θ_i, is

$$p\left(y_i | \Theta_i\right) = \prod_{j=1}^{J} p_{jy_{ij}}(\Theta_i),$$

where $p_{jy}(\Theta_i) = p(Y_{ij} = y | \Theta_i)$ is obtained as

$$p_{jy}^{*}(\Theta_i) - p_{j,y+1}^{*}(\Theta_i),$$

according to Equation 4.4. With binary response variables, $p(y_i | \Theta_i)$ becomes (see Equation 3.1 for the unidimensional case)

$$p\left(y_i | \Theta_i\right) = \prod_{j=1}^{J} p_j(\Theta_i)^{y_{ij}} \left[1 - p_j(\Theta_i)\right]^{1 - y_{ij}}.$$

Finally, adopting a continuous approach, each Θ_i has a multivariate normal distribution with mean $\mathbf{0}$ and variance and covariance matrix

$$\begin{pmatrix} \sigma_{\theta_1}^{2} & \sigma_{\theta_1\theta_2} & \cdots & \sigma_{\theta_1\theta_D} \\ \sigma_{\theta_2\theta_1} & \sigma_{\theta_2}^{2} & \cdots & \sigma_{\theta_2\theta_D} \\ \vdots & \vdots & \ddots & \vdots \\ \sigma_{\theta_D\theta_1} & \sigma_{\theta_D\theta_2} & \cdots & \sigma_{\theta_D}^{2} \end{pmatrix}$$

with elements in the main diagonal denoting the variances of each latent trait and off-diagonal elements being pairwise covariances; see Christensen et al. (2002). Adopting a discrete approach, we assume that the distribution of each

Θ_i has k support points denoted by $\xi_1, \ldots, \xi_k$ and corresponding probabilities $\pi_v = p(\Theta_i = \xi_v)$; see Bartolucci (2007).

Both continuous and discrete approaches may enclose individual covariates that affect the latent traits, along the same lines as those discussed in Sections 6.2.1 and 6.2.2. Similarly, the previous assumptions are compatible with a multilevel structure in which suitable latent variables are used to model the effect of each cluster.

The estimation of multidimensional models is based on algorithms rather similar to those for unidimensional IRT models (Chapter 5); see Skrondal and Rabe-Hesketh (2004) and Reckase (2009) for details under the normal assumption and Bartolucci (2007) and Bacci et al. (2014) for the discrete case.

6.5 Structural Equation Modeling Setting

To conclude the methodological part of this chapter, and then of the book, it is interesting to provide more details on the strong relation between IRT models and SEMs (Wright, 1921; Goldberger, 1972; Duncan, 1975; Lei and Wu, 2007; Bollen et al., 2008; Kline, 2011; Hoyle, 2012). In the following, we first describe the formulation of a very general SEM, and then we illustrate how some of the previously described IRT models may be obtained as special cases.

The motivation of SEMs is to handle latent variables and to model their reciprocal relationships. For this aim, SEMs are typically characterized by a system of multiple equations, where each equation incorporates one or more explanatory latent variables that are assumed to influence a dependent endogenous latent variable. The explanatory variables can be exogenous (i.e., predetermined) or endogenous. In addition, observed variables are usually part of the formulation so that the measurement error may be suitably modeled. The model component that specifies the relations between endogenous and exogenous latent variables is named *structural model*, whereas the *measurement model* links each latent variable to the observed covariates. In the classic SEM, whose best known formulation is the *LInear Structural RELations* (LISREL) model (Jöreskog, 1973, 1977), both latent and observed variables are continuous. The first ones are usually known as common factors (as in the factor analysis), whereas the second ones are the so-called indicators.

Let Θ_i and Ξ_i be the vectors of endogenous and exogenous continuous (usually normal) latent variables for subject $i = 1, \ldots, n$. Also, let y_i and x_i be the vectors of the observed continuous indicators of Θ_i and Ξ_i, respectively. Then, the structural model is formulated as follows:

$$\Theta_i = \kappa_0 + K_\Theta \Theta_i + K_\Xi \Xi_i + \varepsilon_i, \tag{6.12}$$

where κ_0 is the vector of intercepts, K_Θ is the matrix of regression coefficients for the effects of the endogenous variables Θ_i on each other, and K_Ξ is the matrix of regression coefficients for the effects of the latent exogenous variables Ξ_i on the latent endogenous variables Θ_i. Moreover, ε_i is the vector of error terms. Based on the standard assumptions of SEMs, Θ_i and ε_i are normally distributed (with mean 0 to ensure model identifiability) and ε_i and Ξ_i are uncorrelated. The measurement model is formulated as

$$Y_i = \beta_y + \Lambda_y \Theta_i + \varepsilon_{yi}$$
$$X_i = \beta_x + \Lambda_x \Xi_i + \varepsilon_{xi},$$

(6.13)

with β_y and β_x denoting the intercepts and Λ_y and Λ_x denoting the matrices of regression coefficients or *factor loadings* of latent variables Θ_i and Ξ_i on Y_i and X_i, respectively. Besides, ε_{yi} and ε_{xi} are the two error components, also called *unique factors*, and they have mean 0 and are uncorrelated with each other and with Θ_i and Ξ_i.

An important special case of SEM is the *Multiple Indicator Multiple Cause* (MIMIC) model (Zellner, 1970; Hauser and Goldberger, 1971), where the latent variables are regressed on observed covariates and there are no regressions among the latent variables. The measurement model (6.13) for x_i is simplified to $X_i = \Xi_i$ and the structural model follows from Equation 6.12 as

$$\Theta_i = \kappa_0 + K_\Theta \Theta_i + K_1 X_i + \varepsilon_i,$$

(6.14)

where K_Ξ is now denoted by K_1, for coherence with the notation adopted in Section 6.2. Note that in the MIMIC model, the observed covariates in X_i have a direct effect on the endogenous latent variables in Θ_i.

Several generalizations of the LISREL model have been proposed to account for a range of possible extensions. A wide class of models that enclose all the main extensions are provided by the *Generalized Linear Latent and Mixed Model* (GLLAMM), introduced by Skrondal and Rabe-Hesketh (2004) (see also Rabe-Hesketh and Skrondal, 2007) and implemented in the gllamm module of Stata.

One of the main relevant streams of research concerns the treatment of categorical observable variables. Among the first developments, we recall the works of Muthén (1983, 1984); more recently, SEMs with categorical indicators, more properly named items (as in the IRT), are discussed in Edwards et al. (2012) for the dichotomous case and in Bovaird and Koziol (2012) for the ordinal polytomous case. The most common method, used in SEMs to account for the nonlinear relationship between observable categorical responses and latent variables, consists of replacing the observed variables Y_i with their underlying latent continuous (usually normal) responses Y_i^*, as described by Equation 3.5 for the dichotomous case. The relationship between Y_i and Y_i^* can be formalized for any number of categories so that

$$Y_{ij} = y \quad \text{if} \quad v_{jy} < Y_{ij}^* < v_{j,y+1}, \quad y = 0, \ldots, l_j - 1,$$

where v_{jy} denotes the cut point between consecutive item response categories, with $v_{j0} = -\infty$ and $v_{jl_j} = +\infty$. To ensure model identifiability, one cut point has to be constrained to a fixed value, usually $v_{j1} = 0$.

If we assume a single latent variable $\Theta_i = \Theta_i$ and no variable of type Ξ_i and X_i, then the structural model (6.12) is eliminated and the measurement model (6.13) reduces to

$$Y_i^* = \beta_y + \lambda_y \Theta_i + \varepsilon_{yi}, \tag{6.15}$$

where Λ_y becomes the J-dimensional column vector λ_y and Y_i^* is defined as previously specified. If we assume that errors in vector ε_{yi} have a standard logistic distribution and $l_j = 2$ for $j = 1, \ldots, J$, then a SEM based on (6.15) is the same as a 2PL model (Section 3.4), with elements in β_y having an opposite sign. Besides, the Rasch model is obtained if all elements of λ_y are constrained to 1. Along the same lines, the other IRT models also described in the previous chapters may be formulated in terms of SEMs.

The SEM formulation allows us to easily enclose a vector of observed covariates affecting the latent variable Θ_i, as described in Section 6.2.1. In addition to the measurement model (6.15), we specify the structural model (6.14) as follows:

$$\Theta_i = \kappa_0 + x_i' \kappa_1 + \varepsilon_i,$$

where intercept κ_0 becomes the scalar κ_0 and matrix K_1 becomes the vector κ_1. Note that such a structural model is the same as the latent regression model (6.1).

More in general, a multidimensional IRT model (see Section 6.4 for details) with covariates affecting the latent trait is formulated as follows:

$$\Theta_i = \kappa_0 + K_1 X_i + \varepsilon_i \quad \text{(structural model)},$$
$$Y_i^* = \beta_y + \Lambda_y \Theta_i + \varepsilon_{yi} \quad \text{(measurement model)}.$$

Another interesting aspect is the possibility of encompassing clustered data in the SEM setting (Bollen et al., 2008; Rabe-Hesketh et al., 2012). The class of two-level IRT models with covariates presented in Section 6.3.1 may be extended to the multidimensional case and formulated in terms of a *multilevel SEM* as follows:

$$\Theta = \kappa_0 + K_\Theta \Theta + K_1 X + \varepsilon \quad \text{(multilevel structural model)},$$

$$Y_{hi}^* = \beta_y + \Lambda_y \Theta_{hi}^{(1)} + \varepsilon_{yhi} \quad \text{(multilevel measurement model)},$$

where the vector Θ of latent variables is decomposed in a vector $\Theta_{hi}^{(1)}$ containing the latent variables at individual level and in another vector for the latent variables at cluster level. In a similar way, we may decompose the residual vector ε and the covariate vector X in a pair of vectors to account for individual- and cluster-level residuals. In addition, K_Θ is a block upper triangular matrix, where the blocks correspond to the elements of Θ varying at a given level, as the latent variables at lower level are regressed on the latent variables at higher level, but the opposite is not possible. Finally, K_1 denotes a block matrix of regression coefficients of individual- and cluster-level covariates whose observed values are contained in vector X. Note that the elements of K_1 coincide with those of vectors $\kappa_1^{(1)}$ and $\kappa_1^{(2)}$ of Equation 6.10, in the presence of a single latent variable at individual (and cluster) level.

To conclude, all the aforementioned models may also be formulated assuming categorical rather than continuous latent variables. For this, the class of *mixture SEMs* represents the natural extension of classic SEMs, characterized by categorical items and categorical latent variables. For details about this wide class of models, we refer the reader to Jedidi et al. (1997), Dolan and van der Maas (1998), Arminger et al. (1999), Vermunt and Magidson (2005), Bauer and Curran (2004), Bauer (2007), and Dolan (2009). We also recall Bacci and Bartolucci (2015) for an example of a multidimensional LC-IRT model formulated in terms of SEM for the treatment of nonignorable missing item responses.

6.6 Examples

In the following, we illustrate how to estimate in Stata and R the extended IRT models described in this chapter. First, we describe the Stata commands for the estimation of a latent regression 2PL model through the analysis of the dichotomized version of the RLMS data (Section 6.6.1), for the estimation of a multilevel Rasch model through the analysis of the grammar section of the INVALSI data (Section 6.6.2), and for the estimation of a bidimensional graded response model through the analysis of the HADS data (Section 6.6.3). All the proposed analyses are performed by means of function gllamm, under the assumption of normality of the latent traits. We advise the readers that the estimation processes may take a long time.

We also provide two examples in R concerning the estimation of GRM with covariates for the RLMS data (Section 6.6.4) and of a multilevel and multidimensional dichotomous IRT model for the three dimensions of the INVALSI data (Section 6.6.5). In both cases, we assume the discreteness of the latent traits and we perform the analyses through package MultiLCIRT.

6.6.1 RLMS Binary Data: Latent Regression 2PL Model in Stata

In Section 3.8.1, we illustrated how estimating dichotomous IRT models by suitable Stata functions. We considered the dichotomized version of the RLMS data about job satisfaction. Here, we rely on the same data in order to illustrate how including covariates (education, gender, and age) to explain the latent trait; see Section 1.8.2 for a detailed description of the data at issue. The analysis is based on function gllamm.

As usual, the data must be collapsed and reshaped in long form; note that now the reshaping of the data also involves the covariates:

```
. use "RLMS_bin.dta", clear
. gen cons=1
. collapse (sum) wt2=cons, by (education gender age Y1-Y4)
. gen ind=_n
. reshape long Y, i(ind) j(item)

. qui tab item, gen(d)
. forvalues j=1/4{
    generate nd`j'=-d`j'
    }
```

In addition, as gllamm does not accept polytomous qualitative covariates, we generate six dummies, one for each category of education:

```
. qui tab education, gen(ed)
```

The formulation of a latent regression IRT model requires option geqs of function gllamm to be specified. Such an option retrieves a previously defined equation, say f1, which denotes the explanatory variables of the latent trait

```
. eq f1: ed1-ed5 gender age
```

where ed1—ed5 are the dummies referred to the first five categories of the variable education (the last category is taken as reference). Then, the latent regression model is estimated along the same lines illustrated in the examples of previous chapters. For instance, for the 2PL model, we can first estimate the model without covariates to find suitable starting values:

```
. local 11=e(11)
. eq load: nd1 nd2 nd3 nd4
. quietly gllamm Y nd1-nd4, nocons link(logit) fam(bin) i(ind) eqs(load)  w(wt) adapt dots
```

Then, we recall these values through option from in the following latent regression 2PL model, the formulation of which now includes option geqs(f1):

```
* Latent regression 2PL model
. matrix a=e(b)
. gllamm Y nd1-nd4, nocons link(logit) fam(bin) i(ind) eqs(load) geqs(f1) w(wt) from(a) adapt
dots

[...] Output omitted

Regressions of latent variables on covariates
-----------------------------------------------------------------------------

    random effect 1 has 7 covariates:
    ed1: 2.5780788 (2.5252891)
    ed2: 3.7601949 (1.8699718)
    ed3: .49784603 (.57769958)
    ed4: .68138343 (.43528582)
    ed5: .67332745 (.44443921)
    gender: .26921664 (.31440416)
    age: .03953786 (.08444542)
-----------------------------------------------------------------------------
```

Note that covariates may also be directly added in the principal command gllamm, along with dummies nd1–nd4. However, in such a case, we would assume a direct effect of the explanatory variables on the item responses rather than on the latent trait.

In addition to the usual output (here omitted) concerning estimation of the parameters β_j and λ_j, as well as of σ_θ^2 (see Section 3.8.1), for the latent regression model at issue gllamm also shows the estimates of the regression coefficients κ_0 and κ_1 (see Equation 6.1) and the corresponding standard errors (in parentheses) for the explanatory variables.

6.6.2 INVALSI Grammar Data: Multilevel Rasch Model with Covariates in Stata

In Sections 3.8.2 and 5.10.2 we analyzed the INVALSI reduced data through the dichotomous Rasch model and diagnostic tools, respectively. Here, we focus on the grammar section of the complete data (see Section 1.8.1 for a detailed description and Section 2.12.1 for an analysis in the classical test theory framework), which is composed of 10 binary items (denoted by c1–c10). In addition, we consider the nested structure of the data, characterized by students within schools, and the information provided by an individual-level covariate, corresponding to the gender of respondents (gender), and by a school-level covariate, corresponding to the geographical area (dummies area_2, area_3, area_4, and area_5):

```
. use "INVALSI_full.dta", clear
. rename c* Y*
. keep id_school gender area_2 area_3 area_4 area_5 Y1-Y10
```

The most suitable model for such data is a two-level dichotomous Rasch model, which corresponds to a three-level random intercept multivariate logit model with item responses within students and students within schools (see Section 6.3.1 for details). The type of model at issue can be estimated by function gllamm. We first collapse and reshape the data:

```
. gen cons=1
. collapse (sum) wt2=cons, by (id_school gender area_2 area_3 area_4 area_5 Y1-Y10)
. gen ind=_n
. reshape long Y, i(ind) j(item)
. qui tab item, gen(d)
. forvalues j=1/10 {
    generate nd`j'=-d`j'
  }
```

We then estimate a two-level Rasch model without explanatory variables, in order to evaluate the effect of the hierarchical data structure on the variability of the item responses. With respect to a traditional Rasch model, the two-level formulation requires to specify the variable that identifies the highest-level units, that is, id_school, through option i:

```
. * Two-level Rasch model
. sort id_school ind item
. gllamm Y nd2-nd10, i(ind id_school) link(logit) fam(bin) w(wt)  adapt dots

[…] Output omitted

number of level 1 units = 37740
number of level 2 units = 3774
number of level 3 units = 171

Condition Number = 8.9844871

gllamm model

log likelihood = -14190.951
```

```
------------------------------------------------------------------
      Y |    Coef.   Std. Err.     z     P>|z|   [95% Conf. Interval]
--------+---------------------------------------------------------
    nd2 |  1.567659   .0817528   19.18   0.000    1.407427   1.727892
    nd3 |   .9811268   .0852452   11.51   0.000    .8140492   1.148204
    nd4 |  2.366236   .0798491   29.63   0.000    2.209735   2.522737
    nd5 |   .2434562   .0935399    2.60   0.009    .0601213   .4267911
    nd6 |  1.937651    .08053    24.06   0.000    1.779815   2.095487
    nd7 |  1.535851   .0818894   18.76   0.000    1.37535   1.696351
    nd8 |  -.2572915   .1028644   -2.50   0.012   -.4589021   -.055681
    nd9 |  -.6513475   .1131513   -5.76   0.000    -.87312   -.429575
   nd10 |  1.692657   .0812662   20.83   0.000    1.533378   1.851936
   _cons|  3.176355   .1020125   31.14   0.000    2.976414   3.376296
------------------------------------------------------------------

Variances and covariances of random effects
------------------------------------------------------------------
***level 2 (ind)
    var(1): .5045859 (.03714643)

***level 3 (id_school)
    var(1): .87449581 (.11007673)
------------------------------------------------------------------
```

As usual, the output shows the item difficulty estimates, $\hat{\beta}_j$, $j = 2, \ldots, 10$ (column Coef.), and the estimates of the variances of random effects, that is, $\hat{\sigma}^2_{\theta(1)} = 0.505$ and $\hat{\sigma}^2_{\theta(2)} = 0.874$ (Equations (6.7) and (6.8), respectively) with the corresponding standard errors in parentheses. In more detail, we observe that $\hat{\sigma}^2_{\theta(2)}$ is statistically significant, and it contributes to explain a proportion of the total variance of the item responses equal to

$$\frac{0.874}{\pi^2/3 + 0.505 + 0.874} = 0.187,$$

where $\pi^2/3$ denotes the variance of standard logistic distribution at item response level (Snijders and Bosker, 2012).

Parameter estimates of the model without covariates are then used as starting values for the estimation of the model that includes covariates at student and school levels. Covariates are included in a similar way as described in Section 6.6.1, but we need to distinguish individual-level covariates from school-level covariates, specifying two different equations. The second element in each equation name defines which latent variable we refer to; therefore, equation named f1 refers to students' characteristics and equation named f2 refers to schools' characteristics:

```
* Two-level Rasch model with covariates
. matrix a=e(b)
. eq f1: gender
. eq f2: area_2 area_3 area_4 area_5
. gllamm Y nd2-nd10, i(ind id_school) link(logit) fam(bin) geqs(f1 f2) w(wt)  from(a) adapt
 dots

[…] Output omitted

Variances and covariances of random effects
---------------------------------------------------------------------------
***level 2 (ind)
    var(1): .49604004 (.03684619)

***level 3 (id_school)
    var(1): .75678969 (.09617433)

Regressions of latent variables on covariates
---------------------------------------------------------------------------
    random effect 1 has 1 covariates:
    gender: .1742202 (.04103389)
    random effect 2 has 4 covariates:
    area_2: -.25165102 (.26554012)
    area_3: -.00538196 (.24518481)
    area_4: .68495494 (.21775222)
    area_5: .35400985 (.22827129)
---------------------------------------------------------------------------
```

The output concerning item parameters is very similar to what was shown previously. Besides, the first variance component $\hat{\sigma}^2_{\varepsilon(1)}$, instead of $\hat{\sigma}^2_{\theta(1)}$ (see first equation in (6.10)), is substantially unchanged, whereas the second variance component $\hat{\sigma}^2_{\varepsilon(2)}$, instead of $\hat{\sigma}^2_{\theta(2)}$ (see second equation in (6.10)), is slightly smaller. Regression coefficients are also shown, distinguishing between covariates at the student level (gender) and covariates at the school level (area_2-area_5). We can state that female students from schools in the south of Italy have grammar ability levels significantly higher than male students from the schools of northwest Italy.

To conclude the multilevel analysis, it is useful to estimate the levels of ability in grammar for each student and for each school. Through the following command:

```
. gllapred u, u
```

we obtain four new variables: um1 and um2 denote the random intercepts at individual level ($\hat{\varepsilon}^{(1)}_{hi}$) and the random intercepts at school level ($\hat{\varepsilon}^{(2)}_{h}$), respectively, and us1 and us2 are the corresponding standard errors.

By collapsing the data according to variable ind, a new dataset is obtained with one record for each student, so that the empirical distribution of individual random intercepts $\hat{\varepsilon}^{(1)}_{hi}$ may be analyzed through descriptive statistics (see also Figure 6.1a):

```
* describe latent trait at individual level
. collapse u* id_school, by (ind)
. summarize um1, detail
                          (mean) um1
-------------------------------------------------------------
        Percentiles      Smallest
 1%      -1.337733       -2.141185
 5%      -.8500417        -2.045131
10%      -.6503204       -1.880187      Obs               2418
25%      -.3467463       -1.869421      Sum of Wgt.       2418

50%       -.090218                      Mean          -.1066398
                         Largest        Std. Dev.      .4272947
75%       .1800084        .9798151
90%       .4034098        .9961768      Variance       .1825808
95%       .5322558        1.099412      Skewness      -.5942081
99%       .7624114        1.113778      Kurtosis       4.056702

* histogram of random intercepts at student level
. hist um1
```

In addition, collapsing the data again by variable id_school, one record for each school is now provided. The empirical distribution of estimates $\hat{\varepsilon}^{(2)}_{h}$ is shown by the histogram in Figure 6.1b.

```
* describe random intercepts at school level
. collapse u*, by(id_school)
. keep id_school um2 us2
* histogram of random intercepts at school level
. hist um2
```

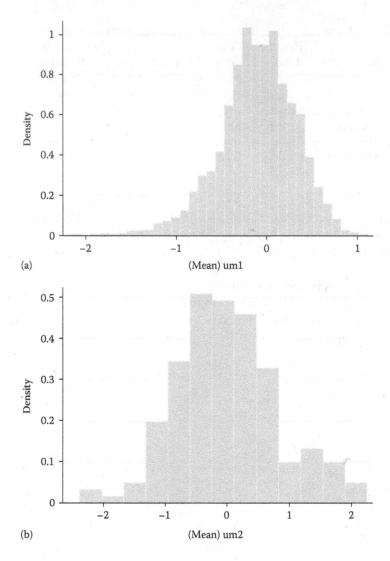

(a)

(b)

FIGURE 6.1
Histograms of random intercept estimates at individual level, $\hat{\varepsilon}_{hi}^{(1)}$ (a), and at school level, $\hat{\varepsilon}_{h}^{(2)}$ (b).

A nice representation of the performance of the schools is given by the so-called caterpillar plot, which allows us to rank the 171 schools as a function of residuals $\hat{\varepsilon}_{h}^{(2)}$ and to perform pairwise comparisons. We first generate a new variable containing the ranks of the schools and, then, we run function `serrbar` as follows:

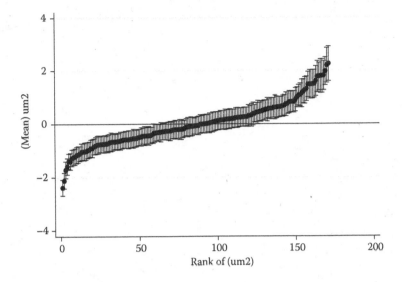

FIGURE 6.2
Random intercept predictions $\hat{\varepsilon}_h^{(2)}$ ordered and endowed with error bars $\left(\pm 1.39 \operatorname{se}\left(\hat{\varepsilon}_h^{(2)}\right)\right)$ for pairwise comparisons between schools.

```
* generate the ranks of the school-level intercept residuals
. egen um2rank = rank(um2)

* caterpillar plot for the school-level random intercepts
. serrbar um2 us2 um2rank, scale(1.39) yline(0)
```

The caterpillar plot in Figure 6.2 consists of a graph that shows on the x-axis the 171 schools ordered by their rank (i.e., variable um2rank) and, on the y-axis, a bar centered on $\hat{\varepsilon}_h^{(2)}$ (i.e., variable um2) whose length is twice 1.39 by the standard error of $\hat{\varepsilon}_h^{(2)}$ (i.e., $2 \cdot 1.39 \cdot$ us2). We remind the reader that, as shown in Goldstein and Healy (1994), two quantities are significantly different at 5% level if and only if their univariate intervals obtained by multiplying the standard errors by 1.39 are disjoint (and not if the confidence intervals at 95% level are disjoint, as a common misconception is). The value 1.39 stems from assumptions of normality, homoscedasticity, and independence of compared quantities, which usually are not satisfied; therefore, the value 1.39 is an approximation and the significance level of 5% has to be understood as an average level.

The caterpillar plot shows that only a few comparisons are significant, due to substantially high standard errors that cause an overlapping of a large number of intervals. However, it is of interest to note the significant difference between schools on the two extreme ends of the plot. The single values of the random intercepts and corresponding schools may be displayed as follows:

```
* the worst 5 schools
. sort um2rank
. list in 1/5
```

```
    +----------------------------------------------+
    | id_sch~1         um2          us2   um2rank |
    |----------------------------------------------|
 1. |      692   -2.380916    .21394149         1 |
 2. |      656   -2.120701     .2099146         2 |
 3. |      997   -1.7136748   .22429723         3 |
 4. |      702   -1.5865829   .22130304         4 |
 5. |      896   -1.3960589   .20543954         5 |
    +----------------------------------------------+
```

```
* the best 5 schools
. list in 167/171
```

```
     +----------------------------------------------+
     | id_sch~1         um2          us2   um2rank |
     |----------------------------------------------|
167. |      269    1.8014514   .45512111       167 |
168. |      105    1.8144193   .39191205       168 |
169. |      841    1.9726434   .37644634       169 |
170. |      349    2.1951311   .47226603       170 |
171. |      313    2.2525741   .46850363       171 |
     +----------------------------------------------+
```

6.6.3 HADS Data: Bidimensional Models in `Stata`

In Section 4.6.1, we illustrated the estimation of several types of models for ordered polytomously scored items, under the assumption of unidimensionality. In particular, we considered the subset of items in the HADS data (Section 1.8.3) that measure depression. Here, we propose a multidimensional analysis, which accounts for both latent traits intended to be measured by HADS, that is, anxiety and depression.

We begin by loading and reshaping the dataset in the usual way but we consider all 14 items:

```
. use "HADS.dta", clear
. gen cons=1
. collapse (sum) wt2=cons, by (Y1-Y14)
. gen ind=_n
. reshape long Y, i(ind) j(item)
. qui tab item, gen(d)
. sort ind
```

The bidimensional version of models illustrated in Section 4.6.1 can be implemented in `Stata` through function `gllamm`. As the main difference with respect to unidimensional models, we need to declare the number of latent traits by using option `nrf`. Moreover, the contribution of every item to each latent trait is specified through option `eqs` on the basis of equations $eq_1 \ldots eq_D$ that are defined before running `gllamm`. Specifically, a rating scale GRM (RS-GRM) with two correlated latent traits is estimated as follows, where equation `load1` specifies that items Y2, Y6, Y7, Y8, Y10, Y11, and Y12 contribute to measure the first latent trait (i.e., anxiety), whereas equation `load2` specifies that the remaining items measure the second latent trait (i.e., depression). Option `nocor` might be introduced in function `gllamm` if we assumed uncorrelated latent traits, which is not the case in this example:

```
* Bidimensional RS-GRM
. eq load1: d2 d6-d8 d10-d12
. eq load2: d1 d3-d5 d9 d13-d14
. gllamm Y d2-d14, i(ind) weight(wt) l(ologit) f(binom)  nrf(2) eqs(load1 load2)  adapt dots

[...] Output omitted

number of level 1 units = 2814
number of level 2 units = 201

Condition Number = 12.900871

gllamm model

log likelihood = -2788.0401
```

response	Coef.	Std. Err.	z	P>|z|	[95% Conf. Interval]	
response						
d2	-.1579996	.2088335	-0.76	0.449	-.5673057	.2513065
d3	.2558534	.2289457	1.12	0.264	-.1928719	.7045787
d4	1.224265	.19901	6.15	0.000	.8342123	1.614317
d5	2.207697	.2012394	10.97	0.000	1.813275	2.602119
d6	.4651402	.2022556	2.30	0.021	.0687265	.861554
d7	.318054	.2028998	1.57	0.117	-.0796223	.7157302
d8	-.029358	.2173985	-0.14	0.893	-.4554513	.3967353
d9	-.9063364	.2288703	-3.96	0.000	-1.354914	-.4577588
d10	.1536432	.2004528	0.77	0.443	-.239237	.5465235
d11	-.3747518	.2127039	-1.76	0.078	-.7916438	.0421401
d12	2.470959	.2054742	12.03	0.000	2.068237	2.873681
d13	-.1557081	.2085806	-0.75	0.455	-.5645186	.2531023
d14	.7982123	.2031209	3.93	0.000	.4001027	1.196322
_cut11						
_cons	-.5952755	.1828742	-3.26	0.001	-.9537024	-.2368486
_cut12						
_cons	2.144171	.1887842	11.36	0.000	1.774161	2.514182
_cut13						
_cons	4.675519	.214515	21.80	0.000	4.255077	5.095961

```
Variances and covariances of random effects
-------------------------------------------------------------------------


***level 2 (ind)

    var(1): 3.3300045 (.76609884)

    loadings for random effect 1
    d2: 1 (fixed)
    d6: .85847637 (.12689189)
    d7: .7088844 (.11338685)
    d8: 1.1576632 (.15914654)
    d10: .68917997 (.10996811)
    d11: 1.0019151 (.14544358)
    d12: .55516287 (.09946501)

    cov(2,1): 2.8577818 (.53753828) cor(2,1): 1
```

```
var(2): 2.4525243 (.59916871)

loadings for random effect 2
d1: 1 (fixed)
d3: 1.58427 (.21852862)
d4: .94854519 (.14127583)
d5: .76556248 (.12491315)
d9: 1.1905761 (.19327513)
d13: 1.0965737 (.1695025)
d14: .86210555 (.13806916)
```

The output concerning the fixed effects is very similar as for the unidimensional model: in particular, we observe that the difficulty estimates for the items related to depression are very similar to those obtained in Section 4.6.1 for the unidimensional RS-GRM (the items in Section 4.6.1 are now renamed). Concerning the random effects, the variances $\sigma^2_{\theta_1}$ and $\sigma^2_{\theta_2}$ of the two latent traits are estimated as well as the correlation $\sigma_{\theta_1\theta_2}$ between them; the corresponding Stata outputs are denoted by var(1), var(2), cov(2,1), respectively. We observe that the very high correlation between anxiety and depression, which arises from the performed analysis, agrees with the results obtained by Bartolucci et al. (2014) on the same dataset under the assumption of discreteness of latent traits.

Estimates of anxiety and depression may be obtained through function gllapred that generates two new variables, um1 and um2, denoting levels of anxiety and depression, respectively, for each individual, and the corresponding standard errors, us1 and us2:

```
. gllapred u, u
. * to display one record for each individual
. collapse u*,by (ind)
. list in 1/10

    +-----------------------------------------------------------+
    | ind        um1          us1          um2          us2 |
    |-----------------------------------------------------------|
 1. |   1   -4.4340656    1.0608946   -3.8052778    .91045084 |
 2. |   2    -3.705636     .94195144   -3.1801456    .80837479 |
 3. |   3   -3.5914413     .93560402   -3.0821447    .80292749 |
 4. |   4   -3.4714904     .90774703   -2.9792039    .77902086 |
 5. |   5   -3.4306958     .87448315   -2.9441942    .75047408 |
    |-----------------------------------------------------------|
 6. |   6   -2.9304389     .79484816    -2.514878     .682132 |
 7. |   7   -3.1297247     .84336593   -2.6859033    .72376954 |
 8. |   8   -3.0514904     .84475413   -2.6187634    .72496088 |
 9. |   9   -2.7933236     .78667534   -2.3972068    .67511815 |
10. |  10   -2.4474273     .72358052   -2.1003615    .62097071 |
    +-----------------------------------------------------------+
```

The other types of bidimensional IRT model are estimated along the same lines, taking into account the peculiarities described in Section 4.6.1 of any of them. For instance, an estimation of models with free item thresholds needs to specify option thresh in function gllamm.

A little more attention is now necessary to implement multidimensional models with constrained item discrimination parameters, because option `constraint` is required. We first use function `constraint define` to fix $\lambda_j = 1$ for each item j (the first item of each dimension may be omitted, because its loading is automatically fixed to 1):

```
constraint define 1 [ind1_11]d6=1
constraint define 2 [ind1_11]d7=1
constraint define 3 [ind1_11]d8=1
constraint define 4 [ind1_11]d10=1
constraint define 5 [ind1_11]d11=1
constraint define 6 [ind1_11]d12=1

constraint define 7 [ind1_21]d3=1
constraint define 8 [ind1_21]d4=1
constraint define 9 [ind1_21]d5=1
constraint define 10 [ind1_21]d9=1
constraint define 11 [ind1_21]d13=1
constraint define 12 [ind1_21]d14=1
```

It should be noted that `ind1_11` is the name assigned by `Stata` to the loading of latent trait 1 at level 1 (i.e., at individual level) and, similarly, `ind1_21` is the name assigned to the loading λ_j of latent trait 2 at level 1. To find out the names needed to correctly specify the constraints, the `gllamm` command must be run with options `noest` and `trace`, such as

```
. gllamm Y d2-d14, i(ind) weight(wt) l(ologit) f(binom)  nrf(2) eqs(load1 load2)  adapt dots
  noest trace
```

The previous constraints are then retrieved in `gllamm` through option `constraint`, specifying the type of constraint. In the case of a doubly constrained GRM, that is, 1P-RS-GRM, the command for the model estimation is given by

```
* Bidimensional 1P-RS-GRM
. gllamm Y d2-d14, i(ind) weight(wt) l(ologit) f(binom)  nrf(2) eqs(load1 load2)
  constraint(1 2 3 4 5 6 7 8 9 10 11 12)  adapt dots
```

which results in the following output for the random effects (the output for fixed effects is here omitted):

```
[…] Output omitted

Variances and covariances of random effects
-----------------------------------------------------------------------
***level 2 (ind)

    var(1): 2.212812 (.30076351)

    loadings for random effect 1
    d2: 1 (fixed)
    d6: 1 (0)
    d7: 1 (0)
    d8: 1 (0)
    d10: 1 (0)
```

```
d11: 1 (0)
d12: 1 (0)

cov(2,1): 2.3610703 (.2880694) cor(2,1): 1

var(2): 2.5192618 (.33593519)

loadings for random effect 2
d1: 1 (fixed)
d3: 1 (0)
d4: 1 (0)
d5: 1 (0)
d9: 1 (0)
d13: 1 (0)
d14: 1 (0)
```

6.6.4 RLMS Data: Models with Covariates in R

In Section 4.6.2, we analyzed the RLMS dataset through polytomous IRT models. Here, we propose an analysis, based on the assumption of discreteness of the latent distribution, which considers some explanatory variables (marital, education, gender, age, and work); see Section 1.8.2 for a detailed description of the data. The analysis is performed through the package MultiLCIRT.

The data are loaded and prepared as follows:

```
> load("RLMS.RData")

# Reverse response categories
> Y = 4-data[ , 6:9]
# Covariates
> X = data[ , 1:5]
```

Note that, different from the analyses performed in Sections 2.12.2 and 4.6.2, we do not remove records now that contain missing responses.

We directly fix the number of latent classes at $k = 3$ and we start from a GRM specification under the multinomial logit parameterization described in Equation 6.3. The model is fitted through function est_multi_poly as follows, where input X is neccessary to estimate the effect of the covariates in this matrix and option output allows us to obtain standard errors for the parameter estimates:

```
> # GRM with covariates
> out1 = est_multi_poly(Y,X=X,k=3,link=1,disc=1,output=TRUE,out_se=TRUE)
Missing data in the dataset, units and items without responses are removed
*-------------------------------------------------------------------------*
Model with multidimensional structure
            [,1] [,2] [,3] [,4]
Dimension 1    1    2    3    4
Link of type =             1
Discrimination index =     1
Constraints on the difficulty = 0
Type of initialization =   0
*-------------------------------------------------------------------------*
```

In order to visualize relevant information about the fitted model and the conditional distribution of the latent classes given the latent trait, we use command summary():

```
> summary(out1)

Call:
est_multi_poly(S = Y, k = 3, X = X, link = 1, disc = 1, output = TRUE)

Log-likelihood:
[1] -7313.1

AIC:
[1] 14692.2

BIC:
[1] 14866.96

Class weights:
[1] 0.3538 0.5311 0.1151

Conditional response probabilities:
, , class = 1

         item
category      1      2      3      4
       0 0.1079 0.1296 0.3418 0.2763
       1 0.3514 0.3471 0.3924 0.4143
       2 0.4488 0.4194 0.1625 0.2120
       3 0.0907 0.1019 0.0939 0.0907
       4 0.0011 0.0019 0.0095 0.0067

, , class = 2

         item
category      1      2      3      4
       0 0.0012 0.0026 0.0722 0.0341
       1 0.0071 0.0132 0.2205 0.1369
       2 0.0806 0.1164 0.2725 0.2903
       3 0.8094 0.7696 0.3747 0.4711
       4 0.1017 0.0982 0.0602 0.0677

, , class = 3

         item
category      1      2      3      4
       0 0.0000 0.0000 0.0073 0.0018
       1 0.0000 0.0001 0.0302 0.0087
       2 0.0003 0.0009 0.0714 0.0317
       3 0.0270 0.0557 0.4857 0.3725
       4 0.9727 0.9433 0.4055 0.5853
```

Results are similar to those shown in Section 4.6.2 with reference to GRM with $k = 3$ latent classes and without covariates. On the basis of the obtained output we conclude that, for all four items, the probabilities of answering with a high response category (denoting a high level of satisfaction) increase from class 1 to class 3, whereas the probabilities of answering with a low response category (denoting a low level of satisfaction) decrease from class 1 to class 3. Consequently, we conclude that the latent classes are ordered according to the job satisfaction level, as also confirmed by the estimated support points $\hat{\xi}_v$ ($v = 1, 2, 3$):

```
> out1$Th
               Class 1   Class 2   Class 3
Dimension 1   2.111929  6.730031  12.48284
```

We also observe that more than one-half (see class weights obtained by command summary()) of the subjects belongs to class 2, which is characterized by an intermediate level of satisfaction. There is 35.81% of subjects in class 1 and they tend to have the lowest level of job satisfaction, whereas the remaining 11.51% of individuals belongs to class 3, which is characterized by the highest level of job satisfaction.

Regarding the item parameters λ_j and β_{jy}, we have:

```
> # Discrimination indices
> out1$gac
[1] 1.0000000 0.8742056 0.4110691 0.5157541
> # Difficulty parameters
> out1$Bec
          [,1]      [,2]     [,3]       [,4]
[1,]  0.0000000  1.949125  4.403137   8.908548
[2,] -0.0661826  2.005466  4.577518   9.266780
[3,]  0.5174592  4.583006  7.367118  13.413814
[4,]  0.2451108  3.668756  6.428818  11.814987
```

The least discriminant and most difficult item is the third one about earnings, followed by the fourth one about opportunities for professional growth.

Finally, the regression parameters κ_{0v} and κ_{1v}, with $v = 2, 3$, for the covariates and the corresponding standard errors are obtained by out1$De and out1$seDe, respectively:

```
>  # Regression coefficients
> out1$De
           logit
                   2            3
 intercept -0.52298646 -0.18414644
 marital   -0.20750239  0.11762631
 education  0.14027079  0.21175586
 gender    -0.06130595 -0.05299865
 age        0.02345263 -0.07965114
 work       0.03388581  0.09087181
> # Standard errors
> out1$seDe
              2          3
[1,] 0.79317682 1.19188948
[2,] 0.13306243 0.20199016
[3,] 0.06465903 0.10017289
[4,] 0.13294886 0.20105779
[5,] 0.03464548 0.05305296
[6,] 0.24831039 0.38074308
> # t-tests
> out1$De/out1$seDe
           logit
                  2           3
 intercept -0.6593567 -0.1544996
 marital   -1.5594364  0.5823368
 education  2.1693923  2.1139038
```

```
gender    -0.4611243 -0.2635991
age        0.6769320 -1.5013513
work       0.1364655  0.2386696
```

Note that the parameterization adopted by default in function est_ multi_poly is the multinomial logit one. It implies that, for each covariate, there are as many estimated regression parameters as latent classes minus 1. More precisely, as the first latent class is the reference class, each regression parameter in column 2 of out1$De detects the effect of the corresponding covariate on the logarithm of the probability of being in class 2 with respect to class 1, whereas each regression parameter in column 3 of out1$De compares the probability of class 3 with that of class 1. Then, to make the interpretation of the covariate effects easier, it is important that the latent classes are increasingly ordered with respect to the latent trait: in such a way, positive (or negative) values of regression parameters denote an increase (or decrease) in the latent trait level, moving from the reference class to the upper class. On the contrary, when the latent classes are not ordered, interpretability problems arise and the parameterization at issue does not make sense. We observe that the most significant covariate is the educational level. Moreover, since the three latent classes of workers are ordered from that with the lowest to that with the highest level of satisfaction and both regression parameters for this covariate are positive (0.1403 and 0.2118), as the educational level increases, the level of job satisfaction also increases.

We now consider the model based on the global logit parameterization specified in Equation 6.4 with the same number of latent classes and GRM formulation of the conditional response probabilities. For this, we have to use the option glob as follows:

```
> out2 = est_multi_poly(Y,X=X,k=3,link=1,disc=1,output=TRUE,out_se=TRUE,glob=TRUE)
Missing data in the dataset, units and items without responses are removed
*------------------------------------------------------------------------------*
Model with multidimensional structure
             [,1] [,2] [,3] [,4]
Dimension 1    1    2    3    4
Link of type =              1
Discrimination index =      1
Constraints on the difficulty = 0
Type of initialization =      0
*------------------------------------------------------------------------------*

> summary(out2)

Call:
est_multi_poly(S = Y, k = 3, X = X, link = 1, disc = 1, output = TRUE,
    glob = TRUE)

Log-likelihood:
[1] -7316.42

AIC:
[1] 14688.85

BIC:
[1] 14837.13

[...] Output omitted
```

The model based on the global logit parameterization has a slightly worse fit with respect to the model based on the multinomial logit parameterization, in terms of log-likelihood (-7316.42 vs -7313.10), but it is more parsimonious. This implies smaller values of the Akaike's information index (AIC) (14688.85 vs 14692.20) and Bayesian information index (BIC) (14837.13 vs 14866.96). The interpretation of the conditional response probabilities, support points, and item parameter estimates is similar (output here omitted), whereas the interpretation of the effect of the covariates is easier, as we now have a single regression parameter for each covariate. In more detail, we have the following results regarding the regression parameters and their standard errors:

```
> # Regression coefficients
> out2$De
                   [,1]
cutoff1      0.53240278
cutoff2     -2.11711160
marital     -0.05027206
education    0.14518340
gender      -0.04565141
age         -0.01972156
work         0.05516895
> # Standard errors
> out2$seDe
                [,1]
[1,]  0.69039312
[2,]  0.69541822
[3,]  0.11451935
[4,]  0.05610286
[5,]  0.11449768
[6,]  0.03012194
[7,]  0.21790125
> # t-tests
> out2$De/out2$seDe
                  [,1]
cutoff1      0.7711589
cutoff2     -3.0443718
marital     -0.4389831
education    2.5878076
gender      -0.3987103
age         -0.6547241
work         0.2531832
```

The only covariate that is clearly significant is the educational level that has a positive effect on the job satisfaction level. The signs of the remaining covariates can be easily interpreted even if these covariates are not significant.

6.6.5 INVALSI Data: Multilevel Multidimensional Models with Covariates in R

Here, we provide an example in R of the estimation of a multilevel and multidimensional dichotomous IRT model for the three dimensions of the INVALSI data (Section 1.8.1), by assuming the discreteness of the latent traits.

In this section, we work with the full sample of 3774 students in 171 schools. The data at issue have a nested structure, characterized by students within schools, and we use an individual-level covariate (gender) and a school-level covariate, corresponding to the geographical area (dummies area_2, area_3, area_4, and area_5).

As usual, we start loading the data, as follows:

```
> load("Invalsi_full.RData")
```

and, then, we define the data structure for the subsequent analyses:

```
# Define a cluster indicator
> clust = as.vector(data[,1])
> clust_unique = unique(clust)
> for(i in 1:length(clust)) clust[i] = which(clust_unique==clust[i])

# Build the matrix for covariates at cluster level
> W0 = as.matrix(data[,3:6])
> W = NULL
> for(cl in 1:max(clust)){
+ W = rbind(W,W0[which(clust==cl)[1],])
+ }

# Build the matrix for covariates at individual level
> X = as.vector(data[,2])

# Item responses matrix
> Y = as.matrix(data[,7:73])
> Y = 1*(Y>0)
> n = nrow(Y)
> r = ncol(Y)
```

In addition, as the model of interest has a multidimensional structure, we associate the items to the corresponding latent traits through matrix multi, having one row for each latent trait:

```
# Define the multidimensional (reading, grammar, and mathematics)
structure of the data
> multi = c(1:30, 31:40, rep(0,20), 41:67,rep(0,3))
> multi = t(matrix(multi,30,3))
```

We perform the analyses through package MultiLCIRT. The function we use to estimate multilevel LC-IRT models is est_multi_poly_clust, which works similarly to est_multi_poly. We now have to specify input W, which stands for the matrix containing the cluster-level covariates, in addition to input X for the individual-level covariates. We also now have a certain number of latent classes at cluster level and a certain number of latent classes at individual level (not necessary the same as for cluster level): options kU and kV allow us to define the values at issue. Finally, option clust identifies the vector of cluster indicators.

Multilevel multidimensional models under the Rasch and the 2PL parameterizations are fitted as follows:

```
> # Set the number of latent classes at cluster and individual levels
> kU = 5
> kV = 3
#  Multilevel multidimensional LC-Rasch model
> out1PL = est_multi_poly_clust(Y, W=W, X=X, kU=kU, kV=kV, clust=clust,
+ link=1,disc=0,multi=multi,fort=TRUE,output=TRUE,disp=TRUE)

[…] Output omitted

#  Multilevel multidimensional LC-2PL model
> out2PL = est_multi_poly_clust(Y, W=W, X=X, kU=kU, kV=kV, clust=clust,
+ link=1,disc=1,multi=multi,fort=TRUE,output=TRUE,disp=TRUE)

[…] Output omitted
```

In the previous code, option `multi` is specified, as the default of function `est_multi_poly_clust` (as well as `est_multi_poly`) is unidimensionality. Moreover, the number of latent classes at cluster level and individual level is set to $k_1 = 3$ and $k_2 = 5$, respectively, in accordance with previous studies carried out with the same INVALSI data (for details, see Gnaldi et al., 2015). To avoid confusion, in the following we denote the latent classes at cluster level as *types of school*, whereas the term latent classes is reserved for the individual level.

To choose the best model, we use the following commands:

```
> out1PL$bic
[1] 219050.1
> out2PL$bic
[1] 217968.8
> out1PL$lk
[1] -109092.7
> out2PL$lk
[1] -108288.5
> out1PL$np
[1] 105
> out2PL$np
[1] 169
```

and we conclude, on the basis of BIC and the likelihood-ratio (LR) test, for the 2PL model with k_1 and k_2 fixed as previously indicated, in comparison to the Rasch model with the same structure of latent classes.

Next, concerning the estimated LC-2PL model with three dimensions and three classes at student level and five classes at school level, we display the ordered latent class abilities and the corresponding average weights, as follows, where `Dimension 1`, `Dimension 2`, and `Dimension 3` stand for reading, grammar, and mathematics, respectively:

```
> # Abilities at student level for each latent class and each dimension
> ind = order(out2PL$Th[1,])
> (Ths = out2PL$Th[,ind])
                Class 1    Class 2   Class 3
Dimension 1 -0.5837673  0.9447017  2.479555
Dimension 2  1.5236052  2.5773002  4.005172
Dimension 3  1.3524168  1.8230297  2.400171

> # Latent class average weights at student level
```

```
> Pivs = out2PL$Piv[,ind,]
> Pivms = matrix(0,n,kV)
> for(h in unique(clust)){
+ ind1 = which(clust==h)
+ for(u in 1:kU) Pivms[ind1,] = Pivms[ind1,]+Pivs[ind1,,u]*out2PL$La[h,u]
+ }
> (pivs = colMeans(Pivms))
[1] 0.1497717 0.4365087 0.4137196
```

An inspection of these estimates shows that students belonging to class 1 tend to have the lowest ability level with respect to all dimensions. Overall, the weight of low attainment students grouped in this class is rather reduced in terms of class proportion, as they account for slightly less than 15%. On the contrary, students grouped in class 2 account for more than 40% of the overall sample and the same holds for class 3.

At a higher level, the distribution of the estimated average abilities for the five chosen latent classes of schools is obtained through the average value of the individual abilities at the school level. In more detail, we first standardize the individual abilities, and, then, we compute the mean values for each type of school, taking into account the weights of each students' latent class, as follows:

```
> # Standardized students' abilities
> mThs = as.vector(Ths%*%pivs)
> vThs = as.vector((Ths-mThs)^2%*%pivs)
> (Ths = (Ths-mThs)*(1/sqrt(vThs)))
              Class 1    Class 2   Class 3
Dimension 1 -1.796467 -0.3770936 1.048209
Dimension 2 -1.639337 -0.4773984 1.097156
Dimension 3 -1.691570  0.4455678 1.082481

> # Compute the mean values of students' abilities for each type of school
> ThU = rep(0,times=5)
> for(u in 1:kU) ThU[u] = sum(Pivs[,,u]%*%t(Ths))/(n*kV)
> ThU
[1] -1.0535877 -0.4435458 -0.3001581  0.3904478  0.9099140
```

The previous output allows us to qualify the schools from those of type 1 that are the worst ones (average ability equals -1.054) to those of type 5 that are the best ones (average ability equals 0.910).

Finally, we can extrapolate the regression parameters (and corresponding standard errors) for the student-level covariate gender directly from the estimated LC-2PL model as follows:

```
> # Regression coefficients and standard errors for gender
> out2PL$DeV
            logit
                  2           3
intercept.1 -0.1976683 -3.28951253
intercept.2  1.9587528 -0.38997127
intercept.3  0.6795607  0.06795358
intercept.4  2.3660613  2.70313188
intercept.5  1.3953187  3.80786442
X.1          0.2018985  0.28854150
```

```
> out2PL$seDeV
            [,1]        [,2]
[1,]  0.1609925  0.9476312
[2,]  0.2545137  0.5895456
[3,]  0.1814492  0.2735874
[4,]  0.4176485  0.4779450
[5,]  0.4523495  0.3884170
[6,]  0.1175724  0.1302895
```

The previous estimates are obtained by taking the first as reference class for the students' ability, which is characterized by the worst level of estimated ability for each of the three involved dimensions (see output Ths), and category males as reference category for the covariate gender. The regression parameters for class 2 and class 3 (0.202 and 0.288, respectively) with respect to class 1, and the corresponding standard errors (0.117 and 0.130, respectively) show that females have a higher tendency to belong to classes 2 and 3, and therefore, they perform better than males at the INVALSI tests.

Similarly, regression parameters for the school-level covariate geographic area are estimated by taking schools of type 1 (i.e., the worst schools) as reference class, and category area_1 (northwest) as reference category for the covariate. For an easier interpretation, the estimated regression parameters and the corresponding standard errors are not shown here, but they are replaced by the estimated a posterior probabilities to belong to each of the five types of schools, given the geographic area. These results are obtained as follows:

```
> ind2 = order(ThU)
> # Posterior probabilities to belong to each type of school
> Las = out2PL$La[,ind2]
> # Conditional posterior probabilities given the geographic area
> Lass = rbind(Las[which(rowSums(W)==0)[1],],Las[which(W[,1]==1)[1],],Las[which(W[,2]==1)[1],],
+ Las[which(W[,3]==1)[1],],Las[which(W[,4]==1)[1],])
> colnames(Lass) = c("Type 1","Type 2","Type 3","Type 4","Type 5")
> rownames(Lass) = c("North-West","North-East","Centre","South","Islands")
> Lass
                Type 1        Type 2      Type 3     Type 4       Type 5
North-West  0.06803550  6.163395e-15  0.4873890  0.4445753  1.259938e-07
North-East  0.06646258  1.744149e-08  0.7805507  0.1183833  3.460340e-02
Centre      0.16489881  8.781523e-02  0.4140284  0.2318233  1.014343e-01
South       0.07949635  2.310756e-01  0.1215358  0.3635172  2.043750e-01
Islands     0.12296530  1.399444e-01  0.1845535  0.2912284  2.613083e-01
```

Overall, the great majority of the Italian schools tend to be classified into average and high attainment schools (type 3 and type 4, respectively). In addition, schools from the northwest and northeast show a very similar profile, as they display a high probability to belong to medium attainment schools and high attainment schools. On the other hand, schools from the south and islands have a relatively high probability to be classified among the best schools and, at the same time, to the worst schools (type 5 and type 1, respectively). The latter apparently inconsistent result may be related to the presence in the southern regions of a few schools with exceptionally positive results, as discussed in Sani and Grilli (2011); see also Battistin et al. (2014).

Exercises

1. Using dataset `mislevy.dat` (downloadable from `http://www.gllamm.org/books/mislevy.dat`), perform the following analyses through `Stata`:

   (a) Reshape the data in long format and generate dummies for gender (male/female) and race (white/nonwhite).

   (b) Estimate a 2PL model including a direct effect of gender and race on the latent variable.

   (c) On the basis of the results at point (b), check the significance of the constant term.

   (d) On the basis of the results at point (b), check the significance of the regression coefficients and compute the corresponding odds ratio.

   (e) Predict the individuals' abilities and discuss the differences between white males and nonwhite males, given the item response patterns.

2. Using the dataset `aggression.dat` (downloadable from `http://www.gllamm.org/aggression.dat`), perform the following analyses through `Stata`:

   (a) After collapsing category 2 and category 1, estimate a bidimensional Rasch model, with items `i1-i12` measuring dimension 1 and items `i13-i24` belonging to dimension 2.

   (b) Estimate the model at point (a) adding the direct effect of covariates `Anger` and `Gender` on the latent traits and discuss the results.

   (c) Estimate the model at point (b) assuming a discrete distribution for the latent traits with three support points (use options `ip(f)` and `nip(#)` of function `gllamm`) and discuss the results.

   (d) On the basis of the model at point (b), predict the individuals' abilities on both dimensions and compare the corresponding empirical distributions (e.g., using descriptive statistics, such as mean, median, variance, and histograms).

   (e) Repeat the exercise using the original data with three ordered categories (0, 1, 2) and estimating an RS-GRM.

3. Consider dataset `verbal` available in the R package `difR` and perform the following analyses through R:

   (a) Using package `MultiLCIRT`, estimate a sequence of bidimensional LC-2PL models with an increasing number k of latent classes from 1 to 5, with `Want`-type items (i.e., items in columns 1–12) belonging to dimension 1 and `Do`-type items (i.e., items in columns 13–24) belonging to dimension 2, and select the optimal value of k on the basis of BIC.

(b) On the basis of the model selected at point (a), discuss the structure of the latent classes in terms of levels of verbal aggression and corresponding weights.

(c) On the basis of the model selected at point (a), include the effects of gender and anger score through a multinomial logit parameterization.

(d) On the basis of the results at point (c), evaluate the significance of the regression coefficients and interpret their meaning.

(e) Repeat the analyses at points (c) and (d) by using a global logit parameterization.

4. Consider dataset Anxiety available in the R package lordif and perform the following analyses through R:

(a) After collapsing category 2 and category 1 together and category 5 with category 4 together, estimate the unidimensional models GRM, 1P-GRM, and 1P-RS-GRM with $k = 3$ latent classes, using package MultiLCIRT.

(b) On the basis of the results at point (a), select the model with the best fit.

(c) Apply function class_item to the model selected at point (b) and investigate the dimensionality of the set of items at issue.

(d) Estimate the multidimensional graded response type model selected at points (b) and (c), including the effects of age, gender, and education through a global logit parameterization.

(e) On the basis of the results at point (d), evaluate the significance of the regression coefficients and interpret their meaning.

References

Abramowitz, M. and Stegun, I. A. (1965). *Handbook of Mathematical Functions with Formulas, Graphs, and Mathematical Tables*. Dover, New York.

Acock, A. C. (2008). *A Gentle Introduction to Stata*. Stata Press, College Station, TX.

Adams, R. J., Wilson, M., and Wu, M. L. (1997a). Multilevel item response models: An approach to errors in variables regression. *Journal of Educational and Behavioral Statistics*, 22:47–76.

Adams, R. J., Wilson, M., and Wang, W. (1997b). The multidimensional random coefficients multinomial logit. *Applied Psychological Measurement*, 21:1–24.

Agresti, A. (1993). Computing conditional maximum likelihood estimates for generalized Rasch models using simple loglinear models with diagonals parameters. *Scandinavian Journal of Statistics*, 20:63–71.

Agresti, A. (2002). *Categorical Data Analysis*. John Wiley & Sons, Hoboken, NJ.

Agresti, A., Booth, J. G., Hobert, J. P., and Caffo, B. (2000). Random-effects modeling of categorical response data. *Sociological Methodology*, 30:27–80.

Akaike, H. (1973). Information theory and an extension of the maximum likelihood principle. In Petrov, B. N. and Csaki, F., eds., *Second International Symposium of Information Theory*, pp. 267–281, Budapest, Hungary. Akademiai Kiado.

Allen, M. J. and Yen, W. M. (1979). *Introduction to Measurement Theory*. Waveland Press, Lond Grove, IL.

American Educational Research Association, American Psychological Association, and National Council on Measurement in Education. (1999). *Standards for Educational and Psychological Testing*. American Educational Research Association, Washington, DC.

Andersen, E. B. (1970). Asymptotic properties of conditional maximum-likelihood estimators. *Journal of the Royal Statistical Society-Series B*, 32:283–301.

Andersen, E. B. (1972). The numerical solution of a set of conditional estimation equations. *Journal of the Royal Statistical Society-Series B*, 34:42–54.

Andersen, E. B. (1973). A goodness of fit test for the Rasch model. *Psychometrika*, 38:123–140.

Andersen, E. B. (1977). Sufficient statistics and latent trait models. *Psychometrika*, 42:69–81.

Andrich, D. (1978a). Application of a psychometric rating model to ordered categories which are scored with successive integers. *Applied Psychological Measurement*, 2:581–594.

Andrich, D. (1978b). A rating formulation for ordered response categories. *Psychometrika*, 43:561–573.

Andrich, D. (1988). *Rasch Models for Measurement*. Sage Publications, Inc., Newbury Park, CA.

Andrich, D. (2010). Understanding the response structure and process in the polytomous Rasch model. In Nering, M. L. and Ostini, R., eds., *Handbook of Polytomous Item Response Theory Models*, pp. 123–152. Routledge, Taylor & Francis Group, New York.

Arminger, G., Stein, P., and Wittenberg, J. (1999). Mixtures of conditional mean- and covariance-structure models. *Psychometrika*, 64:475–494.

Bacci, S. and Bartolucci, F. (2015). A multidimensional finite mixture SEM for non-ignorable missing responses to test items. *Structural Equation Modeling: A Multidisciplinary Journal*, DOI: 10.1080/10705511.2014.937376.

Bacci, S., Bartolucci, F., and Gnaldi, M. (2014). A class of multidimensional latent class IRT models for ordinal polytomous item responses. *Communication in Statistics— Theory and Methods*, 43:787–800.

Bacci, S. and Caviezel, V. (2011). Multilevel IRT models for the university teaching evaluation. *Journal of Applied Statistics*, 38:2775–2791.

Bacci, S. and Gnaldi, M. (2015). A classification of university courses based on students' satisfaction: an application of a two-level mixture Item Response Theory (IRT) model. *Quality & Quantity*, 49:927–940.

Baetschmann, G., Staub, K. E., and Winkelmann, R. (2011). Consistent estimation of the fixed effects ordered logit model. Technical Report 5443, IZA, Bonn, Germany.

Baker, F. (2001). *The Basics of Item Response Theory*. ERIC Clearinghouse on Assessment and Evaluation, University of Maryland, College Park, MD.

Baker, F. B. and Kim, S.-H. (2004). *Item Response Theory: Parameter Estimation Techniques*, 2nd edn. Marcel Dekker, New York.

Barndorff-Nielsen, O. (1978). *Information and Exponential Families in Statistical Theory*. John Wiley & Sons, Chichester, UK.

Bartolucci, F. (2007). A class of multidimensional IRT models for testing unidimensionality and clustering items. *Psychometrika*, 72:141–157.

Bartolucci, F., Bacci, S., and Gnaldi, M. (2014). MultiLCIRT: An R package for multidimensional latent class item response models. *Computational Statistics & Data Analysis*, 71:971–985.

Bartolucci, F., Colombi, R., and Forcina, A. (2007). An extended class of marginal link functions for modelling contingency tables by equality and inequality constraints. *Statistica Sinica*, 17:691–711.

Bartolucci, F., Farcomeni, A., and Pennoni, F. (2013). *Latent Markov Models for Longitudinal Data*. Chapman & Hall/CRC Taylor & Francis Group, Boca Raton, FL.

Bartolucci, F. and Forcina, A. (2000). A likelihood ratio test for MTP2 within binary variables. *The Annals of Statistics*, 28:1206–1218.

Battistin, E., De Nadai, M., and Vuri, D. (2014). Counting rotten apples: Student achievement and score manipulation in Italian elementary schools. Technical Report 8405, IZA, Bonn, Germany.

Bauer, D. J. (2007). Observations on the use of growth mixture models in psychological research. *Multivariate Behavioral Research*, 42:757–786.

Bauer, D. J. and Curran, P. J. (2004). The integration of continuous and discrete latent variable models: Potential problems and promising opportunities. *Psychological Methods*, 9:3–29.

Bertoli-Barsotti, L. (2005). On the lack of the co-monotonicity between Likert scores and Rasch-based measures. *Journal of Applied Measurement*, 6:71–79.

Birnbaum, A. (1968). Some latent trait models and their use in inferring an examinee's ability. In Lord, F. M. and Novick, M. R., eds., *Statistical Theories of Mental Test Scores*, pp. 395–479. Addison-Wesley, Reading, MA.

Bock, R. D. (1972). Estimating item parameters and latent ability when responses are scored in two or more nominal categories. *Psychometrika*, 37:29–51.

Bock, R. D. and Aitkin, M. (1981). Marginal maximum likelihood estimation of item parameters: Application of an EM algorithm. *Psychometrika,* 46: 443–459.

Bock, R. D. and Lieberman, M. (1970). Fitting a response model for *n* dichotomously scored items. *Psychometrika*, 35:179–197.

Bolck, A., Croon, M., and Hagenaars, J. (2004). Estimating latent structure models with categorical variables: One-step versus three-step estimators. *Political Analysis*, 12:3–27.

Bollen, K. A., Rabe-Hesketh, S., and Skrondal, A. (2008). Structural equation models. In Box-Steffensmeier, J. M., Brady, H., and Collier, D., eds., *Oxford Handbook of Political Methodology*, chapter 18, pp. 432–455. Oxford University Press, Oxford, NY.

Bollen, K. A. and Curran, P. J. (2006). *Latent Curve Models: A Structural Equation Perspective*, vol. 467. John Wiley & Sons, Hoboken, NJ.

Bond, T. G. and Fox, C. M. (2007). *Applying the Rasch Model. Fundamental Measurement in the Human Sciences*. Lawrence Erlbaum Associates, Mahwah, NJ.

Bovaird, J. A. and Koziol, N. A. (2012). Measurement models for ordered-categorical indicators. In Hoyle, R. H., ed., *Handbook of Structural Equation Modeling*, chapter 29, pp. 495–511. The Guilford Press, New York.

Brennan, R. L. (2001). *Generalizability Theory*. Springer-Verlag, New York.

Briggs, D. and Wilson, M. (2003). An introduction to multidimensional measurement using Rasch models. *Journal of Applied Measurement*, 4:87–100.

Camilli, G. (1992). A conceptual analysis of differential item functioning in terms of a multidimensional item response model. *Applied Psychological Measurement*, 16:129–147.

Casella, G. and Berger, R. L. (2006). *Statistical Inference*. Thomson Learning, Pacific Grove, CA.

Chalmers, R. P. (2012). mirt: A multidimensional item response theory package for the R environment. *Journal of Statistical Software*, 48:1–29.

Chamberlain, G. (1980). Analysis of covariance with qualitative data. *Review of Economic Studies*, 47:225–238.

Cheong, Y. F. and Raudenbush, S. W. (2000). Measurement and structural models for children's problem behaviours. *Psychological Methods*, 5:477–495.

Choi, S. W., Gibbons, L. E., and Crane, P. K. (2011). lordif: An R package for detecting differential item functioning using iterative hybrid ordinal logistic regression/item response theory and Monte Carlo simulations. *Journal of Statistical Software*, 39:1–30.

Christensen, K., Bjorner, J., Kreiner, S., and Petersen, J. (2002). Testing unidimensionality in polytomous Rasch models. *Psychometrika*, 67:563–574.

Christensen, K., Bjorner, J., Kreiner, S., and Petersen, J. (2004). Latent regression in loglinear Rasch models. *Communications in Statistics. Theory and Methods*, 33:1295–1313.

Costantini, M., Musso, M., Viterbori, P., Bonci, F., Del Mastro, L., Garrone, O., Venturini, M., and Morasso, G. (1999). Detecting psychological distress in cancer patients: Validity of the Italian version of the hospital anxiety and depression scale. *Support Care Cancer*, 7:121–127.

Crane, P. K., Gibbons, L. E., Jolley, L., and van Belle, G. (2006). Differential item functioning analysis with ordinal logistic regression techniques: DIFdetect and difwithpar. *Medical Care*, 44:S115–S123.

Crocker, L. and Algina, J. (1986). *Introduction to Classical and Modern Test Theory*. Holt, Rinehart and Winston, Inc., Toronto, Ontario, Canada.

Cronbach, L. J., Gleser, G. C., Nanda, H., and Rajaratnam, N. (1972). *The Dependability of Behavioral Measurements: Theory of Generalizability for Scores and Profiles*. John Wiley & Sons, New York.

Cronbach, L. J. and Meehl, P. E. (1955). Construct validity in psychological tests. *Psychological Bulletin*, 51:181–301.

Cronbach, L. J., Rajaratnam, N., and Gleser, G. C. (1963). Theory of generalizability: A liberalization of reliability theory. *British Journal of Statistical Psychology*, 16:137–163.

Dalgaard, P. (2008). *Introductory Statistics with R*. Springer-Verlag, New York.

Dayton, C. M. and Macready, G. B. (1988). Concomitant-variable latent-class models. *Journal of the American Statistical Association*, 83:173–178.

De Boeck, P. and Wilson, M. (2004a). *Explanatory Item Response Models. A Generalized Linear and Nonlinear Approach*. Springer-Verlag, New York.

De Boeck, P. and Wilson, M. (2004b). A framework for item response models. In De Boeck, P. and Wilson, M., eds., *Explanatory Item Response Models. A Generalized Linear and Nonlinear Approach*, pp. 3–41. Springer-Verlag, New York.

de Gruijter, D. N. M. and van der Kamp, L. J. T. (2008). *Statistical Test Theory for the Behavioral Sciences*. Chapman & Hall/CRC, Boca Raton, FL.

Dempster, A. P., Laird, N. M., and Rubin, D. B. (1977). Maximum likelihood from incomplete data via the EM algorithm (with discussion). *Journal of the Royal Statistical Society-Series B*, 39:1–38.

Dolan, C. and van der Maas, H. (1998). Fitting multivariate normal finite mixtures subject to structural equation modeling. *Psychometrika*, 63:227–253.

Dolan, C. V. (2009). Structural equation mixture modeling. In Millsap, R. E. and Maydeu-Olivares, A., eds., *The Sage Handbook of Quantitative Methods in Psychology*, pp. 568–591. Sage Publications, Inc., Thousand Oaks, CA.

Dorans, N. and Kulick, E. (1986). Demonstrating the utility of the standardization approach to assessing unexpected differential item performance on the Scholastic Aptitude Test. *Journal of Educational Measurement*, 23:355–368.

Downing, S. M. and Haladyna, T. M. (2006). *Handbook of Test Development*. Lawrence Erlbaum Associates, Inc., Mahwah, NJ.

Drasgow, F. (1988). Polychoric and polyserial correlations. In Kotz, L. and Johnson, N. L., eds., *Encyclopedia of Statistical Sciences*, vol. 7, pp. 69–74. John Wiley & Sons, New York.

Drasgow, F. and Lissak, R. (1983). Modified parallel analysis: A procedure for examining the latent dimensionality of dichotomously scored item responses. *Journal of Applied Psychology*, 68:363–373.

Duncan, O. D. (1975). *Introduction to Structural Equation Models*. Academic Press, New York.

Duncan, O. D. and Stenbeck, M. (1987). Are Likert scales unidimensional? *Social Science Research*, 16:245–259.

Edgeworth, F. Y. (1908). On the probable errors of frequency-constants. *Journal of the Royal Statistical Society*, 71:651–678.

Edwards, M. C., Wirth, R. J., Houts, C. R., and Xi, N. (2012). Categorical data in the structural equation modeling framework. In Hoyle, R. H., ed., *Handbook of Structural Equation Modeling*, chapter 12, pp. 195–208. The Guilford Press, New York.

Embretson, S. E. (1991). A multidimensional latent trait model for measuring learning and change. *Psychometrika*, 56:495–515.

Fawcett, T. (2006). An introduction to ROC analysis. *Pattern Recognition Letters*, 27:861–874.

Feldt, L. S. and Brennan, R. L. (1989). Reliability. In Linn, R. L., ed., *Educational Measurement*, pp. 105–146. American Council on Education/Macmillan Publisher, Washington, DC.

Ferguson, G. A. (1942). Item selection by the constant process. *Psychometrika*, 7:19–29.

Fieuws, S., Spiessens, B., and Draney, K. (2004). Mixture models. In De Boeck, P. and Wilson, M., eds., *Explanatory Item Response Models. A Generalized Linear and Nonlinear Approach*, pp. 317–340. Springer-Verlag, New York.

Fischer, G. H. (1981). On the existence and uniqueness of maximum-likelihood estimates in the Rasch model. *Psychometrika*, 46:59–77.

Fischer, G. H. (1995). Derivations of the Rasch model. In Fischer, G. H. and Molenaar, I. W., eds., *Rasch Models. Foundations, Recent Developments, and Applications*, pp. 15–38. Springer-Verlag, New York.

Formann, A. K. (1986). A note on the computation of the second-order derivatives of the elementary symmetric functions in the Rasch model. *Psychometrika*, 51:335–339.

Formann, A. K. (1995). Linear logistic latent class analysis and the Rasch model. In Fischer, G. H. and Molenaar, I. W., eds., *Rasch Models. Foundations, Recent Developments and Applications*, pp. 239–255. Springer-Verlag, New York.

Formann, A. K. (2007a). (Almost) equivalence between conditional and mixture maximum likelihood estimates for some models of the Rasch type. In von Davier, M. and Carstensen, C., eds., *Multivariate and Mixture Distribution Rasch Models*, pp. 177–189. Springer-Verlag, New York.

Formann, A. K. (2007b). Mixture analysis of multivariate categorical data with covariates and missing entries. *Computational Statistics and Data Analysis*, 51:5236–5246.

Fox, J. P. (2004). Applications of multilevel IRT modeling. School effectiveness and school improvement. *International Journal of Research Policy Practice*, 15:261–280.

Fox, J. P. (2005). Multilevel IRT using dichotomous and polytomous response data. *British Journal of Mathematical and Statistical Psychology*, 58:145–172.

Fox, J. P. and Glas, C. A. W. (2001). Bayesian estimation of a multilevel IRT model using Gibbs sampling. *Psychometrika*, 66:271–288.

Furr, R. and Bacharach, V. R. (2008). *Psychometrics. An Introduction*. Sage Publications, Inc., Los Angeles, CA.

Glas, C. (1988). The derivation of some tests for the Rasch model from the multinomial distribution. *Psychometrika*, 53:525–546.

Glas, C. and Verhelst, N. (1995a). Testing the Rasch model. In Fischer, G. H. and Molenaar, I. W., eds., *Rasch Models. Foundations, Recent Developments, and Applications*, pp. 69–95. Springer-Verlag, New York.

Glas, C. and Verhelst, N. (1995b). Tests of fit for polytomous Rasch models. In Fischer, G. H. and Molenaar, I. W., eds., *Rasch Models. Foundations, Recent Developments, and Applications*, pp. 325–352. Springer-Verlag, New York.

Gnaldi, M. and Bacci, S. (2015). Joint assessment of the latent trait dimensionality and observable differential item functioning of students' national tests. *Quality & Quantity*. DOI 10.1007/s11135-015-0214-0.

Gnaldi, M., Bacci, S., and Bartolucci, F. (2015). A multilevel finite mixture item response model to cluster examinees and schools. *Advances in Data Analysis and Classification*. DOI: 10.1007/s11634-014-0196-0.

Goldberger, A. (1972). Structural equation models in the social sciences. *Econometrica: Journal of Econometric Society*, 40:979–1001.

Goldstein, H. (1980). Dimensionality, bias, independence and measurement scale problems in latent trait test score models. *British Journal of Mathematical and Statistical Psychology*, 33:234–260.

Goldstein, H. (2003). *Multilevel Statistical Models*. Arnold, London, UK.

Goldstein, H. and Healy, M. J. R. (1994). The graphical presentation of a collection of means. *Journal of the Royal Statistical Society-Series A*, 158:175–177.

Goodman, L. A. (1974). Exploratory latent structure analysis using both identifiable and unidentifiable models. *Biometrika*, 61:215–231.

Gorsuch, R. L. (1983). *Factor Analysis*. Lawrence Erlbaum, Hillsdale, NJ.

Gulliksen, H. (1950). *Theory of Mental Tests*. John Wiley & Sons, New York.

Gustafson, J. E. (1980). A solution of the conditional estimation problem for long tests in the Rasch model for dichotomous items. *Educational and Psychological Measurement*, 40:377–385.

Guttman, L. (1945). A basis for analyzing test-retest reliability. *Psychometrika*, 10:255–282.

Haley, J. A. and McNeil, B. J. (1982). The meaning and use of the area under a receiver operating characteristic (ROC) curve. *Radiology*, 143:29–36.

Hambleton, R. K., Linden van der, W. J., and Wells, C. S. (2010). IRT models for the analysis of polytomously scored data. In Nering, M. L. and Ostini, R., eds., *Handbook of Polytomous Item Response Theory Models*, pp. 21–42. Routledge, Taylor & Francis Group, New York.

Hambleton, R. K. and Swaminathan, H. (1985). *Item Response Theory: Principles and Applications*. Kluwer Nijhoff, Boston, MA.

Hambleton, R. K., Swaminathan, H., and Rogers, H. J. (1991). *Fundamentals of Item Response Theory*. Sage Publications, Inc., Newbury Park, CA.

Hand, D. J. and Till, R. J. (2001). A simple generalization of the area under the ROC curve to multiple class classification problems. *Machine Learning*, 45:171–186.

Hardouin, J. (2007). Rasch analysis: Estimation and tests with `raschtest`. *The Stata Journal*, 7:22–44.

Hauser, R. M. and Goldberger, A. S. (1971). The treatment of unobservable variables in path analysis. In Costner, H. L., ed., *Sociological Methodology*, pp. 81–117. Jossey-Bass, San Francisco, CA.

Heinen, T. (1996). *Latent Class and Discrete Latent Traits Models: Similarities and Differences*. Sage Publications, Inc., Thousand Oaks, CA.

Hemker, B. T., Sijtsma, K., Molenaar, I. W., and Junker, B. W. (1996). Polytomous IRT models and monotone likelihood ratio of the total score. *Psychometrika*, 61:679–693.

Hemker, B. T., Sijtsma, K., Molenaar, I. W., and Junker, B. W. (1997). Stochastic ordering using the latent trait and the sum score in polytomous IRT models. *Psychometrika*, 62:331–347.

Hemker, B. T., van der Ark, L. A., and Sijtsma, K. (2001). On measurement properties of continuation ratio models. *Psychometrika*, 66:487–506.

Hoijtink, H. (1995). Linear and repeated measures models for the person parameters. In Fischer, G. H. and Molenaar, I. W., eds., *Rasch Models. Foundations, Recent Developments, and Applications*, pp. 203–214. Springer-Verlag, New York.

Holland, P. W. (1990). On the sampling theory foundations of item response theory models. *Psychometrika*, 55:577–601.

Holland, P. W. and Thayer, D. T. (1988). Differential item performance and the Mantel-Haenszel procedure. In Wainer, H. and Braun, H. I., eds., *Test Validity*, pp. 129–145. Erlbaum, Hillsdale, NJ.

Hosmer, D. W. and Lemeshow, S. (1980). A goodness-of-fit test for the multiple logistic regression model. *Communications in Statistics*, A10:1043–1069.

Hosmer, D. W. and Lemeshow, S. (2000). *Applied Logistic Regression*. John Wiley & Sons, New York.

Hoyle, R. H. (2012). *Handbook of Structural Equation Modeling*. The Guilford Press, New York.

Huang, G.-H. and Bandeen-Roche, K. (2004). Building an identifiable latent class model with covariate effects on underlying and measured variables. *Psychometrika*, 69:5–32.

INVALSI. (2009). Esame di stato di primo ciclo. a.s. 2008–2009. Technical Report, INVALSI.

Jedidi, K., Jagpal, H., and DeSarbo, W. (1997). STEMM: A general finite mixture structural equation model. *Journal of Classification*, 14:23–50.

Jensen, A. R. (1980). *Bias in Mental Testing*. Free Press, New York.

Jöreskog, K. G. (1973). A general method for estimating a linear structural equation system. In Goldberger, A. S. and Duncan, O. D., eds., *Structural Equation Models in the Social Sciences*, pp. 85–112. Seminar Press, New York.

Jöreskog, K. G. (1977). Structural equation models in the social sciences: Specification, estimation and testing. In Krishnaiah, P. R., ed., *Applications of Statistics*, pp. 265–287. North-Holland, Amsterdam, the Netherlands.

Kamata, A. (2001). Item analysis by the hierarchical generalized linear model. *Journal of Educational Measurement*, 38:79–93.

Kamata, A. and Cheong, Y. F. (2007). Multilevel Rasch models. In von Davier, M. and Carstensen, C., eds., *Multivariate and Mixture Distribution Rasch Models*, pp. 217–232. Springer-Verlag, New York.

Kelderman, H. (1996). Multidimensional Rasch models for partial-credit scoring. *Applied Psychological Measurement*, 20:155–168.

Kelderman, H. (1997). Loglinear multidimensional item response models for polytomously scored items. In van der Linden, W. and Hambleton, R., eds., *Handbook of Modern Item Response Theory*, pp. 287–304. Springer-Verlag, New York.

Kelderman, H. and Rijkes, C. P. M. (1994). Loglinear multidimensional IRT models for polytomously scored items. *Psychometrika*, 59:149–176.

Kim, S.-H., Cohen, A. S., and Park, T.-H. (1995). Detection of differential item functioning in multiple groups. *Journal of Educational Measurement*, 32:261–276.

Kline, R. B. (2011). *Principles and Practice of Structural Equation Modeling*. The Guilford Press, New York.

Kline, T. (2005). *Psychological Testing: A Practical Approach to Design and Evaluation*. Sage Publications, Inc., Thousand Oaks, CA.

Kuder, G. F. and Richardson, M. W. (1937). The theory of the estimation of test reliability. *Psychometrika*, 2:151–160.

Langheine, R. and Rost, J. (1988). *Latent Trait and Latent Class Models*. Plenum, New York.

Lazarsfeld, P. F. and Henry, N. W. (1968). *Latent Structure Analysis*. Houghton Mifflin, Boston, MA.

Lee, S.-Y., Poon, W. Y., and Bentler, P. M. (1995). A two-stage estimation of structural equation models with continuous and polytomous variables. *British Journal of Mathematical and Statistical Psychology*, 48:339–358.

Lei, P.-W. and Wu, Q. (2007). Introduction to structural equation modeling: Issues and practical considerations. *Educational Measurement: Issues and Practice*, 26:33–43.

Linacre, J. M. and Wright, B. D. (1994). Dichotomous mean-square chi-square fit statistics. *Rasch Measurement Transactions*, 8:360.

Lindsay, B., Clogg, C., and Greco, J. (1991). Semiparametric estimation in the Rasch model and related exponential response models, including a simple latent class model for item analysis. *Journal of the American Statistical Association*, 86:96–107.

Liou, M. (1994). More on the computation of higher-order derivatives of the elementary symmetric functions in the Rasch model. *Applied Psychological Measurement*, 18:53–62.

Lord, F. M. (1952). *A Theory of Test Scores (Psychometric Monograph No. 7)*. Psychometric Corporation, Richmond, VA.

Lord, F. M. (1984). *Maximum Likelihood and Bayesian Parameter Estimation in IRT*. Educational Testing Service, Princeton, NJ.

Lord, F. M. (1980). *Applications of Item Response Theory to Practical Testing Problems*. Erlbaum, Hillsdale, MI.

Lord, F. M. and Novick, M. R. (1968). *Statistical Theories of Mental Test Scores*. Addison-Wesley Publishing Company, Inc., Reading, MA.

Luecht, R. M. and Miller, R. (1992). Unidimensional calibrations and interpretations of composite traits for multidimensional tests. *Applied Psychological Measurement*, 16:279–293.

Magis, D., Béland, S., Tuerlinckx, F., and De Boeck, P. (2010). A general framework and an R package for the detection of dichotomous differential item functioning. *Behavior Research Methods*, 42:847–862.

Maier, K. S. (2001). A Rasch hierarchical measurement model. *Journal of Educational and Behavioral Statistics*, 26:307–330.

Maier, K. S. (2002). Modeling incomplete scaling questionnaire data with a partial credit hierarchical measurement model. *Journal of Educational and Behavioral Statistics*, 27:271–289.

Maij-de Meij, A. M., Kelderman, H., and van der Flier, H. (2008). Fitting a mixture item response theory model to personality questionnaire data: Characterizing latent classes and investigating possibilities for improving prediction. *Applied Psychological Measurement*, 32:611–631.

Mair, P., Hatzinger, R., and Maier, M. J. (2013). eRm. R package version 0.15-3, http://CRAN.R-project.org/package=eRm, accessed April 8, 2015.

Mair, P., Reise, S., and Bentler, P. (2008). IRT goodness-of-fit using approaches from logistic regression. Technical Report, UCLA Department of Statistics, Los Angeles, CA.

Mantel, N. and Haenszel, W. (1959). Statistical aspects of the analysis of data from retrospective studies of disease. *Journal of the National Cancer Institute*, 22:719–748.

Martin-Löf, P. (1973). *Statistika modeller*. Institütet för Försäkringsmatemetik och Matematiks Statistiks vid Stockholms Universitet, vol. 31, pp. 223–264, Anteckningar från seminarier lasåret 1969–1970, utarbetade av Rolf Sundberg Obetydligt ändrat nytryck October 1973. Stockholm, Sweden.

Masters, G. (1982). A Rasch model for partial credit scoring. *Psychometrika*, 47:149–174.

Maydeu-Olivares, A., Drasgow, F., and Mead, A. (1994). Distinguishing among parametric item response models for polychotomous ordered data. *Applied Psychological Measurement*, 18:245–256.

McCullagh, P. and Nelder, J. A. (1989). *Generalized Linear Models*, 2nd edn. Chapman & Hall/CRC, London, UK.

McCullagh, P. and Tibshirani, R. (1990). A simple method for the adjustment of profile likelihoods. *Journal of the Royal Statistical Society-Series B*, 52:325–344.

McCulloch, C. and Searle, S. (2001). *Generalized, Linear, and Mixed Models*. John Wiley & Sons, New York.

McDonald, R. P. (1999). *Test Theory: A Unified Treatment*. Lawrence Erlbaum Associates, Mahwah, NJ.

McFadden, D. (1974). Conditional logit analysis of qualitative choice behavior. In Zarembka, P., ed., *Frontiers in Econometrics*, pp. 105–142. Academic Press, New York.

McLachlan, G. and Peel, D. (2000). *Finite Mixture Models*. John Wiley & Sons, New York.

Mellenbergh, G. J. (1995). Conceptual notes on models for discrete polytomous item responses. *Applied Psychological Measurement*, 19:91–100.

Mesbah, M. (2004). Measurement and analysis of health related quality of life and environmental data. *Environmetrics*, 15:471–481.

Messick, S. (1989). Validity. In Linn, R. L., ed., *Educational Measurement*, 3rd edn. Macmillan, New York.

Metz, C. E. (1978). Basic principles of ROC analysis. *Seminars in Nuclear Medicine*, 8:283–298.

Mislevy, R. J. (1987). Exploiting auxiliary information about examinees in the estimation of item parameters. *Applied Psychological Measurement*, 11:81–91.

Mislevy, R. J. and Bock, R. D. (1989). A hierarchical item-response model for educational testing. In Bock, R., ed., *Multilevel Analysis of Educational Data*, pp. 57–74. Academic Press, San Diego, CA.

Molenaar, I. W. (1983a). Item steps (Heymans Bulletin 83-630-EX). University of Groningen, Groningen, the Netherlands.

Molenaar, I. W. (1983b). Some improved diagnostics for failure of the Rasch model. *Psychometrika*, 48:49–72.

Molenaar, I. W. (1997). Logistic mixture models. In Van der Linden, W. and Hambleton, R. K., eds., *Handbook of Modern Item Response Theory*, pp. 449–463. Springer-Verlag, New York.

Mood, A. M., Graybill, F. A., and Boes, D. C. (1974). *Introduction to the Theory of Statistics*, 3rd edn. McGraw-Hill, New York.

Muraki, E. (1990). Fitting a polytomous item response model to Likert-type data. *Applied Psychological Measurement*, 14:59–71.

Muraki, E. (1992). A generalized partial credit model: Application of an EM algorithm. *Applied Psychological Measurement*, 16:159–176.

Muraki, E. (1993). Information functions of the generalized partial credit model. *Applied Psychological Measurement*, 17:351–363.

Muraki, E. and Carlson, J. E. (1995). Full-information factor analysis for polytomous item responses. *Applied Psychological Measurement*, 19:73–90.

Muthén, B. O. (1983). Latent variable structural modelling with categorical data. *Journal of Econometrics*, 22:43–65.

Muthén, B. O. (1984). A general structural equation model with dichotomous, ordered categorical and continuous latent indicators. *Psychometrika*, 49:115–132.

Netemeyer, R. G., Bearden, W. O., and Sharma, S. (2003). *Scaling Procedures*. Sage, Publications, Inc., Thousand Oaks, CA.

Novick, M. R. (1966). The axioms and principal results of classical test theory. *Journal of Mathematical Psychology*, 3:1–18.

Ostini, R. and Nering, M. L. (2010). New perspectives and applications. In Nering, M. L. and Ostini, R., eds., *Handbook of Polytomous Item Response Theory Models*, pp. 3–20. Routledge, Taylor & Francis Group, New York.

Partchev, I. (2014). irtoys: Simple interface to the estimation and plotting of IRT models. R package version 0.1.7, http://CRAN.R-project.org/package=irtoys., accessed April 8, 2015.

Pastor, D. A. (2003). The use of multilevel IRT modelling in applied research: An illustration. *Applied Measurement in Education*, 16:223–243.

Pastor, D. A. and Beretvas, S. N. (2006). Longitudinal Rasch modelling in the context of psychotherapy outcomes assessment. *Applied Psychological Measurement*, 30:100–120.

Pearson, K. and Filon, L. N. G. (1898). Mathematical contributions to the theory of evolution IV. On the probable errors of frequency constants and on the influence of random selection on variation and correlation. *Philosophical Transactions of the Royal Society-A*, 191:229–311.

Penfield, R. D. (2001). Assessing differential item functioning among multiple groups: A comparison of three Mantel-Haenszel procedures. *Applied Measurement in Education*, 14:235–259.

Pfanzagl, J. (1993). A case of asymptotic equivalence between conditional and marginal maximum likelihood estimators. *Journal of Statistical Planning and Inference*, 35:301–307.

Pinheiro, J. C. and Bates, D. M. (1995). Approximation to the log-likelihood function in the nonlinear mixed effects models. *Journal of Computational and Graphical Statistics*, 4:12–35.

Ponocny, I. (2001). Nonparametric goodness-of-fit tests for the Rasch model. *Psychometrika*, 66:437–460.

Provost, F. and Domingos, P. (2001). Well-trained PETs: Improving probability estimation trees. Technical Report, CeDER Working Paper #IS-00-04, Stern School of Business, New York University, New York.

Rabe-Hesketh, S. and Skrondal, A. (2007). Latent variable modelling: A survey. *Scandinavian Journal of Statistics*, 34:712–745.

Rabe-Hesketh, S., Skrondal, A., and Pickles, A. (2004). *GLLAMM Manual*. University of California, Berkeley, CA.

Rabe-Hesketh, S., Skrondal, A., and Zheng, X. (2012). Multilevel structural equation modeling. In Hoyle, R. H., ed., *Handbook of Structural Equation Modeling*, chapter 30, pp. 512–531. The Guilford Press, New York.

Raju, N. (1988). The area between two item characteristic curves. *Psychometrika*, 53:495–502.

Rasch, G. (1960). Probabilistic models for some intelligence and attainment tests. Danish Institute for Educational Research, Copenhagen, Denmark.

Rasch, G. (1961). On general laws and the meaning of measurement in psychology. *Proceedings of the IV Berkeley Symposium on Mathematical Statistics and Probability*, 4:321–333.

Rasch, G. (1967). An informal report on a theory of objectivity in comparisons. In Van der Kamp, L. J. T. and Vlek, C. A. J., eds., *Psychological Measurement Theory. Proceedings of the NUFFIC International Summer Session in Science at Het Oude Hof*, pp. 1–19. University of Leyden, Leyden, MA.

Rasch, G. (1977). On specific objectivity: An attempt at formalizing the request for generality and validity of scientific statements. *The Danish Yearbook of Philosophy*, 14:58–94.

Raudenbush, S. W. and Sampson, R. (1999). Assessing direct and indirect effects in multilevel designs with latent variables. *Sociological Methods and Research*, 28:123–153.

Raykov, T. and Marcoulides, G. A. (2011). *Introduction to Psychometric Theory*. Routledge, Taylor & Francis Group, New York.

Reckase, M. D. (2009). *Multidimensional Item Response Theory*. Springer-Verlag, New York.

Richardson, M. W. (1936). The relationship between difficulty and the differential validity of a test. *Psychometrika*, 1:33–49.

Rijmen, F. and Briggs, D. (2004). Multiple person dimensions and latent item predictors. In Wilson, M. and De Boeck, P., eds., *Explanatory Item Response Models: A Generalized Linear and Nonlinear Approach*, pp. 247–265. Springer-Verlag, New York.

Rijmen, F., Tuerlinckx, F., De Boeck, P., and Kuppens, P. (2003). A nonlinear mixed model framework for item response theory. *Psychological Methods*, 8: 185–205.

Rizopoulos, D. (2006). ltm: An R package for latent variable modeling and item response theory analyses. *Journal of Statistical Software*, 17:1–25.

Robinson, W. S. (1950). Ecological correlations and the behavior of individuals. *American Sociological Review*, 15:351–357.

Rost, J. (1990). Rasch models in latent classes: An integration of two approaches to item analysis. *Applied Psychological Measurement*, 14: 271–282.

Rost, J. and von Davier, M. (1995). Mixture distribution Rasch models. In Fischer, G. H. and Molenaar, I. W., eds., *Rasch Models. Foundations, Recent Developments and Applications*, pp. 257–268. Springer-Verlag, New York.

Rudner, L. M., Getson, P. R., and Knight, D. L. (1980). Biased item detection techniques. *Journal of Educational Statistics*, 5:213–233.

Samejima, F. (1969). *Estimation of Ability Using a Response Pattern of Graded Scores*. Psychometrika Monograph, 17. Psychometric Society, Richmond, VA.

Samejima, F. (1972). *A General Model for Free-Response Data*. Psychometrika Monograph, 18. Psychometric Society, Richmond, VA.

Samejima, F. (1995). Acceleration model in the heterogeneous case of the general graded response model. *Psychometrika*, 60:549–572.

Samejima, F. (1996). Evaluation of mathematical models for ordered polychotomous responses. *Behaviormetrika*, 23:17–35.

Sani, C. and Grilli, L. (2011). Differential variability of test scores among schools: A multilevel analysis of the fifth-grade INVALSI test using heteroscedastic random effects. *Journal of Applied Quantitative Methods*, 6:88–99.

Schilling, S. and Bock, R. D. (2005). High-dimensional maximum marginal likelihood item factor analysis by adaptive quadrature. *Psychometrika*, 70: 533–555.

Schwarz, G. (1978). Estimating the dimension of a model. *Annals of Statistics*, 6:461–464.

Sijtsma, K. and Hemker, B. T. (1998). Nonparametric polytomous IRT models for invariant item ordering, with results for parametric models. *Psychometrika*, 63:183–200.

Sijtsma, K. and Hemker, B. T. (2000). A taxonomy of IRT models for ordering persons and items using simple sum scores. *Journal of Educational and Behavioral Statistics*, 25:391–415.

Simonoff, J. S. (1998). Logistic regression, categorical predictors, and goodness-of-fit: It depends on who you ask. *The American Statistician*, 10:10–14.

Sing, T., Sander, O., Beerenwinkel, N., and Lengauer, T. (2013). ROCR: Visualizing the performance of scoring classifiers. R package version 1.0–5, http://CRAN.R-project.org/package=ROCR, accessed April 8, 2015.

Skrondal, A. and Rabe-Hesketh, S. (2004). *Generalized Latent Variable Modeling. Multilevel, Longitudinal and Structural Equation Models*. Chapman & Hall/CRC, London, UK.

Smit, A., Kelderman, H., and van der Flier, H. (1999). Collateral information and mixed Rasch models. *Methods of Psychological Research (Online)*, 4:19–32.

Smit, A., Kelderman, H., and van der Flier, H. (2000). The mixed Birnbaum model: Estimation using collateral information. *Methods of Psychological Research Online*, 5:31–43.

Snijders, T. A. B. and Bosker, R. J. (2012). *Multilevel Analysis. An Introduction to Basic and Advanced Multilevel Modeling*. Sage Publications, Inc., London, UK.

Spaan, M. (2007). Evolution of a test item. *Language Assessment Quarterly*, 4:279–293.

Spearman, C. (1904). General intelligence objectively determined and measured. *The American Journal of Psychology*, 15:201–292.

Swaminathan, H. and Rogers, H. (1990). Detecting differential item functioning using logistic regression procedures. *Journal of Educational Measurement*, 27:361–370.

Swets, J. A. and Pickett, R. M. (1982). *Evaluation of Diagnostic Systems: Methods from Signal Detection Theory*. Academic Press, New York.

Tay, L., Newman, D., and Vermunt, J. (2011). Using mixed-measurement item response theory with covariates (MM-IRT-C) to ascertain observed and unobserved measurement equivalence. *Organizational Research Methods*, 14:147–176.

Thissen, D. (1982). Marginal maximum likelihood estimation for the one-parameter logistic model. *Psychometrika*, 47:175–186.

Thissen, D. and Steinberg, L. (1984). A response model for multiple choice items. *Psychometrika*, 49:501–519.

Thissen, D. and Steinberg, L. (1986). A taxonomy of item response models. *Psychometrika*, 51:567–577.

Thissen, D., Steinberg, L., and Wainer, H. (1988). Use of item response theory in the study of group differences in trace lines. In Wainer, H. and Braun, H. I., eds., *Test Validity*, pp. 147–169. Erlbaum, Hillsdale, NJ.

Thissen, D. and Wainer, H. (2001). *Test Scoring*. Lawrence Erlbaum, Mahwah, NJ.

Thurstone, L. L. (1925). A method of scaling psychological and educational tests. *Journal of Educational Psychology*, 16:433–451.

Timminga, E. and Adema, J. J. (1995). Test construction from item banks. In Fischer, G. H. and Molenaar, I. W., eds., *Rasch Models. Foundations, Recent Developments and Applications*, pp. 111–127. Springer-Verlag, New York.

Tuerlinckx, F. and Wang, W.-C. (2004). Models for polytomous data. In De Boeck, P. and Wilson, M., eds., *Explanatory Item Response Models. A Generalized Linear and Nonlinear Approach*, pp. 75–109. Springer-Verlag, New York.

Tutz, G. (1990). Sequential item response models with an ordered response. *British Journal of Mathematical and Statistical Psychology*, 43:39–55.

Van den Noortgate, W. and Paek, I. (2004). Person regression models. In Wilson, M. and De Boeck, P., eds., *Explanatory Item Response Models: A Generalized Linear and Nonlinear Approach*, pp. 167–187. Springer-Verlag, New York.

Van den Wollenberg, A. (1982). Two new test statistics for the Rasch model. *Psychometrika*, 47:123–140.

Van der Ark, L. A. (2001). Relationships and properties of polytomous item response theory models. *Applied Psychological Measurement*, 25:273–282.

Van der Linden, W. and Hambleton, R. K. (1997). *Handbook of Modern Item Response Theory*. Springer-Verlag, New York.

Verhelst, N. D. and Eggen, T. J. H. M. (1989). Psychometric and statistical aspects of assessment research. Technical Report, PPON-rapport 4, CITO, Arnhem, the Netherlands.

Verhelst, N. D., Glas, C. A. W., and Vries de, H. H. (1997). A steps model to analyze partial credit. In Van der Linden, W. J. and Hambleton, R. K., eds., *Handbook of Modern Item Response Theory*, pp. 123–138. Springer-Verlag, New York.

Verhelst, N. D., Glas, C. A. W., and Vries de, H. H. (2005). IRT models for rating scale data. In Everitt, B. and Howell, D., eds., *Encyclopedia of Statistics in Behavioral Science*, pp. 995–1003. John Wiley & Sons, West Sussex, UK.

Vermunt, J. K. and Magidson, J. (2005). Structural equation models: Mixture models. In Everitt, B. and Howell, D., eds., *Encyclopedia of Statistics in Behavioral Science*, pp. 1922–1927. John Wiley & Sons, West Sussex, UK.

Vermunt, J. K. (2008). Multilevel latent variable modeling: An application in education testing. *Austrian Journal of Statistics*, 37:285–299.

von Davier, M. and Rost, J. (1995). Polytomous mixed Rasch models. In Fischer, G. H. and Molenaar, I. W., eds., *Rasch Models. Foundations, Recent Developments, and Applications*, pp. 371–379. Springer-Verlag, New York.

Wang, W.-C. and Su, Y.-Y. (2004). Factors influencing the Mantel and generalized Mantel-Haenszel methods for the assessment of differential item functioning in polytomous items. *Applied Psychological Measurement*, 28:450–480.

Wang, W.-C., Wilson, M. R., and Adams, R. J. (1997). Rasch models for multidimensionality between items and within items. In Wilson, M., Draney, K., and Eglehard, G., eds., *Objective Measurement: Theory into Practice*, pp. 139–155. Ablex Publishing, Greenwich, CT.

Westen, D. and Rosenthal, R. (2003). Quantifying construct validity: Two simple measures. *Journal of Personality and Social Psychology*, 84:608–618.

Wilson, E. B. and Hilferty, M. M. (1931). The distribution of chi-square. *Proceedings of the National Academy of Sciences of the United States of America*, 17:684–688.

Wilson, M. and De Boeck, P. (2004). Descriptive and explanatory item response models. In Wilson, M. and De Boeck, P., eds., *Explanatory Item Response Models: A Generalized Linear and Nonlinear Approach*, pp. 43–74. Springer-Verlag, New York.

Wooldridge, J. M. (2002). *Econometric Analysis of Cross Section and Panel Data*. MIT Press, Cambridge, MA.

Wright, B. and Masters, G. (1982). *Rating Scale Analysis*. Mesa Press, Chicago, IL.

Wright, B. and Stone, M. (1999). *Measurement Essentials*. Wide Range, Inc., Wilmington, DE.

Wright, S. (1921). Correlation and causation. *Journal of Agricultural Research*, 20:557–585.

Yao, L. and Schwarz, R. D. (2006). A multidimensional partial credit model with associated item and test statistics: An application to mixed-format tests. *Applied Psychological Measurement*, 30:469–492.

Zellner, A. (1970). Estimation of regression relationships containing unobservable variables. *International Economic Review*, 11:441–454.

Zheng, X. and Rabe-Hesketh, S. (2007). Estimating parameters of dichotomous and ordinal item response models with gllamm. *The Stata Journal*, 7:313–333.

Zigmond, A. and Snaith, R. (1983). The hospital anxiety and depression scale. *Acta Psychiatrika Scandinavica*, 67:361–370.

Zimmerman, D. W. (1975). Probability spaces, Hilbert spaces, and the axioms of test theory. *Psychometrika*, 40:395–412.

Zimmerman, D. W. and Williams, R. H. (1982). Gain scores in research can be highly reliable. *Journal of Educational Measurement*, 19:149–154.

Zwinderman, A. H. (1991). A generalized Rasch model for manifest predictors. *Psychometrika*, 56:589–600.

Zwinderman, A. H. (1997). Response models with manifest predictors. In Van der Linden, W. J. and Hambleton, R. K., eds., *Handbook of Modern Item Response Theory*, pp. 245–257. Springer-Verlag, New York.

List of Main Symbols

In the following text, we present the list of the principal symbols used in the book and listed in the same order as they are introduced in the book chapters.

Symbol	Description
i	individual (or sample unit) number
n	sample size
j	item number
J	number of items
s_i	true score of individual i
θ_i	latent trait level of individual i
y_{ij}	response of individual i to item j
l_j	number of categories of item j
$\mathbf{y}_i$	vector of all responses provided by subject i
S_i	random variable for the true score
Θ_i	random variable for the latent trait level
$\mathbf{Y}_i$	random vector of response variables
$y_{i\cdot}$	total score of examinee i
$y_{\cdot j}$	total score for item j
$\bar{y}_{i\cdot}$	average score of examinee i
$\bar{y}_{\cdot j}$	average score for item j
$\bar{y}_{\cdot\cdot}$	mean individual score
$v_{i\cdot}$	variance score of examinee i
$v_{\cdot j}$	variance score of item j
$v_{\cdot\cdot}$	variance of individual scores
$\mathrm{cov}_{j_1 j_2}$	sample covariance between items j_1 and j_2
$\mathrm{cor}_{j_1 j_2}$	sample correlation between items j_1 and j_2
t	test number
T	number of tests
$J^{(t)}$	number of items in test t
$y_{ij}^{(t)}$	response to item j of test t by subject i
$\mathbf{y}_i^{(t)}$	vector of responses to test t by subject i
$y_{i\cdot}^{(t)}$	total score attained at test t by subject i
$\bar{y}_{i\cdot}^{(t)}$	mean score attained at test t by subject i
$y_{\cdot j}^{(t)}$	total score for item j of test t
$\bar{y}_{\cdot j}^{(t)}$	average score for item j of test t

$\bar{y}_{..}^{(t)}$	mean score at test t
$\mu_X = E(X)$	expected value of random variable X
$\sigma_X^2 = V(X)$	variance of random variable X
$\text{Cov}(X, Y)$	covariance between random variables X and Y
$\rho_{XY} = \text{Cor}(X, Y)$	correlation between random variables X and Y
η_i	error term affecting the score of individual i
$Y_{i.}^{(t)}$	random variable for the observed score for individual i at test t
$S_i^{(t)}$	true score for individual i at test t
$\eta_i^{(t)}$	error term affecting the score of individual i at test t
ρ_{ys}^2	reliability
$\hat{\rho}_{ys}^2$	estimate of reliability based on the alternate form method
$\tilde{\rho}_{ys}^2$	estimate of reliability based on the split-half method
α	Cronbach's α
$\hat{\alpha}$	estimate of Cronbach's α
KR20	Kuder–Richardson 20 coefficient
α^{st}	generalized Spearman–Brown coefficient
$\overline{\text{Cor}}$	average of the correlations between any pair of items
$\hat{\alpha}^{st}$	estimate of the generalized Spearman–Brown coefficient
$\overline{\text{cor}}$	average of the sample correlations between each pair of items
$\hat{s}_i$	true score estimate for individual i
$1 - \alpha$	confidence level of interval estimates
$z_{\alpha/2}$	quantile at level $1 - \alpha/2$ of the standard normal distribution
$\hat{se}(\cdot)$	estimated standard errors of a certain random variable or estimator
$\hat{s}_i^{adj}$	adjusted true score estimate
$\text{cor}_{pm}^{(j)}$	item-to-total correlation
$\text{cor}_{pbis}^{(j)}$	point-biserial correlation coefficient
cor_s	Spearman's rank-order correlation
$p_j(\theta_i)$	item characteristic curve
$p(y_{1j_1}, y_{2j_2} | \theta_i)$	joint probability of a pair of responses given θ_i
$p(y_i | \theta_i)$	joint conditional probability of y_i given θ_i
$p(y_i)$	manifest probability of y_i
$f(\theta_i)$	a priori distribution of Θ_i
Y	$n \times J$ matrix of observed responses
θ	vector of all individual abilities
$p(Y | \theta)$	conditional probability of Y given θ
$p(Y)$	manifest probability of Y
β_j	difficulty level of binary item j
λ_j	discriminating parameter of item j

δ_j	pseudo-guessing parameter of item j
$1\{\cdot\}$	indicator function
v	support point number for the latent discrete distribution
k	number of support points for the latent discrete distribution
ξ_v	support point for the latent discrete distribution
π_v	mass probability for the latent discrete distribution
$p_{jy}(\theta_i)$	item response category characteristic curve
$\boldsymbol{p}_j(\theta_i)$	vector with conditional probabilities of each category given θ_i
$p^*_{jy}(\theta_i)$	cumulative operating characteristic curve
A_y, B_y	sets of item categories in a logit definition
$g_y(\cdot)$	link function
β_{jy}	difficulty level for response category y of item j
τ_y	cutoff point for category y
ψ	vector of model parameters
$L_J(\psi)$	joint likelihood function
$\ell_J(\psi)$	joint log-likelihood function
$\omega_r(e)$	elementary symmetric function
$p_C(\boldsymbol{y}_i)$	conditional probability of the response configuration $\boldsymbol{y}_i$ given y_i.
$L_C(\psi)$	conditional likelihood function
$\ell_C(\psi)$	conditional log-likelihood function
$\boldsymbol{I}_C(\psi)$	observed information matrix
$L_M(\psi)$	marginal likelihood function
$\ell_M(\psi)$	marginal log-likelihood function
$\ell^*_M(\psi)$	complete marginal log-likelihood function
z_{iv}	indicator function of individual i belonging to latent class v
$I_{jy}(\theta_i)$	information associated with response category y of item j
$I_j(\theta_i)$	information associated with item j
$I(\theta)$	test information
$m_{y_1 y_2}$	number of predicted responses equal to y_1 and observed responses equal to y_2
AUC_y	area under the ROC curve for response category y
AUC_{y_1, y_2}	area under the ROC curve for categories y_1 and y_2
AIC	Akaike information criterion index
BIC	Bayesian information criterion index
$\#par$	number of free model parameters
R^2_{MF}	pseudo-R^2
LR	likelihood ratio test statistic
h	cluster of individuals
H	number of clusters of individuals
$\hat{\ell}_{Ch}$	conditional maximum log-likelihood estimated for cluster h
$d^{(j)}_h$	difference between the observed number and the expected number of persons belonging to cluster h and giving response 1 to item j

$o_{yh}^{(j)}$	observed number of persons belonging to cluster h and giving response y to item j
$\hat{o}_{yh}^{(j)}$	expected value under a given model of the number of persons belonging to cluster h and giving response y to item j
r	test score
n_r	number of subjects with score r
$q_j(r)$	probability of endorsing item j for a subject with raw score equal to r
$\hat{q}_j(r)$	probability of endorsing item j for a subject with raw score equal to r estimated through the CML method
T_h	sum for all items of the squared standardized differences between the observed and the expected number of persons belonging to cluster h and giving response 1 to item j
R_{1c}	test statistic proposed by Glas (1988)
Q_1	test statistic proposed by Van den Wollenberg (1982)
S_j	test statistic proposed by Verhelst and Eggen (1989)
U_j	test statistic proposed by Molenaar (1983b)
ML	test statistic for the Martin-Löf (1973) test
d	latent trait identification number
D	total number of latent traits
$\mathcal{J}_d$	subset of items measuring dimension d
T_1, T_{11}, T_2, T_5	test statistics proposed by Ponocny (2001)
$\overline{\text{cor}}_{j_1 j_2}$	average of the correlations between items j_1 and j_2 computed on simulated data
$\hat{y}_{ij}$	response expected value
e_{ij}	score residual
out_j	outfit statistic or unweighted mean square statistic for item j
out_j^*	standardized outfit statistic or standardized unweighted mean square statistic for item j
in_j	infit statistic or weighted mean square statistic for item j
in_j^*	standardized infit statistic or standardized weighted mean square statistic for item j
out_i	outfit statistic or unweighted mean square statistics for person i
in_i	infit statistic or weighted mean square statistic for person i
$\beta_j^{(h)}$	difficulty parameter for binary item j in case person i belongs to cluster h
MH_j	test statistic for the Mantel–Haenszel test
$o_{yhr}^{(j)}$	number of persons belonging to cluster h, giving response y to item j, and having test score r
$ST-p-DIF$	test statistic for the standardized p difference test

Q_j	test statistic for Lord's test
T_4	test statistic for the T_4 test of Ponocny (2001)
x_i	vector of individual covariates
κ_0	constant term in a latent regression model
κ_1	vector of the regression parameters in a latent regression model
ε_i	error term affecting the latent trait level of individual i
π_{iv}	mass probability for the latent discrete distribution, depending on the individual-specific covariates
κ_{0v}	constant term in a concomitant variable latent class model
κ_{1v}	vector of the regression parameters in a concomitant variable LC model
n_h	number of individuals within cluster h
Y_{hij}	response provided by subject i in cluster h to item j
$x_{hi}^{(1)}$	vector of individual-specific covariates
$x_h^{(2)}$	vector of cluster-specific covariates
$\Theta_{hi}^{(1)}$	individual-level residual for person i belonging to cluster h
$\Theta_h^{(2)}$	cluster-level residual for cluster h
$\varepsilon_{hi}^{(1)}$	error term affecting the individual-level residuals
$\varepsilon_h^{(2)}$	error term affecting the cluster-level residuals
β_{00j}	difficulty of item j at the first time occasion
β_{01j}	variation of difficulty of item j along time
$V_{hi}^{(1)}$	discrete latent variable at individual level
k_1	number of support points for the latent discrete distribution at individual level
$V_h^{(2)}$	discrete latent variable at cluster level
k_2	number of support points for the latent discrete distribution at cluster level
$v^{(1)}$	support point number for the latent discrete distribution at individual level
$\xi_{v^{(1)}}$	support point for the latent discrete distribution at individual level
$\pi_{hi,v^{(1)}|v^{(2)}}^{(1)}$	mass probability for the latent discrete distribution at individual level
$v^{(2)}$	support point number for the latent discrete distribution at cluster level
$\kappa_{0v^{(1)}v^{(2)}}^{(1)}$	constant term at individual level in a multilevel LC model
$\kappa_{1v^{(1)}v^{(2)}}^{(1)}$	vector of regression coefficients of individual-level covariates
$\pi_{hv^{(2)}}^{(2)}$	mass probability for the latent discrete distribution at cluster level

$\kappa^{(2)}_{0v^{(2)}}$ constant term at cluster level in a multilevel LC model

$\kappa^{(2)}_{1v^{(2)}}$ vector of regression coefficients of cluster-level covariates

Θ_i vector of latent traits for subject i

θ_{id} level of latent trait d for subject i

$p_j(\Theta_i)$ item characteristic curve in the multidimensional case

$p^*_{jy}(\Theta_i)$ cumulative operating characteristic curve in the multidimensional case

$p(y_i|\Theta_i)$ joint conditional probability of y_i given Θ_i

$p_{jy}(\Theta_i)$ item response category characteristic curve in the multidimensional case

Index